THE LEGACY OF BELLEAU WOOD

THE LEGACY OF BELLEAU WOOD

100 Years of Making Marines and Winning Battles

AN ANTHOLOGY

Edited by
Paul Westermeyer

Art Editor
Breanne Robertson, PhD

History Division
United States Marine Corps
Quantico, Virginia
2018

LIBRARY OF CONGRESS CATALOGING-IN-PUBLICATION DATA

Names: Westermeyer, Paul W., editor. | Robertson, Breanne, editor. | Marine Corps University (U.S.). History Division, issuing body.
Title: The Legacy of Belleau Wood : 100 years of making Marines and winning battles, an anthology / edited by Paul Westermeyer ; art editor: Breanne Robertson.
Other titles: 100 years of making Marines and winning battles, an anthology
Description: Quantico, Virginia : History Division, United States Marine Corps, 2018. | Includes bibliographical references.
Identifiers: LCCN 2018000255 | ISBN 9780997317480 (paperback)
Subjects: LCSH: United States. Marine Corps. | United States. Marine Corps—History.
Classification: LCC VE23 .L39 2018 | DDC 359.9/60973—dc23 | SUDOC D 214.511:M 33
LC record available at https://lccn.loc.gov/2018000255

Disclaimer

The views expressed in this publication are solely those of the authors. They do not necessarily reflect the opinions of the organizations for which they work, Marine Corps University, the U.S. Marine Corps, the Department of the Navy, or the U.S. government. The information contained in this book was accurate at the time of printing. Every effort has been made to secure copyright permission on excerpts and artworks reproduced in this volume. Please contact the editors to rectify inadvertent errors or omissions.

Published by
Marine Corps History Division
2044 Broadway Drive
Quantico, VA 22134
www.mcu.usmcu.edu/historydivision

1st Printing, 2018

For sale by the Superintendent of Documents, U.S. Government Publishing Office
Internet: bookstore.gpo.gov Phone: toll free (866) 512-1800; DC area (202) 512-1800
Fax: (202) 512-2104 Mail: Stop IDCC, Washington, DC 20402-0001

ISBN 978-0-16-094412-3

CONTENTS

CONTENTS

FOREWORD

The title of this work, *The Legacy of Belleau Wood: 100 Years of Making Marines and Winning Battles*, is a bit of a misnomer. Certainly, the U.S. Marines of today benefit from the legacy of the Corps' initial, legendary battle in the First World War, but today's Marines also have inherited the legacy of Château-Thierry, Saint-Mihiel, and the Meuse-Argonne, not to mention those who served in the Corps' first aerial units, behind the lines, and aboard U.S. Navy ships throughout the conflict. The Corps emerged from the First World War as an elite, modern fighting organization capable of transferring the determination and skill of its prewar professionals to the mass of enthusiastic volunteers who flocked to proudly wear the eagle, globe, and anchor. Beyond that, its officers were consummate professionals capable of command and organization at the tactical and operational levels of war.

In the century since Belleau Wood, the Corps has continued to make Marines and win battles, maintaining the highest standards while continuing to innovate across the spectrum of conflict. Marines developed the techniques of amphibious warfare that enabled success during the drive across the Pacific during World War II and instilled the *esprit de corps* that allowed the 1st Marine Division to overcome adversity despite all odds during the battle at Chosin Reservoir. The same legacy was at work as Marines responded to the challenges of the all-volunteer force in the 1970s and fought conventional, counterinsurgency, and counterterrorism conflicts in the three decades since the end of the Cold War.

At the heart of our Corps' effectiveness is its commitment to the basics of infantry tactics and small unit leadership. As General Alfred M. Gray Jr. said, "Every Marine is, first and foremost, a rifleman. All other conditions are secondary." The legacy of Belleau Wood is what makes Marines who are dedicated to no single method or concept of warfare and committed to no specific strategy or technological advance. Marines are devoted to the idea of being Marines, willing and able to adapt to new technologies and strategies as required, while maintaining fidelity to our core beliefs.

I would like to personally thank our university for putting this piece together. We must never forget our history and always strive to learn more about the legacy of the Marines who have gone before us. The fight for Belleau Wood was the birthplace of the modern day Marine Corps. The lessons learned from that fight, most importantly that we must continue to adapt to change, still resonates with our Corps today. As we train for each mission, we must remember to educate for the future—Belleau Wood, and the Marines who fought there, are an everlasting memory pointing us in the right direction as we march toward the future.

General Robert B. Neller
Commandant
United States Marine Corps

PREFACE

In the summer of 2017, the newly arrived president of Marine Corps University, Brigadier General William J. Bowers, ordered a lecture series, "The Legacy of Belleau Wood: 100 Years of Making Marines and Winning Battles." The series would include four lectures, and it was to be supported by an anthology produced by History Division, providing readings to the students on the topics each lecture would cover. The intent was to produce an anthology of lasting worth to Marines, broadly depicting keystone moments in the history of the Corps during the century following the Battle of Belleau Wood.

This volume presents a collection of 40 extracts, articles, letters, orders, interviews, and biographies. The work is intended to serve as a general overview and provisional reference to inform both Marines and the general public of the broad outlines of notable trends and controversies in Marine Corps history.

Additional support for this work came from the *Journal of Military History*, *Naval History*, *MCU Journal*, *Sea Power*, *Proceedings*, *Leatherneck Magazine*, and the *Marine Corps Gazette*, all of whom gave permission to reprint their articles. Their cooperation made this anthology possible.

The Legacy of Belleau Wood could not have been published without the professional efforts of the History Division staff. The editors would like to thank the former director of Marine Corps History, Dr. Charles P. Neimeyer, and Deputy Director Paul J. Weber. Colleagues Douglas E. Nash, Dr. Fred H. Allison, Dr. Alexandra Kindell, Joan C. Thomas, Alisa M. Whitley, Kara Newcomer, and Annette D. Amerman provided unflagging professional advice and support. Our Editing and Design Branch, led by Angela J. Anderson, was instrumental in transforming the manuscript into a finished product, editing the manuscript, and overseeing the production process.

This anthology is organized into four chapters covering the rise of the modern Marine Corps, the lessons of the Second World War and the Korean War, the Wilson/Barrow renaissance of the 1970s and 1980s, the modern Marine Corps and the future, and an appendix detailing military historiography as a subject of professional study for Marines. This work is not meant to be an authoritative history, but rather it is intended to be used as a starting point for Marines, the general public, and academic researchers.

THE LEGACY OF BELLEAU WOOD

American illustrator N. C. Wyeth's painting, *It was after this attack that the High Command published to the German army: "The moral effect of our own gunfire can not seriously impede the advance of the American Infantry,"* appeared in *Redbook* magazine in the early 1930s.
Brandywine River Museum of Art

CHAPTER ONE

The Rise of the Early Modern Marine Corps and World War I

by Paul Westermeyer

As the twentieth century approached, thoughtful Marines had cause to be concerned for the future of their Corps. Founded in 1775 during the American War of Independence almost as an afterthought, the Corps' served throughout the nineteenth century in the traditional Marine role of ships detachments, maintaining shipboard discipline and providing trained soldiery for boarding and landing parties. In addition, the budget conscious American government added naval yard security to the traditional duties of Marines in the Age of Sail (ca. 1570–1860). The early Corps performed these duties honorably, occasionally with distinction, and also often served ashore to augment the small United States Army, most notably in the Seminole Wars, the Mexican War, and the Civil War.

Despite the Corps' demonstrated usefulness, obsolescence threatened at the end of the nineteenth century. Naval technology and changing demographics were quickly rendering the Corps' traditional duties moot. Its future as a combat organization was bleak; it seemed likely its ranks would be absorbed by the Army or reduced to the ceremonial band in the capital and to watchmen at various naval yards along the coasts.

Instead, during the next decades, the Marine Corps transformed itself, increasing dramatically the professionalism of its officers and finding a *raison d'etre* as the nation's expeditionary force in being. The hard realities of wind and sail had provided the need for Marines in the Age of Sail, and the technological necessities of the Age of Steam provided the same impetus for Marines. Navies required overseas bases to project their power, and the Corps would (occasionally over its own objections) take the mission of seizing and defending advanced naval bases.

The Spanish-American War foreshadowed this future when a battalion of Marines seized Guantánamo Bay as a forward base from which the U.S. Navy could blockade the island of Cuba. The Corps' success at Guantánamo compared favorably to the more haphazard embarkation and ship of the Army expedition, as did the deployment of Marines in the Far East as "colonial infantry." The Army's own increased professionalism and sense of purpose added another danger to the Corps' existence, however. Army officers saw the volunteer regiments and state militias that comprised a large percentage of American ground forces through the Spanish-American War as a primary cause for the chaos and unprofessionalism of mobilization in that conflict, and lobbied to formalize the state forces into the "National Guard." This put the Corps' secondary purpose as an augmentation to American land forces at risk.

America's entry into World War I provided an opportunity for the Corps to prove its increased professionalism and its value as a military Service, but there were significant difficulties. The Army was equally determined to prove it was a modern, professional organization on par with European armies; as such, it was not eager to grant the Navy Department's land forces a chance to show it up. Organizationally, the Corps would have to field regiments and brigades to act on the western front, and the Army questioned whether the Corps had any officers capable of commanding and staffing large land forces of that nature. And like the Army, the Marine Corps would be forced to expand rapidly, going from 10,397 officers and men in 1916 to 75,101 in 1918.

The Corps responded to these challenges with vigor, producing an enviable battlefield record in the war despite its relatively tiny size. Meanwhile, General John A. Lejeune proved Marine officers were just as professionally capable as their Army and Navy counterparts, rising to command the Army's 2d Division. And in the Corps' first major action in the Great War, the Battle of Belleau Wood, Marine tenacity and media savvy catapulted the Corps into even greater public consciousness, cementing the Marine Corps' self-proclaimed reputation as an elite force into reality.

MILITARY PROFESSIONALISM
The Case of the U.S. Marine Officer Corps, 1880–1898

by Jack Shulimson
Journal of Military History, 1996

In December 1880, Marine Captain Henry Clay Cochrane, recently promoted after 18 years of military service, caustically entered in his diary, "a great year in the M.C. [Marine Corps]."[1] He then recorded a litany of misfortunes that had befallen the officer corps: one lieutenant had died in a riding accident; another had been sent "home insane;" while still another was dismissed for cause. A Navy court-martial convicted a senior Marine colonel for behavior unbecoming an officer and the Philadelphia police arrested a Marine major for drunkenly accosting women in the street. Reform legislation that would modestly expand the officer corps and open up promotions remained stalled in Congress. The president's appointment of a well-connected junior lieutenant to a plump staff position of assistant quartermaster over more than 30 of his seniors caused even more rancor. There was little to encourage that "small band of officers" who desired to reform the Corps.[2]

The Marine Corps was an organizational anomaly and in some disarray. Dispersed into small ship's detachments and navy yard guards of usually 100 men or less, the Marine Corps had no formal company, battalion, or regimental structure. Marine reformers among the younger officers, called for either a "funeral or a resuscitation" for the Corps.[3]

During the three decades following the Civil War, the officers of the U.S. Marine Corps underwent a metamorphosis from an almost moribund state to a near professional status. At the beginning of the period, influence was the

[1] The original article came from Jack Shulimson, "Military Professionalism: The Case of the U.S. Marine Officer Corps, 1880–1898," *Journal of Military History* 60, no. 2 (1996): 231–42. Minor revisions were made to the text based on current standards for style, grammar, punctuation, and spelling.

[2] Henry Clay Cochrane Diary, General Entry, 1880, Henry Clay Cochrane Papers, PCS 1, Marine Corps Historical Center (MCHC); Commandant of the Marine Corps, *Annual Report, 1880,* 529; Capt Woodhull S. Schenck to SecNav, 10 March 1880, Letters Received, 1880, RG 80, General Records of the Department of the Navy, National Archives and Records Administration, Washington, DC (NARA); *New York Times,* 12 February 1880, 5; and *Army and Navy Journal,* 3 July 1880, 978.

[3] Henry Clay Cochrane, "A Resuscitation or a Funeral," 1 October 1875, Cochrane Papers, MCHC; *New York Times,* 29 December 1875, 5; and Maj James Forney, "The Marines: With an Account of Life at a Marine Post," in *United Service: A Monthly Review of Military and Naval Affairs*, vol. 1 (Philadelphia: L. R. Hamersly, April 1889), 89, 94–95.

Group of Marine officers, 1st Battalion Marines, Navy Yard, Portsmouth, NH, 1898.
Naval History and Heritage Command, NH119980

usual avenue of entry. Older officers were apathetic while junior and company grade officers faced long frustrating years with little prospect for promotion. By the turn of the century, nevertheless, a complete reversal had nearly occurred. Both entry into and promotion in the officer corps depended upon passing relatively stringent examinations. Many officers were graduates of the [United States] Naval Academy, and still others had attended advanced schools of both the Army and Navy. The Spanish-American War and internal reform account only in part for this transformation.

My study of the Marine officer corps would confirm [American political scientist] Samuel P. Huntington's periodization of military professionalism but differs markedly with his concept of the "isolation" of the American military. While accepting Huntington's general definition of the military professional: "expertise in the management of violence, responsibility, and corporateness," I would maintain that such descriptive traits are much less important than "professional jurisdiction."[4]

The Marine officer experience was a result of the several larger trends that altered all aspects of American life during this period. Among the most important of these impulses were technological change, advances in formal knowledge, an expanding industrialism, a restructuring both of private and public organizations, and a per-

[4] Samuel P. Huntington, *The Soldier and the State: The Theory and Politics of Civil-Military Relations* (Cambridge, MA, 1957), 7–10, 19–54.

Typical tent quarters for Marines at Camp Sampson, Key West, FL.
Marine Corps History Division

vasive professionalization of American society. Whether called a "search for order," the "visible hand," or the "organizational revolution," the dominant feature of this entire process was its avowed emphasis upon rationality and control.[5]

It was in the industrialization and urbanization processes of the late nineteenth century that the professions, like other aspects of American life, took on their modern cast. There was a virtual explosion of professional groups and organizations. Of the 58 professional organizations formed during the nineteenth century, 37 or more than 60 percent came into existence between 1870 and 1900. These newer associations reflected an increasing specialization and tendency toward technology in response to the needs of the new age.[6]

While there is general agreement about the centrality of the professions to the social and economic structure of modern civilization, there are wide differences in the scholarly com-

[5] Daniel Beaver, "The American Military and the 'New' Institutional History" (paper delivered at the Annual Meeting of the Organization of American Historians, Indianapolis, IN, May 1985), 2–4; James L. Abrahamson, *American Arms for a New Century: The Making of a Great Military Power* (New York: Free Press, 1981), xii; Kenneth E. Boulding, *The Organizational Revolution: A Study in the Ethics of Economic Organization* (Chicago, IL: Quadrangle Books, 1968), 49, 202; Robert H. Wiebe, *The Search for Order, 1877–1920* (New York: Hill and Wang, 1967), vii–ix, xiii–xiv, 11–43; Alfred D. Chandler Jr., *The Visible Hand: The Managerial Revolution in American Business* (Cambridge, MA: Harvard University Press, 1977), 1–14; and Thomas L. Haskell, *The Emergence of Professional Social Science: The American Social Science Association and the Nineteenth Century Crisis of Authority* (Urbana: University of Illinois Press, 1977), 3–4, 234.

[6] Burton J. Bledstein, *The Culture of Professionalism: The Middle Class and the Development of Higher Education in America* (New York: Norton, 1976), 80; and Ralph S. Bates, *Scientific Societies in the United States* (Cambridge, MA: Pergamon Press, 1965), 85, 105, 121.

munity about the nature of professionalism. Some would restrict the discussion to such professions as the law and medicine, while others write articles about the professionalization of nearly everyone. There is even disagreement over whether there is a standard definition of professionalism, or whether there is even a need for such a definition. As one author observed, the debate has resulted in a "confusion so profound that there are even disagreements about the existence of the confusion."[7]

While much of the discussion of professionalism remains mired in disputation about traits, functions, and power, sociologist Andrew Abbott has suggested that scholars have left unexamined the actual work that professions do.[8] For

U.S. Marine Corps recruitment poster showing Marines firing artillery from the deck of a ship, 1914.
Library of Congress Prints and Photographs Division

[7] A. P. Carr-Saunders and P. A. Wilson, *The Professions* (Oxford, UK: Clarendon Press, 1933), 3; Harold L. Wilensky, "The Professionalization of Everyone?," *American Journal of Sociology* 70, no. 2 (1964): 137–58; Terence J. Johnson, ed., *Professions and Power* (London: Routledge, 1972), 23; Eliot Freidson, *Professional Powers: A Study of the Institutionalization of Formal Knowledge* (Chicago, IL: University of Chicago Press, 1986), 30–32; Geoffrey Millerson, *The Qualifying Associations: A Study in Professionalization* (London: Routledge, 1964), 4–5; Geoffrey Millerson, "Dilemmas of Professionalism," *New Society* (June 1964): 15–16; and W. J. Goode, "The Theoretical Limits of Professionalization," in A. Etzioni, ed., *The Semi-professions and Their Organization* (New York: Free Press, 1969), 266–313. The quotation is from Johnson, *Professions and Power,* 22.

[8] Among sociologists, who have written most extensively on the subject, there are at least three identifiable schools on the subject of professionalism. For convenience, they can be designated the functionalists, the structuralists or "traitists," and the "monopolists" or power theorists. The functionalists, largely associated with [American sociologist] Talcott Parsons, submit that the professions provide the vital functions of society. With an emphasis on "cognitive rationality" and the social responsibility of the professions, Parsons and his adherents focused largely on the expert-client relationship. For discussion of the functionalist perspective in the literature, see Talcott Parsons, "Professions," *International Encyclopedia of the Social Sciences*, ed. David L. Sills (New York: Macmillan, 1968), 12, 536–47; Johnson, *Professions and Power,* 23, 32–47; Bernard Barber, "Some Problems in the Sociology of the Professions," *Daedalus* (Fall 1963): 669–88; Michael F. Winter, *The Culture and Control of Expertise: Toward a Sociological Understanding of Librarianship* (Westport, CT: Greenwood Press, 1988), 42–44; Andrew Abbott, *The System of Professions: An Essay on the Division of Expert Labor* (Chicago, IL: University of Chicago Press, 1988), 15; and Andrew Abbott, "Perspectives on Professionalism" (paper delivered at the Annual Meeting of the Organization of American Historians, Indianapolis, IN, May 1985), 2–3. In contrast to the functionalists, the structuralists concentrated on the structure or attributes of the professions. They attempted to identify the defining characteristics of any given profession. The difficulty, however, was that there were probably more differences among the structuralists than there were between the structuralists and the functionalists. For discussion on the literature of the structuralists, see Abbott, *The System of Professions*, 15; Abbott, "Perspectives on Professionalism," 2–3; Johnson, *Professions and Power*, 23–30; Winter, *Culture and Control of Expertise*, 21–23, 26–27, 37, 119; Millerson, "Dilemmas of Professionalism," 15; Bengt Abrahamsson, *Military Professionalism and Political Power* (Beverly Hills, CA: Sage Publications, 1972), 14–15; and Wilensky, "The Professionalization of Everyone?," 137–58. Beginning in the 1960s, more recent writers have questioned the validity of both the structuralist and functionalist positions. With such scholars as Eliot Freidson, Magali S. Larson, and Terence Johnson in the forefront, they contended that both schools emphasized the positive while ignoring the negative aspects of professionalism. Criticizing the older scholarship for its over concern with form and structure, Freidson, Johnson, and Larson argued that the new scholarship should concentrate upon the concept of professions as ideologies to control the work place. Rather than accepting the older view of collegiality and mutual trust as the trademark of professional life, they insisted

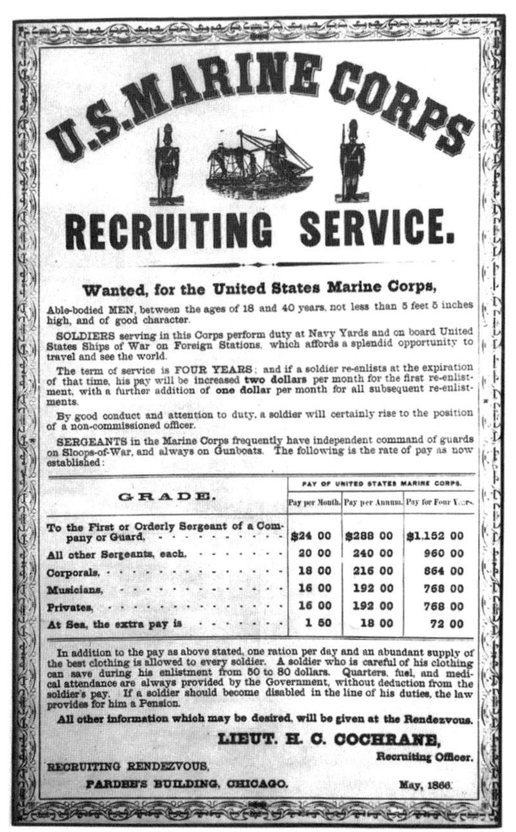

U.S. MARINE CORPS

RECRUITING SERVICE.

Wanted, for the United States Marine Corps,

Able-bodied MEN, between the ages of 18 and 40 years, not less than 5 feet 5 inches high, and of good character.

SOLDIERS serving in this Corps perform duty at Navy Yards and on board United States Ships of War on Foreign Stations, which affords a splendid opportunity to travel and see the world.

The term of service is FOUR YEARS; and if a soldier re-enlists at the expiration of that time, his pay will be increased **two dollars** per month for the first re-enlistment, with a further addition of **one dollar** per month for all subsequent re-enlistments.

By good conduct and attention to duty, a soldier will certainly rise to the position of a non-commissioned officer.

SERGEANTS in the Marine Corps frequently have independent command of guards on Sloops-of-War, and always on Gunboats. The following is the rate of pay as now established:

GRADE.	PAY OF UNITED STATES MARINE CORPS.		
	Pay per Month.	Pay per Annum.	Pay for Four Years.
To the First or Orderly Sergeant of a Company or Guard,	$24 00	$288 00	$1.152 00
All other Sergeants, each,	20 00	240 00	960 00
Corporals,	18 00	216 00	864 00
Musicians,	16 00	192 00	768 00
Privates,	16 00	192 00	768 00
At Sea, the extra pay is	1 50	18 00	72 00

In addition to the pay as above stated, one ration per day and an abundant supply of the best clothing is allowed to every soldier. A soldier who is careful of his clothing can save during his enlistment from 50 to 80 dollars. Quarters, fuel, and medical attendance are always provided by the Government, without deduction from the soldier's pay. If a soldier should become disabled in the line of his duties, the law provides for him a Pension.

All other information which may be desired, will be given at the Rendezvous.

LIEUT. H. C. COCHRANE,
Recruiting Officer.

RECRUITING RENDEZVOUS,
PARDEE'S BUILDING, CHICAGO.

May, 1866.

U.S. Marine Corps Recruiting Service broadside, 1866.
Official U.S. Marine Corps image

Abbott, it is not the characteristic or structure of professionalism that is significant but rather the work the profession performs, how it controls that work, and its relationship with similar or competing professions. The essential element is the link between the profession and its work, which Abbott labels "jurisdiction." From the perspective of intersecting professional jurisdictions, he argues, one can determine how professions develop, how they relate to one another, and what is the basis of the work they do.[9]

From this perspective, professions neither stand alone nor are encompassed by any overarching theory of professionalism: "They exist in a system." The system, however, is a complex one. Changes in one area may work their way through the entire system of professions, but more likely this "ripple effect" is confined to "one general task area, through the rearrangement of jurisdictions, the strengthening of some, the weakening of others." Thus, the interaction of professions in any given instance is usually limited to such general task areas. Assuming that the profession of military officer is one such task area or system within the general system of professionalism, in effect a system within a system, Abbott's construct provides a most useful framework.[10]

Some scholars have questioned whether military officership, itself, is a profession. As [historian] Allan R. Millett suggests, the viewpoint depends somewhat upon one's concept about the morality of war and the need of society for military force. In the early 1930s, British sociologists A. M. Carr-Saunders and P. A. Wilson dismissed discussion of the military officer "because the service which soldiers are trained to render is one which it is hoped they will never be called upon to perform."[11] Yet, by the end of the nineteenth century, both the U.S. Army and the Navy officer corps had taken on attributes of professionalism common to all the modern professions.[12]

that "dominance and autonomy" were its chief characteristics. For extended discussion of the power theorists, see Eliot Freidson, "Professions and the Occupational Principle," in Freidson, ed., *The Professions and Their Prospects* (Beverly Hills, CA: Sage Publications, 1973), 19–38, 29–31; Freidson, *Professional Powers*, 28–29, 211; Abbott, *The System of Professions*, 5; Johnson, *Professions and Power*, 32, 37–39; Larson, *Rise of Professionalism*, xvi–xviii; and Winter, *Culture and Control*, 44–45, 50.

[9] Abbott, "Perspectives on Professionalism," 10; and Abbott, *System of Professions*, 8–9, 18–20, 86–91, 318, 320–21.

[10] Abbott, "Perspectives on Professionalism," 13–15; and Abbott, *System of Professions*, 2–3, 33, 35, 59, 69, 86–91, 111–12.

[11] Allan Millett, *Military Professionalism and Officership in America* (Columbus, OH: Mershon Center, Ohio State University, 1977), 12–13; and Carr-Saunders and Wilson, *The Professions*, 3.

[12] Morris Janowitz and Roger W. Little, *Sociology and the Military Establishment*, 3d ed. (Beverly Hills, CA: Sage Publications, 1974),

The literature on military professionalism has largely concentrated on such attributes as education, inculcation of military ethics, and socialization of the officer corps. Much of the discussion has dealt with the professionalization of both the individual officer and of the officer corps in general. A large concern is the correlation of the professionalization of the military officer corps to its relationship, both political and professional, to the civilian community.[13]

Military professionalism did not evolve in a single line. Both the Navy and Army were complex, multifaceted organizations. The broad rubric of military officer contained various specialties with varied interests and ambitions. Within the Navy, for example, there were line officers, engineers, surgeons, paymasters, and naval constructors, not to mention the Marines, each with his own specific skills and each competing for part of the naval budget. Large independent bureaus, mostly made up of staff officers, wielded great power within both the Army and Navy and especially with Congress. In both Services, often divisive differences existed between line and staff. This occasionally acrimonious debate between staff and line was more than a simple struggle for power; it was in actuality a struggle for recognition and the establishment of jurisdictions by competing military professionals. Both Navy and Army line officers interpreted military officership in the narrow sense of command on the battlefield or on board ship. They viewed other officers and the bureaus as merely support to their own activities. Although not differing with the line officers over the general functions of the Army and Navy, the bureau and staff officers wanted to retain control of their own discrete organizations. While willing to cooperate and consult with the line, the bureaus and staffs "used all their resources at the intersects of power" within the Services and with Congress "to gain acceptance as valuable soldiers of the Republic."[14]

As part of the federal government, the military could not hope to escape the political realities of the time. Political considerations permeated the entire military structure. The president appointed his secretaries of Army and Navy for their political strengths, not because of

127; William B. Skelton, "Professionalization in the U.S. Army Officer Corps during the Age of Jackson," *Armed Forces and Society* 1, no. 4 (August 1975): 443–71; William B. Skelton, *An American Profession of Arms: The Army Officer Corps, 1784–1861* (Lawrence: University Press of Kansas, 1992), 361–62; Timothy K. Nenninger, "The Fort Leavenworth Schools: Post-Graduate Military Education and Professionalization in the U.S. Army, 1880–1920" (PhD dissertation, University of Wisconsin, 1974), 2–3; and Timothy K. Nenninger, *The Leavenworth Schools and the Old Army: Education, Professionalism, and the Officer Corps of the United States Army, 1881–1918* (Westport, CT: Greenwood Press, 1978), 6. Skelton would argue that the actual professionalization of the Army officer corps occurred much earlier, during the antebellum period, and that the Civil War and its aftermath only disrupted this process.

[13] Huntington, *Soldier and the State*, 1–10, 19–54; Abrahamsson, *Military Professionalism*, 12–13, 15–17, 19, 36, 59–60, 69; Millett, *Military Professionalism*, 2, 4–5, 12–13, 18–22; Allan R. Millett, "Professional Military Education and Marine Officers," *Marine Corps Gazette*, November 1989, 46–56; Janowitz and Little, *Sociology and the Military Establishment*, 123–24, 127, 142; Skelton, "Professionalization in the U.S. Army," 443–71; Skelton, *An American Profession of Arms*, 359–60; Nenninger, *Leavenworth Schools*, 3–20; Abrahamson, *America Arms for a New Century*, xiii–xv, 33–36, 40, 48, 147–48, 150; John Gates, "The Alleged Isolation of U.S. Army Officers in the Late 19th Century," *Parameters* 10, no. 3 (September 1980): 32–45; John Gates, "The 'New' Military Professionalism," *Armed Forces and Society* 11, no. 3 (Spring 1985): 427–36; Edward M. Coffman, *The Old Army: A Portrait of the American Army in Peacetime, 1784–1898* (New York: Oxford University Press, 1986), 96, 270; Sam C. Sarkesian, "Moral and Ethical Foundations of Military Professionalism," in *Military Ethics and Professionalism: A Collection of Essays*, ed. James Brown and Michael J. Collins (Washington, DC: National Defense University, 1981), 1–22; Sam C. Sarkesian, *Beyond the Battlefield: The New Military Professionalism* (Cambridge, MA: Pergamon Press, 1981), ix–xi, 7, 9–13, 42, 254; Ronald H. Spector, *The Naval War College and the Development of the Naval Profession* (Newport, RI: Naval War College, 1977), 1–26; and Carol Ann Reardon, "The Study of Military History and the Growth of Professionalism in the U.S. Army before World War I" (PhD dissertation, University of Kentucky, 1987), 1–5, 11, 15, 40. Like this author, most of the above would accept Huntington's definition of the military professional, but would reject the concept of his alleged isolation from the rest of society.

[14] Beaver, "The American Military and the 'New' Institutional History," 12–14.

their military expertise. Military officers were dependent upon the patronage system for their initial appointments to West Point or Annapolis or for their direct commissions into the Service. Political sponsorship often assured officers of career-enhancing positions and assignments. Individual congressmen and senators generally were not especially interested in broad questions of military policy and organization, but were very concerned with questions of patronage. As one writer observed, "Perhaps nothing caused so many irritations to so many members of the House and Senate as their failure to have their way in the thousands of small appointments that were required across the country from year to year." Military professionals, such as Emory Upton, who challenged the political system directly, often ended up defeated and frustrated. Most successful military officers became adept at bureaucratic politics. Remarking on the political adroitness of the military, one historian concluded, "military services acted much as pressure groups, exploiting their contacts with those in positions of influence and using public-relations techniques to create general support for military reform."[15]

Using Abbott's construct and carrying it one step further, one can make the case that for the military profession, jurisdiction is where the outside forces of society, the individual military organization or Service, and professionalism all come together. During the nineteenth century, the outside driving societal force was the new technology and industrialism, which revolutionized warfare on both land and sea. For the naval officer, it resulted in a professional jurisdiction based upon organization, technology, and the strategic concept of sea power; for the Army officer, it created a jurisdiction based upon organization, new weaponry, and the study of land warfare. In contrast to both the Navy and Army officer, the Marine officer had no such clear demarcation of jurisdictional responsibility. While sharing many of the attributes of both the Army and Navy officer, the Marine officer's jurisdiction revolved around organization and mission.[16]

The Marine officers of the last decades of the nineteenth century accommodated slowly and with difficulty to the sudden challenges in organization and mission that the modern era posed. In the decade following the Civil War, despite the appearance of a nascent reform movement, Marine officers failed to agree among themselves about their own role or that of their Service.[17]

During the 1880s, naval reformers and professionals inaugurated an intellectual ferment

[15] Huntington, *Soldier and the State,* 163–92; Gates, "New Military Professionalism," 432–33; Leonard D. White, *The Republican Era: A Study in Administrative History, 1861–1901* (New York: Macmillan, 1958), 26–27; Ambrose, *Upton,* 135; Millett and Maslowski, *For the Common Defense,* 256–58; and Abrahamsson, *America Arms for a New Century,* 150. The first quote is from White while the second is from Abrahamsson.

[16] Much of the examination of military professionalization has ignored the question of professional military jurisdictions, except for the differences between line and staff officers in the Army and line and engineers in the Navy. Historian Peter Karsten perhaps skirts the fringes of professional jurisdictions when he argues that it was "career anxiety" that accounted for the professionalization of the so-called "naval aristocracy." Carol Ann Reardon touches upon a jurisdictional dispute, not between rival groups of military officers, but between professional historians and Army officers in the use and understanding of military history. For the most part, however, most writers on military professionalization have largely ignored the jurisdictional areas that both differentiate and link together the various military professionals. See Peter Karsten, *The Naval Aristocracy: The Golden Age of Annapolis and the Emergence of Modern American Navalism* (New York, 1972), 292–93; and Reardon, "Growth of Professionalism in the U.S. Army before World War I," 8, 34–36, 38. It is interesting to contrast the confrontation between Army officers and historians to the accommodation reached by naval officers and the profession of history. RAdm Alfred Thayer Mahan, the proponent of history and seapower, served a term as president of the American Historical Association, and RAdm French E. Chadwick, the chronicler of the Spanish-American War, was also a respected member of the historical community.

[17] This and the following paragraphs are based on Jack Shulimson, *The Marines' Search for a Mission, 1880–1898* (Lawrence: University Press of Kansas, 1993).

within the Navy that led to new thinking about naval warfare that tied the Marine Corps and its officers closer to the Navy. Curiously, Marine officers and Marine Commandant Colonel Charles G. McCawley played only a peripheral role in these developments. The Marine Commandant saw reform in much narrower terms. He and other Marine reformers sought to make the Marine Corps an elite guard force for the Navy, concentrating on raising officer standards, improving military discipline, and cutting desertion rates. Their efforts to push reform measures through Congress, however, proved fruitless. It was only a legislative fluke resulting from the efforts of certain naval progressives that finally led to new Marine officers coming from the Naval Academy, the only major innovation during this period. While these naval professionals had no great love for the Marine Corps, they were willing to use the Marine Corps to further their own interests. It was only a few Marine nonconformists, such as Captain Daniel Pratt Mannix, who established a school of application in China, who sought out new directions for their Corps.

Richard N. Brooke, *Col Charles G. McCawley, Eighth Commandant of the Marine Corps* (1876–91).
Art Collection, National Museum of the Marine Corps

The years from 1885 to 1889 continued to be a transitional period for Marine officer professionalism. Although the 1885 U.S. intervention in Panama foreshadowed the employment of future Marine expeditionary forces, it resulted in the first rift between the Navy progressives and the Marine leadership. The Marine Commandant rejected any change in Marine structure, including the establishment of a permanent Marine landing force. Still, naval professionals in their literature continued to discuss landing operations and at the same time hinted at limiting the role of Marine officers on board ship. On the other hand, Marine officers demonstrated little initiative, although the new second lieutenants, graduates of the Naval Academy, infused some fresh blood and a certain elan into the Marine officer corps.

During the tumultuous years from 1889 to 1893 for the Marine Corps, the new battleship Navy became a reality and the United States shifted to a naval policy of control of the seas, at least in American coastal waters and in the Caribbean. These developments led to new opportunities for Marine officer professionalism as well as new hazards. A Navy board recommended the removal of Marines from naval warships, while on the other hand, in the Navy's first contingency planning effort, Captain Alfred Thayer Mahan called for Marines to "constitute a most important re-enforcement, nay, backbone, to any force landing on the enemy's coast."[18]

A new Marine Commandant, Colonel Charles

[18] "Contingency Plan of Operations in Case of War with Great Britain, New York, December 1890," reprinted in *Letters and Papers of Alfred Thayer Mahan*, ed. Robert Seager II and Doris D. Maguire, 3 vols. (Annapolis: Naval Institute Press, 1975), 3, 559–82.

Richard N. Brooke, *MajGen Charles Heywood, Ninth Commandant of the Marine Corps* (1891–1903).
Art Collection, National Museum of the Marine Corps

Heywood, influenced by Captain Mannix and other Marine reformers, introduced several measures to raise Marine officer professionalism. Among other reforms, these included more stringent promotion standards, including examinations and the establishment of a School of Application. At the same time, the retirement of several senior officers opened up promotion opportunities. Some Naval Academy cadets sought out the Marine Corps because it actually offered faster immediate promotion than the Navy.

Heywood also attempted to identify the mission of the Marine Corps with that of the Navy. He tried to assign Marine ship detachments under their own officers to the secondary gun batteries of the new steel armored ships. In a sense, this endeavor backfired in that it brought the Marines into conflict with the Navy progressives. In what amounted to a jurisdictional dispute, younger naval professionals pushed openly to remove Marine detachments from warships. This issue about the secondary battery and the Marine ship detachments would continue to color and blur all Marine and Navy professional relationships.

From 1893 to the Spanish-American War, minor skirmishes continued to revolve around the secondary battery and Marine ship detachment issues. An aborted attempt by Army artillery officers to form Marine artillery organization under the Navy complicated the situation even further. At the same time, the Marine Corps School of Application flourished, Marine officer standards continued to be raised, and Marine officers attended the Naval War College and served on the faculty of that institution.

During these years, reform merged with officer professionalism and institutional and individual self-interest. These factors combined with patronage politics in Congress and in the Navy Department to provide the background for the playing out of the Marine and Navy officer relationships during this period. The criticism of the naval progressives, however, had little to do with the quality of the Marine officer or enlisted men, but rather with the relevance of the Marine Corps to the Navy. They simply did not want Marines on board Navy warships. At the end of 1897, the secretary of the Navy created a Navy Personnel Board, headed by Assistant Secretary of the Navy Theodore Roosevelt, to study the complete reorganization of officers in the Navy Department. When the Navy Personnel Board examined the possible amalgamation of the Marine officer corps with the Navy, doubt and confusion shook the Marine officer corps. Many Marine officers wondered aloud whether they had any role with the Navy and even suggested half-seriously the incorporation of the

Marine Corps into the coast artillery regiments of the Army. Obviously, despite the advances in the attainment of the outward professional attributes of the military officer, the Marine officer had failed to achieve a clear jurisdictional link as a professional in the Navy. In a sense, it was a period of missed connections.

The Spanish-American War proved to be a defining period for the Marine Corps. While not fully knowing how they would use it, naval authorities immediately ordered the establishment of a Marine battalion with its own transport. Although numbering less than a quarter of the active Marine Corps, this battalion's activities not only received public approbation, but also had implications for the future relationship of the Marine Corps with the Navy. Despite a somewhat rocky start at Guantánamo, the Marine 1st Battalion proved itself in combat. By seizing the heights on Guantánamo, it provided a safe anchorage for Navy ships. In effect, the Marines seized and protected an advanced base for the fleet blockading Santiago.

Navy strategists and planners also learned another lesson from the war. They quickly realized that Army and Navy officers may have very different and even possibly conflicting goals in a military campaign. The dispute between the Army and Navy at Santiago reflected the separate approaches of professional Army and Navy officers. For the Army, the vital objective was the capture of the Spanish garrison and the city of Santiago. On the other hand, the Navy's aim was the destruction of the Spanish fleet. For its part, the Army designed an overland campaign to capture the city and was unwilling to sacrifice men to take the heights overlooking the narrow channel into Santiago Bay. At the same time, the Navy refused to chance the loss of any of its ships by running the channel. Although both attained their desired ends, their basic conflict remained unresolved. For the Navy, the message was that it could not depend upon the Army to secure land-based sites for naval purposes. The Navy required its own land force. It had this in the Marine Corps.

This drawing by Col John W. Thomason Jr. depicts a Marine company commander in World War I.
Art Collection, National Museum of the Marine Corps

In 1900, the newly formed Navy General Board assigned to the Marine Corps the establishment of advanced bases in support of the fleet. This advanced base mission finally gave the Marine officer that clear connection to the Navy that had eluded him over the previous two decades. The resulting dynamics created a tension that influenced both the professionalization of the Marine officer corps and the institutional survival of the Marine Corps itself.

As important, however, as all of this may have been for Marine professionalism, the basic factor that stands out was the inherent insecurity of the Marine officer. Throughout this entire period, it seemed that no matter how much the Marine officer tried to improve himself, he was rebuffed by his naval colleagues. Although a small minority of naval officers may have looked to the Marine Corps as a possible landing force for the Navy, most looked upon it largely as a relic of the Age of Sail. The circumstances of the Spanish-American War and its consequences, however, forced the Navy progressives to reexamine their views about the Marine Corps. This still left the Marine officer in an innocuous position and resulted in a certain amount of institutional and professional schizophrenia and paranoia. For example, although Colonel Heywood and other Marine officers accepted the new advanced base mission in 1900, they continued to resist abandoning any of their traditional relationships with the Navy. Heywood and his successors continued to argue the viability of the assignment of Marines on board Navy warships. They constantly defined, redefined, and justified their roles and missions both to themselves and to everyone else.

What then can be concluded about Marine officer professionalism? From one aspect, it was self-directed and simulated the features of any other professional military officer group. In a very real way, the Marine officer's concerns with military discipline, tactics, and questions of enlisted morale and desertion rates were very similar to those of Army officers. These resulted in a certain elitism and esprit de corps. They had almost nothing, however, to do with mission and, most important, with the establishment of an area of professional jurisdiction. If the Marine officer had to depend only on these acquired professional traits, he and his Corps would have disappeared entirely or been absorbed completely by the Army. The intersection of the jurisdiction and responsibilities of the Marine officer with those of his naval counterpart, therefore, provides the best approach for the analysis of Marine officer professionalism.

Marine officer professionalism, thus, consisted of two separate but related strains. In the first instance, it consisted of the outward traits that characterized most professionals. The second strain related to the professional jurisdiction that the Marine officer had to carve out for himself within the Navy. After the Spanish-American War, while the rhetoric of the secondary battery and Marine guards on board Navy warships occasionally ruffled the Marine and Navy connection, there developed during the next decades a "continuity and consensus" about the Marine mission with the Navy. The consensus centered around the advanced base and expeditionary roles of the Marine Corps.[19]

The Marine officer participated in, affected, and was affected by the society and forces around him. This included the basic search for structure that characterized much of American life during the last decades of the nineteenth century. As shown here, this search for structure

[19] Graham A. Cosmas and Jack Shulimson, "Continuity and Consensus: The Evolution of the Marine Advance Base Force, 1900–1922," in *Proceedings of the Citadel Conference on War and Diplomacy, 1977*, ed. David H. White and John W. Gordon (Charleston, SC: Citadel Academy, 1979), 31–36.

Col Donald L. Dickson produced this drawing as part of a series illustrating U.S. Marine Corps uniforms throughout history. This scene, created in 1936, depicts two majors in full dress, a first lieutenant, and a private at the turn of the twentieth century.
Art Collection, National Museum of the Marine Corps

took place in a confused arena where the old forms continued to have full play, and competed with the new. The new professionalism coexisted with partisan politics, competing interest groups, personal and institutional self-interest, advancing technology, and an America beginning to look outward.

TEDDY ROOSEVELT AND THE CORPS' SEA-GOING MISSION

by Jack Shulimson and Graham A. Cosmas

Marine Corps Gazette, 1981

President Theodore Roosevelt's attempt in November 1908 to remove Marine guards from the warships of the U.S. Navy resulted in a noisy congressional and public controversy.[1] This episode is often depicted as a simple melodrama in which Marines heroically and effectively rose to save their Corps from a cabal of naval officers bent on its destruction. In fact, the issues were more complex and were related to the effort to redefine Marine Corps roles and missions in the twentieth century steam-and-steel Navy. In the larger context, the controversy illustrates both the complex bureaucratic infighting that shaped so much of Progressive Era reform and the growing estrangement between the lame duck Roosevelt and the Old Guard Republican congressional leadership.

In November 1908, the Marine Corps consisted of 267 officers and 9,100 enlisted men. Approximately one-third of this force was stationed afloat, mostly as guard detachments on warships. Another third was on shore duty outside the continental United States with the largest contingent in the Philippines. The remaining one-third served within the United States as navy yard guards and constituted a reserve from which expeditionary forces could be organized. Since the Spanish-American War, Marine Corps strength had expanded threefold. In the latest increase, in 1908, Congress had added almost 800 officers and men and had advanced the Commandant of the Corps to the rank of major general.

While operating under the Navy Department, the Marine Corps enjoyed the legal status of a separate Service. Its staff in Washington, headed by the Commandant, was closely allied with the powerful Navy Department bureaus and had a reputation for skillful and effective congressional lobbying. Despite this reputation, Headquarters Marine Corps, in the words of one Marine officer, was "not altogether a happy family." Major General Commandant George

[1] The original article came from Jack Shulimson and Graham A. Cosmas, "Teddy Roosevelt and the Corps' Sea-Going Mission," *Marine Corps Gazette* 65, no. 11 (1981): 54–61. Minor revisions were made to the text based on current standards for style, grammar, punctuation, and spelling.

Editorial cartoon by John L. De Mar addressing the controversy over Marines on Navy ships.
The Washington Post, *26 February 1909*

F. Elliott, known for his blunt and often hasty speech, was partially deaf and rumored to be overly fond of the bottle. His staff was riddled with intrigue as ambitious, politically connected officers pursued their own bureaucratic aggrandizement. Field Marines often regarded the Washington staff with suspicion. Lieutenant Colonel John A. Lejeune denounced "the politicians stationed at Headquarters" and declared, "Fortunately the real Marine Corps is elsewhere and consists of the 10,000 officers and men who are scattered around the world."

Within the Navy, sharp divisions had emerged between the so-called progressive reformers and the largely conservative bureau chiefs. The reformers, mostly young commanders and captains, favored establishing a Navy general staff, modeled on that recently created for the Army. President Roosevelt generally sympathized with the reformers and had as his personal naval aide one of the most aggressive of them, Commander William S. Sims, yet the reformers usually met frustration at the hands of the bureau chiefs who enjoyed strong congressional support. The reformers generally viewed the Marine Corps, or at least its Washington headquarters, which usually sided with the bureau chiefs, as an obstacle to their plans. One of the more vociferous Navy progressives, Commander William F. Fullam, claimed that "the Marines and the bureau system are twins. Both must go before our Navy . . . can be properly prepared for war."

Since the early 1890s, Fullam had been in the forefront of a movement among naval officers to take Marine guard detachments off the Navy's fighting ships. Fullam and his cohorts especially objected to the use of Marines as ships' policemen, on the grounds that it was an anachronistic holdover from the days of the press gang and was detrimental to the training, discipline, and status of the modern bluejacket.

The Fullamites envisioned a new mission for the Marine Corps within the Navy, once the Corps was freed from its obsolete tasks and was properly organized. The reformers urged that the Marines be formed into permanent battalions and given their own transports, so that they could accompany the fleet either as an expeditionary force or to seize and fortify advance bases. While many Marine officers eagerly embraced the advance base mission, all Marines insisted that the ships' guards be retained. They claimed that service on board warships kept Marines in close day-to-day association with the Navy and provided them with many of the skills needed for expeditionary and advance base duty. By 1908, Fullam's position had gained many adherents among Navy line officers, but Headquarters Marine Corps, with its allies in Congress and the bureaus had defeated repeated efforts to remove the detachments from capital ships.

By mid-1908, naval reform was in the air. The reformers proposed to a sympathetic President Roosevelt the formation of an independent civilian-military commission to study Navy Department reorganization, specifically the breakup of the bureau system. As key instigators of the commission proposal, Fullam, in command of the Navy training station at Newport, and Commander Sims tried to use Sims' influence with the resident to have the Marines removed from ships. Fullam saw success on the Marine question as "an entering wedge" to break the power of the bureaus. "No legislation and no Congressional action are needed," he told Sims, "but it prepares the way for the new gospel that the men and officers who go to sea and make the ship—the Navy—efficient must control the ship."

On 16 September, Sims, in a long memorandum to the president, outlined the case against the Marines. He reviewed the 20-year history of the issue, emphasizing Fullam's arguments that the use of Marines as ships' policemen undermined the discipline and morale of the bluejackets. Sims cited the fact that the Bureau of Navigation had twice recommended the removal of the Marines, but that "General Elliott goes to the Secretary and successfully combats the proposition." Sims urged Roosevelt to cut through this political tangle by using his executive authority to order the Marines off the ships. He stated: "The effect of removing the Marines from the ships would be electrical, because the demand is universal."

Besides Sims, Fullam used a number of other formal and informal channels to reach the president and [the] secretary of the Navy. On 31 August, W. D. Walker, editor of *Army and Navy Life* and a close associate of the naval reformers, urged Roosevelt to remove the Marine guards, employing essentially the same arguments as Fullam and Sims. More important, a close Fullam associate, Commander William R. Shoemaker, in the Bureau of Navigation, convinced the bureau chief, Rear Admiral John E. Pillsbury, to revive the bureau's earlier removal recommendation. On 16 October, Pillsbury wrote to Secretary of the Navy Victor H. Metcalf that "the time has arrived when all marine detach-

"Inspection of Marines of the United States Man-of-War 'Chicago'," wood-engraved plate after Thure de Thulstrup, from *Harper's Weekly*, 15 March 1890.
Prints, Drawings, and Watercolors from the Anne S. K. Brown Military Collection, Brown Digital Repository, Brown University Library

ments should be removed from . . . naval vessels." Secretary Metcalf brought up the proposal at a cabinet meeting, and President Roosevelt approved it. On 23 October, Metcalf formally concurred in Pillsbury's recommendation and directed that it be carried out.

Up to this point, all those involved in making the decision had carefully avoided consulting or informing General Elliott. Elliott, however, had received hints that the Marines' shipboard position again was under attack. Earlier in October, Admiral Pillsbury had issued an order reducing the size of the Marine guard on one of the battleships. Although Elliott had persuaded Metcalf to rescind this order, he realized that the struggle was far from over. On 30 October, he discussed the issue with Sims and stated that he planned to ask Roosevelt directly to "have the pressure stopped." Before Elliott could meet with the president, however, Secretary Metcalf informed the Commandant that the Marines were to come off the ships. Elliott at once counterattacked. After an unsatisfactory meeting with Admiral Pillsbury, Elliott, on 7 November, made a final appeal to Metcalf. He presented the secretary a long memorandum, prepared by his staff, which declared that:

> *The proposed removal of Marines from vessels of the Navy is . . . contrary to the long established and uninterrupted custom of the service, contrary to all precedents and rulings . . . contrary to the wishes of Congress, and is based upon no argument which is cogent or potent.*

Metcalf rejected the Marine plea and informed

the Commandant that the president already had decided on removal. Elliott then requested permission to take his case directly to Roosevelt.

On 9 November, in his meeting with the president, Elliott found Roosevelt sympathetic to the Marines but firmly committed to their removal. In the course of the conversation, Elliott emphasized that many Marine officers viewed abolition of the ships' guards as the "death knell" of the Corps. Roosevelt asked whether Elliott shared this opinion. Candidly, the Commandant replied that he did not. Roosevelt then instructed the general to draw up a statement of the Marine Corps mission once the guards were removed from the ships.

Elliott entrusted the preparation of the proposed order to three officers of his personal staff: Lieutenant Colonel James [E.] Mahoney, Lieutenant Colonel Eli K. Cole, and Major Charles G. Long. All three were Naval Academy graduates who had been closely associated with the emerging advance base mission. Their draft order avoided mention of the ships' guards and provided that Marines were to garrison navy yards and naval stations within and beyond the continental limits of the United States. Marines were to "furnish the first line of . . . mobile defense" for overseas naval stations, and they were to help man the fortifications of such bases. The Corps was to garrison the Panama Canal Zone and furnish other such garrisons and expeditionary forces for duties beyond the seas as necessary. In an enclosure to the memorandum, the three officers recommended organization of the Marine Corps, once the ships' guards were withdrawn, into nine permanent 1,100-man regiments. Elliott and his staff obviously were making a virtue out of necessity by trying to stake a firm claim to the advanced base and expeditionary role, as well as making an expandable expeditionary organization, while conceding the loss of the ships' detachments.

On 12 November, President Roosevelt incorporated the exact wording of Elliott's memorandum in his executive order. The order did not mention ships' guards or call for their removal, although all those concerned understood that to be its intent. During the next several months, the Bureau of Navigation gradually began the removal of the ships' detachments. By early 1909, about 800 of the 2,700 ships' guards had come off.

The immediate reaction to the executive order was predictable. Naval officers generally approved. Upon hearing the news of Roosevelt's decision, Fullam exclaimed: "Hurrah for the President! God Bless him!" and compared the executive order to [Abraham] Lincoln's Emancipation Proclamation.

Marine officers looked upon the executive order with misgivings at best, and most saw it as a first step toward the elimination of their Corps. One Marine officer stated: "The President's order . . . in effect reduces the Marine Corps to the status of watchmen." Rumors circulated in Washington that Marine officers were organizing to lobby Congress for reversal of Roosevelt's decision. Despite the unhappiness among his officers, General Elliott loyally supported the executive order in public, claiming that it would be "the making of the Marine Corps." On 16 November, in response to the reported Marine lobbying efforts, Elliott issued a special order forbidding such activity as "contrary to the motto of the Corps—for 'Semper Fidelis' would be but a meaningless term if it shone only on the sunny side of life or duty."

Even as Elliott publicly looked toward a new role for the Marine Corps within the Navy, Major General Leonard Wood, a confidant of

Marine Guard, War College at Coasters Harbor, Newport, RI, between 1890–1891.
Detroit Publishing Company Photograph Collection, Library of Congress Prints and Photographs Division

Roosevelt and a leading Army progressive, saw the removal of Marines from ships as an opportunity to incorporate the Corps into the Army. Wood and most other senior Army officers were looking for a way to expand the Army's infantry. The Marine Corps had a prominent place in Army proposals for achieving this objective. During 1907, the Army Chief of Staff, Lieutenant General J. Franklin Bell, floated as a trial balloon a plan to transfer the Army's large coast artillery corps to the Navy (and incorporate it in the Marine Corps). This would leave room in the Army for more infantry regiments. Wood, then commanding general, Division of the Philippines, offered as a counterproposal the simple incorporation of the Marines into the Army. Wood, who had a wide circle of acquaintances within the Navy and Marine Corps, respected Marine military efficiency but had gained the impression that the Navy no longer needed the Corps. Late in 1907, he wrote in a letter intended for Roosevelt's eye that the Marine Corps:

> *is an able body, but its desire for enlargement is productive of unrest. A large portion of the navy are in favor of dispensing with Marines on board ship. . . . Their numbers are . . . far in excess of the actual needs of the navy. We need them in the army.*

Neither of these plans had gone beyond the talking stage when Roosevelt's executive order reopened the entire issue of the Marines' future. Wood had just returned to the United States to take over the Department of the East. He was already regarded as the leading candidate to succeed Bell as Army chief of staff. At Roosevelt's invitation, Wood spent several days in mid-November as a houseguest at the Executive Mansion. During this visit, Wood pressed upon Roosevelt this view that the Marines should be incorporated into the Army. He argued that Elliott, through the executive order, was aiming to establish an expanded Marine infantry under the Navy Department. Wood pointed out that the president, under his executive authority, could order the Marines to duty with the Army, as had been done temporarily several times in the past. Having established such a fait accompli, Roosevelt, at a later time, could work out with Congress and the Service departments the legal details of the transfer. Roosevelt was receptive to Wood's proposal. Already irritated with Marine lobbying, he told his military aide, Captain Archibald [W.] Butt, that the Marines "should be absorbed into the Army, and no vestige of their organization should be allowed to remain."

While in Washington, Wood informally discussed his ideas with General Bell and other high-ranking Army officers. He also made an ill-fated overture to two key Marine Corps staff officers, Colonel Frank L. Denny and Lieutenant Colonel Charles L. McCawley. Both officers were well known in Washington social circles, and both had strong political connections. Denny, the son of a prominent Indiana Republican, had many Army acquaintances and nursed ambitions to become Commandant of the Marine Corps. McCawley was the son of a former Commandant and had been the military social aide to Presidents [William] McKinley and [Theodore] Roosevelt. In a chance encounter with the two men on the street in front of the White House, Wood told them that he personally favored transfer of the Marine Corps to the Army and confided that the president was inclined to such a course of action. He asked Denny and McCawley to sound out Marine officer sentiment.

On 23 November, Denny and McCawley told the Commandant, who had just returned to Washington, about the proposed merger with the Army and the president's tentative support for the idea. Much to their surprise, General Elliott angrily denounced such a move. In a letter of protest to General Wood, Elliott claimed that neither he nor the secretary of the Navy had been told of this proposal and declared: "I would as soon believe there was a lost chord in Heaven" as to believe the president, after redefining the Corps' mission, would contemplate separating the Marines from the Navy. Replying to Elliott, Wood reiterated his own support for [an] Army-Marine amalgamation but denied that he spoke for the president.

In a further exchange of letters, Elliott declared that Wood, as an Army general, had no right to discuss disposition of the Marine Corps, which was a separate Service. The Commandant insisted that "the entire Army and Marine Corps, with the exception of the general officers, would be bitterly opposed to such amalgamation." Wood apologized to Roosevelt for bringing his name into the discussion and forwarded all his correspondence on the subject. On 28 November, Roosevelt, in a letter addressed "Dear Leonard," committed himself on the amalgamation issue. He wrote, "You are quite welcome to quote me on that matter. I think the Marines should be incorporated with the Army." Wood, on 2 December, flatly informed Elliott that the president supported the transfer. The entire incident convinced Elliott, who up to now had publicly defended removal of the Marine guards, that he and the Marine Corps were being double-crossed. As he later stated, "While we had been following quietly our duties, elimination and absorption were casting unknown to us their shadows at our heels."

Elliott was among the last to learn about Wood's scheme. Almost as soon as Wood had arrived in Washington, the future of the Marine Corps had become a matter of public and private speculation. Fairly accurate accounts of Wood's proposals and Roosevelt's reaction appeared in newspapers and journals. While few Marines expressed any enthusiasm about going into the Army, many thought such a course of action inevitable as a result of the removal of ships' guards. In an extreme expression of this point of view, one officer declared: "It is imperative that we immediately sever every possible connection with the Navy by transfer to some branch of the Army."

The regular House Naval Affairs Committee hearings on the annual Navy Department appropriation provided the scene for the first political skirmish over both removal of the Marine detachments and the merger of the Marines with the Army. On 9 December, in his testimony, Admiral Pillsbury flatly stated the Navy Department position: "I think that it will be a very great mistake to put them [the Marines] in the Army. We want them in the Navy. We do not want them on board ship." Although the Marine officers, including General Elliott, made no mention of the subject in their public testimony, Elliott informed the committee off the record that he now opposed removal of the ships' detachments. In perhaps the shrewdest maneuver of the hearing, Lieutenant Colonel George E. Richards, assistant paymaster of the Corps, responding to a prearranged question from a committee member, presented a memorandum estimating that it would cost the Navy Department an additional $425,000 to replace Marines with sailors on board ships. At the end of the session, the committee voted to hold supplementary hearings by a subcommittee on the entire Marine issue.

In the period between the conclusion of the full House committee hearings in December and the opening of the subcommittee hearings in January, the Marine Corps and its allies mobilized for the struggle. Marine staff officers prepared several detailed memoranda supporting their position. On 20 December, a group of Marine officers from several East Coast navy yards met privately [in] Boston to discuss "the new status of the Marine Corps." While they publicly denied that their meeting had anything to do with attempts to reverse the president's executive order, few observers believed they met for any other purpose. Sims and Fullam exchanged rumors and warnings about the Marines' organizing and lobbying efforts. The Army question, meanwhile, faded into the background. Although Wood continued to discuss the subject privately, neither he nor Roosevelt took any overt action. They and the War Department were apparently unwilling to directly challenge Navy control of the Marines if the Navy wanted to retain the Corps.

When the subcommittee began its hearings on 9 January 1909, it was obvious that pro-Marine forces were in control. Representative Thomas H. Butler, who presided over most of the sessions, had a son in the Marine Corps and was on the record as opposing Roosevelt's executive order. The clerk of the subcommittee was a former Marine officer. General Elliott and his staff attended almost the entire hearing, and the subcommittee permitted them to cross-examine witnesses. Commander Fullam described the atmosphere of the proceedings: "The Marine colonels were ever present. A stranger

Adrian Lamb, *Theodore Roosevelt*, 1967. Copy of 1908 original by Philiop Alexius de László.
National Portrait Gallery, Smithsonian Institution, gift of the Theodore Roosevelt Association

could not have distinguished them from members of the Committee. They rose at will to exhort, object, and cross-examine." Although one-sided, Fullam's observations were in the main correct. He and the other reformers faced a rigged jury and a hanging judge.

Before the hearings ended on 15 January, a parade of 34 witnesses testified. All of the

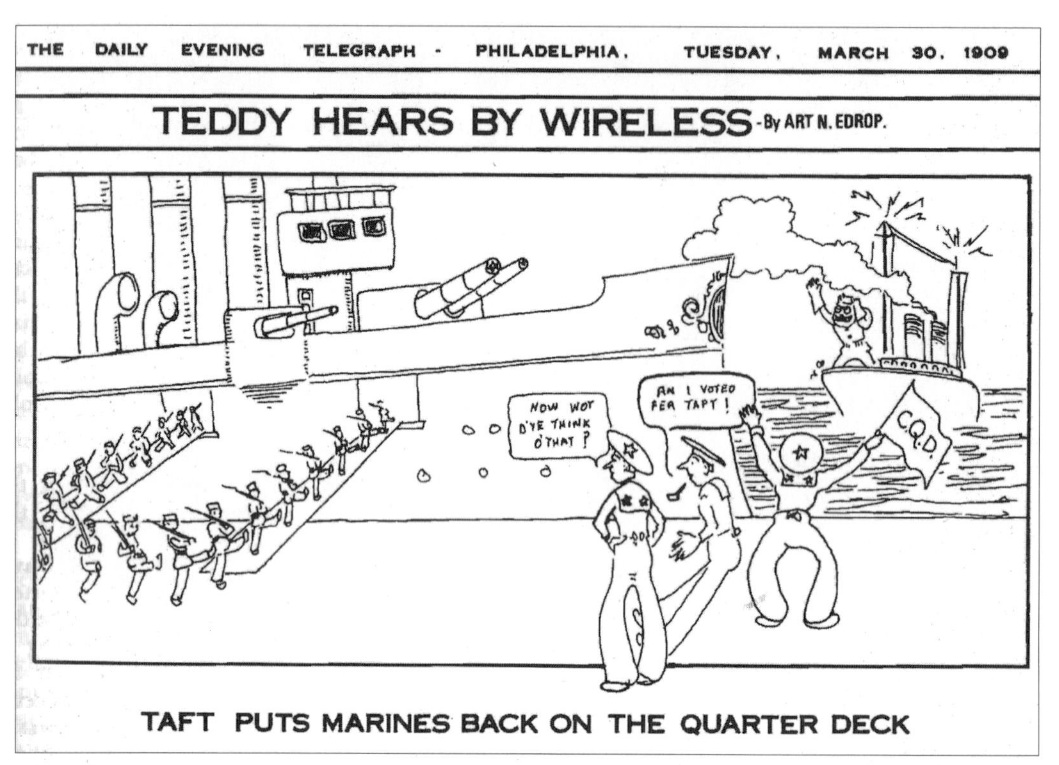

Editorial cartoon by Arthur N. Edrop.
Philadelphia Daily Evening Telegraph, *30 March 1909*

Marines opposed withdrawal of the guard detachments from ships, while the Navy officers split evenly for and against. Both sides reiterated their traditional arguments for and against keeping Marines on warships. Using rudimentary cost-effectiveness analysis, they presented conflicting estimates of the expense involved in replacing Marines with sailors.

While the subcommittee focused on the cost issue, the question of transferring the Marine Corps to the Army was never far from the surface. Several Marine and Navy opponents of the executive order warned that removal of the guard detachments might lead to the Navy losing the Marine Corps, while supporters of the order affirmed their desire to keep the Marines in the Navy. Fullam, for example, declared: "If I were king here tomorrow, I would preserve the Marine Corps . . . as a splendidly organized mobile force, to serve with the Navy." Secretary [Truman H.] Newberry testified that if it were a choice between losing the Marines and putting them back on ship, "I would rather put them back aboard ship." The prospect of absorption of the Marines by the Army was also a stumbling block to congressional supporters of Roosevelt. Representative John W. Weeks wrote to Fullam: "My mind now inclines to leave in the hands of the Executive the question of where the Marines shall serve, but takes a positive stand against action which will tend to amalgamate the Corps with the Army."

When the full Naval Affairs Committee reported the naval appropriation bill to the House on 16 January, it was clear that the Marine point of view had prevailed. The committee recommended insertion in the bill of a provision that:

> *Hereafter officers and enlisted men of the Marine Corps shall serve . . . on board all battleships and armored cruisers . . . in detachments of not less than eight per centum of the strength of the enlisted men of the Navy on said vessels.*

When the appropriation bill came up for consideration before the House, administration forces, assisted by vigorous Navy Department and White House lobbying, turned the tables on the Marines. On 21 January, the House passed the bill without the proposed amendment to keep Marines on board ships.

The fight now shifted to the Senate Naval Affairs Committee, where the Marine Corps could depend on the support of the powerful chairman, Senator Eugene Hale of Maine. Hale, a staunch Roosevelt opponent, was at loggerheads with the president over Navy Department reorganization in general and specifically had come out against taking the Marines off ships. Without bothering to hold hearings on the question of Marine removal, Kale's committee on 10 February reported the appropriation bill to the Senate with numerous amendments, including reinsertion of the House committee's original provision overturning Roosevelt's executive order.

On the Senate floor, the administration made a major effort to defeat the amendment. Massachusetts Senator Henry Cabot Lodge, a personal friend of Roosevelt and long-time supporter of a big Navy, led the fight, liberally supplied with argument and documents by Sims and Fullam. During the Senate debate on 16 and 17 February, Lodge restated the reformers' arguments about the need to restructure the Marine Corps, but significantly disavowed any intention to put the Marines into the Army and stated that he himself would oppose any such effort. Senator Hale, on the other hand, kept hammering at the point that Congress had equal authority with the president over the Navy Department and warned that "the underlying purpose [of removal] is to take these people away from the navy and in the end turn them over to the army." When the amendment came up for final approval on the 17th [of February], it passed by a vote of 51 to 12. This result reflected more personal and political hostility to Roosevelt than conviction about the status of the Marine Corps. Among the supporters of the amendment were most of the Democrats and a strong contingent of conservative Republicans. All of the opponents of the amendment were either Roosevelt loyalists, such as Lodge, or Republican progressives, including William E. Borah and Robert M. LaFollette.

After Senate passage of the entire bill on the 17th [of February], the legislation went to a conference committee headed by Senator Hale and Representative George E. Foss, chairman of the House Naval Affairs Committee. As part of the complex bargaining over dozens of amendments, the House initially refused to accept the Senate provision on the Marines. Roosevelt, however, now was willing to surrender on the Marine issue in order to obtain favorable consideration on the other naval issues. On 18 February, he wrote to Representative Foss: "The bill as it passed the Senate will, as regards this point, do a little damage [but] it does not do very much." Roosevelt made no mention of putting the Marines in the Army and declared that he had issued his executive order "with the explicit object of retaining the marines for the purpose of an expeditionary force." With this signal from the president, the House conferees gave way on the Marine issue. On 1 March, both houses passed the naval appropriation bill with the amendment requiring return of the Marine guards to the ships of the fleet.

During the remaining days of his administration, Roosevelt and Secretary Newberry attempted to find loopholes in the language of the appropriation act, which would permit the president to keep the Marines off the ships. Newberry declared: "I have issued no orders about the return of Marines to the ships and will not do so."

The new president, William Howard Taft, was not about to challenge Congress and immediately took steps to reverse Roosevelt's final measures. As early as 25 January, the president-elect had taken a conciliatory tone, writing to Senator Hale:

> *I intend, so far as possible, to do nothing without full consultation with you managers of the Senate, and while of course it is not expected that we may always agree, it may be asserted that we shall never surprise each other.*

On 5 April, Taft's attorney general, at the Navy Department's request, declared that in his opinion the congressional requirement that Marines make up 8 percent of a ship's crew was constitutional. Very soon thereafter, Marines began marching up the gangplanks of Navy warships, and the controversy was over.

The participants reacted predictably to the outcome. For the Army, it was a case of very little ventured and nothing gained, since Wood's negotiations had been entirely confidential and informal, although quite serious in intent. Some Army officers, nevertheless, believed that "a great opportunity has been lost by the restoration of the Marines to the ships." Navy reformers such as Fullam railed against the decision, denouncing the "parlor and club colonels" of the Marine Corps and grumbling that the entire Navy was "at the mercy of the shore-staying staff and their political friends." More moderate reformers, for example the respected Rear Admiral Stephen B. Luce, founder of the Navy War College, warned that withdrawal of the ships' guards would have led to the "obliteration" of the Marine Corps. Taking Luce's lead, the Navy's General Board in later years would refuse to support the Fullamites in their agitation for removal of the Marine guards on the grounds that such action would lead to the loss of the Corps to the Army. Marines breathed a sigh of relief over what they considered their narrow escape and would cling ever more tenaciously to what was in effect a relatively minor mission. They viewed Fullam and his henchmen with suspicion and often outright hostility and believed they were continually vulnerable to power grabs by ambitious Army and Navy officers. On the occasion of renewed agitation by Fullam in 1913, Major Smedley D. Butler exploded in a letter to his Quaker father, Representative Thomas Butler, who had chaired the special subcommittee in 1909: "I wish somebody would beat the S.O.B. to death. Please try to help us, Father," he pleaded, "for the Lord only knows what will become of our little Corps."

Despite Butler's alone-against-the-world outlook, the Marines in 1908–9 owed their success against Roosevelt's executive order only partially to their own political action. The Marine Corps approached the removal issue with divided councils. General Elliott, obviously influenced by the advance base-oriented members of his informal staff, initially tried to trade acquiescence in the removal of the detachments for a reinforced and expanded Corps designed around the advance base and expeditionary missions. There was much justice in the accusation, made by both Admiral Luce and General Wood, that the Major General Commandant was trying to take advantage of Roosevelt's order to establish an army of his own. Probably a majority of Marine officers in the field, as well as key members of the Headquarters staff, adamantly opposed removal of the guards from the beginning. Still other Marines, typified by Denny and McCawley, simply sought to turn the situation to their own personal advantage and flirted, more or less seriously, with amalgamation into the Army. Whether Elliott was simply swayed by the conflicting currents within the Corps or acting from firm conviction is not entirely clear from the evidence. What is certain is that he swung into active opposition to removal of the Marine guards only after becoming convinced that the president had betrayed him.

President Roosevelt did a great deal to frustrate his own order by, in effect, double-crossing both the Marine Corps and the Navy reformers through his dealings with Wood. Even these factors and the Marine lobbying would not have been enough to reverse Roosevelt's order, had it not been for the general anti-Roosevelt hostility of the conservative Republican Senate leadership and the particular enmity of Senator Hale for all manifestations of naval reform. Taft's retreat from Roosevelt's policy toward the Marines foreshadowed the new President's gradual drift into alliance with the conservative faction of the Republican Party. In the end, then, the ships' detachments owed their salvation at least as much to the cross-purposes of their enemies as to the efforts of their friends. Perhaps a newspaper's amateur poet had the last word:

The guard they stood at attention,
Like they didn't give a damn,
To hear the word of the overlord
The original great I am
And he tells us we ain't wanted,
That the jackies will go it alone.
But I thought I heard an under word
From a power behind the throne.

Join the U.S. Marine Corps – Soldiers of the Sea! U.S. Marine Corps recruitment poster showing a soldier holding a Lewis machine gun and standing in the bow of a small boat, ca. 1914–18.
Library of Congress Prints and Photographs Division

DEFINING THE DUTIES OF THE UNITED STATES MARINE CORPS

by President Theodore Roosevelt
12 November 1908

In accordance with the power vested in me by section 1619, Revised Statutes of the United States, the following duties are assigned to the United States Marine Corps:

(1) To garrison the different navy yards and naval stations, both within and beyond the continental limits of the United States.

(2) To furnish the first line of the mobile defense of naval bases and naval stations beyond the continental limits of the United States.

(3) To man such naval defenses, and to aid in manning, if necessary, such other defenses, as may be erected for the defense of naval bases and naval stations beyond the continental limits of the United States.

(4) To garrison the Isthmian Canal Zone, Panama.

(5) To furnish such garrisons and expeditionary forces for duties beyond the seas as may be necessary in time of peace.

Theodore Roosevelt
THE WHITE HOUSE
12 November 1908
(No. 969)[1]

[1] The original content came from Exec. Order No. 969, 3 C.F.R. (12 November 1908). Minor revisions were made to the text based on current standards for style, grammar, punctuation, and spelling.

MARINE CORPS OFFICERS' PHYSICAL FITNESS

by President Theodore Roosevelt

EXECUTIVE ORDER[1]

1. Officers of the United States Marine Corps, of whatever rank, will be examined physically and undergo the tests herein prescribed at least once in every two years; the time of such examinations to be designated by the Commandant of the Corps so as to interfere as little as possible with their regular duties, and the tests to be carried out in the United States between May 1st and July 1st, as the Commandant of the Corps may direct, and on foreign stations between December 1st and February 1st.

2. All field officers will be required to take a riding test of 90 miles, this distance to be covered in three days. Physical examinations before and after riding, and the riding tests, to be the same as those prescribed for the United States Army by General Orders, No. 79 (paragraph 3), War Department, May 14, 1908.

3. Line officers of the Marine Corps in the grade of captain or lieutenant will be required to walk 50 miles, this distance to be divided into three days, actual marching time, including rests, 20 hours. In battle, time is essential and ground may have to be covered on the run; if these officers are not equal to the average physical strength of their companies the men will be held back, resulting in unnecessary loss of life and probably defeat. Company officers will, therefore, be required, during one of the marching periods, to double-time 200 yards, with a half minute's rest; then 300 yards, with one minute's rest; and then complete the test in a 200 yard dash, making in all 700 yards on the double-time, with one-and-one-half minutes' rest. The physical examinations before and after the tests to

[1] The original content came from Executive Order No. 989, 3 C.F.R. (9 December 1908). Minor revisions were made to the text based on current standards for style, grammar, punctuation, and spelling.

Pvt C. LeRoy Baldridge, *With the Second Division*, ca. 1918.
Art Collection, National Museum of the Marine Corps

be the same as provided for in paragraph 2 of this order.

4. The Commandant of the Marine Corps will be required to make such of the above tests as the Secretary of the Navy shall direct.

5. Field officers of the permanent staff of the Marine Corps who have arrived at an age and rank, which renders it highly improbable that they will ever be assigned to any duty requiring participation in active military operations in the field, may, upon their own application, be excused from the physical test, but not from the physical examination, prescribed above. Such a request, however, if granted, will be regarded by the executive authority as conclusive reason for not selecting the applicant for any future promotion in volunteer rank, or for assignment, selection or promotion to a position involving participation in operations of the line of the Marine Corps, or in competition with officers of the line of the Marine Corps for any position.

Theodore Roosevelt
THE WHITE HOUSE
9 December 1908
(No. 989.)

A PLEA FOR A MISSION AND DOCTRINE

by Major John H. Russell

As information from war-torn Europe gradually drifts across the Atlantic, we learn of the use of new implements of war and the consequent changes to modern tactics. In all of this intelligence, the one point that stands out clearly is the high degree of "efficiency" of the opposing armies of Germany and France. These forces serve as a "standard of efficiency" to which military organizations can and should be trained.[1]

It is therefore but natural that we, of the Marine Corps, should turn to our own organization and compare its efficiency, as we know or believe it to be, with the standard set for us. Such a comparison shows that, while in recent years great strides have been made in improving the efficiency of the Corps, there are some factors that go to make efficiency that have been overlooked or a sufficient amount of stress not laid on them. It is for the purpose of succinctly pointing out these deficiencies and suggesting remedies that this article has been undertaken.

EFFICIENCY

Efficiency is often defined as "the quality of producing results." It is of high or low standard according to the results produced. To reach its maximum all the factors that enter into it must be developed to their maximum and thoroughly harmonized. Then, and only then, can an organization, either public or private, be said to be efficient.

While the necessity for a high degree of efficiency in a private organization is great and is usually stimulated by competition and money greed; in a public organization, especially in a military or naval organization, the necessity for the maximum efficiency becomes peremptory, while the suscitating influences [that] assist the private concern are lost.

To be truly efficient, a military or naval organization must be prepared to place at the command of its government and in the shortest possible time, all its power.

[1] The original article came from Maj John H. Russell, USMC, "A Plea for a Mission and Doctrine," *Marine Corps Gazette* 1, no. 2 (June 1916). Minor revisions were made to the text based on current standards for style, grammar, punctuation, and spelling.

Color drawing by Col Donald L. Dickson.
Art Collection, National Museum of the Marine Corps

The governing factors of such efficiency may be stated as follows:

(a) Organization
(b) Materiel
(c) Personnel
(d) Policy
(e) Leadership
(f) Discipline
(g) Morale
(h) Doctrine

The value of some of these factors is not as great as the value of others, but each and every factor is important. Lacking any one [of these,] the maximum degree of efficiency can never be attained.

It is, accordingly, of the utmost consequence that every military organization carefully develop each factor and include the coordination of all. Such an organization then would become a multiple of the factors or an organic mass. A healthy, sound organization that is capable, in the shortest possible time, of placing all its power behind its blow.

ORGANIZATION

To accomplish the exchange of commodities private business organizations are necessary. The transfer of goods from producer to consumer is thus affected. Formerly, it was the custom for business to create the demand for goods but a scientific investigation of the subject induced, in part, by numerous failures, soon established the general principle that the demand or necessity creates business. This is the only logical assumption and, at the present time, no great business is undertaken without a careful and exhaustive study that clearly demonstrates the necessity for its establishment. Such an investigation conducted along modern lines, ensures as well as can be ensured, a lucrative profit [that] is the final object of all private enterprises. In other words it may be said that "business, like Government, is an evolution and grows out of general economic conditions."

The necessity for a certain undertaking having once been shown the next step is to outline,

John A. Coughlin, *First in France*, ca. 1917.
Art Collection, National Museum of the Marine Corps

in general terms, the "task" to be accomplished. For example, wheat raised in the [Midwest] may ultimately be destined for England or some other nonwheat-producing area but the definite task of the fanner is to raise the largest possible amount of wheat in the most economical manner. His work is then accomplished. The transporting to the mill, the milling, the storing in elevators and the final shipment form separate and complete tasks with which the farmer is only indirectly concerned. The above principle of the division of labor applies, equally well, to nearly every form of human activity.

Public or governmental business, like private business, is created by demand. It is a fact that the final object is not the same, for while in private business it is financial gain, in public business it is social betterment. The underlying principles, however, are the same and the analogy may be carried to many points of similarity in both organization and methods.

As already stated the determination of the task or "mission" is the second step. What is to be accomplished must be clearly and definitely understood by everyone charged with the direction of a business, either public or private. In many cases, especially in public undertakings, the mission can only be stated in very general terms and in the accomplishment of it many "special" or "sub-missions" may be found necessary, but the "general mission" will always be found to stand out clearly above them all. It represents the purpose for which the organization was created and exists and never, for a moment, must it be permitted to become smothered by the introduction of "minor missions." The trail once lost is hard to regain.

Organization may be defined as the act of bringing together related or interdependent parts into one organic whole so that each part is, at once, [an] end and [a] means. In other words the cooperation between the various units must be perfect.

It is generally asserted that the success of certain private undertakings, over others, is due to their more efficient organization. The fact that German business firms have been successful competitors with those of other nations, in all parts of the world, has been stated to be due to their more perfect organization.

The analogy between a great business and a military organization is especially close. Each has its mission, each is divided into various branches or units, which must be separately officered and united into a perfectly disciplined, controlled, and efficient organization. In each case, the organization must be such as will best suit the ful-

U.S. Marines – First to Fight for Democracy. U.S. Marine Corps recruitment poster designed by Leon Alaric Shafer, 1917.
Art Collection, U.S. Navy

This recruitment photograph shows a group of U.S. Marines as the "first to fight" in France during WWI.
American Unofficial Collection of World War I Photographs, compiled 1917–1918 (Record Group 165), Still Picture Records Section, National Archives and Records Administration

fillment of the general mission. This is the prime factor of organization for which all others must be laid aside. Furthermore, it is a fact that a military organization must be perfected in time of "peace" for after "war" has been decided on it will be too late.

The writer believes that the general mission of the Marine Corps is: to cooperate with the Navy in peace and war to the end that in the event of a war the Marine Corps could be of greatest value to the Navy.

But is this the general mission? How many officers of the Marine Corps, if interrogated separately, would give the same answer? What then is our "great work"? No matter how well an organization is organized, if it does not know its mission how can it reach the highest degree of efficiency? It must necessarily lack a concerted action to accomplish its work.

In performing its task the Marine Corps will, naturally, have many special missions presented to it, in fact in years of peace, they are apt to become so numerous that the impression is likely to prevail that such subsidiary work is not at all subsidiary but is, in reality, the master work of the Marine Corps. Such an impression is worse than misleading, it is dangerously false, and if allowed to permeate the Service would result in its failure to properly prepare itself for the real issue and cause it to fight at an enormous and perhaps decisive disadvantage.

It is believed that the general mission of the Marine Corps should be drawn up by a board of Marine officers appointed for that purpose. The result of this board's work to be submitted to a conference of the field officers of the Corps, or as many as might be available, for discussion, amendment, if necessary, and ratification. The conference to be presided over by the Major General Commandant of the Marine Corps. Every officer on entering the Corps would be at once instructed in the mission of the Marine

Corps and commanding officers would preach it to all their subordinates.

PERSONNEL

The importance of this factor is paramount with poor personnel, no matter how well organized and equipped, an organization will, in short order, deteriorate. In fact, in general terms, the efficiency of an organization may be gauged by its personnel.

MATERIEL

This factor depends, to a large extent, on the organization and personnel. If the organization is excellent and the personnel alert to its necessities the materiel should, in a well-governed nation, be brought to a standard equal to or better than a similar organization belonging to any other power.

If, on the other hand, the organization is defective and the personnel of poor quality the materiel is certain to be correspondingly in poor condition and obsolete.

First in the Fight – Always Faithful – Be a U.S. Marine! U.S. Marine Corps recruitment poster designed by James Montgomery Flagg.
Library of Congress Prints and Photographs Division

POLICY

After the organization of a public or private undertaking has been perfected management begins.

The "policy" of an organization may be defined as the system of management necessary to accomplish the mission. It is the conduct of the affairs of the organization. For governmental organizations, to a great extent, policy is governed by regulations but nevertheless a great deal is left and must necessarily be left to commanding officers permitting them to initiate a policy of their own covering their particular commands.

LEADERSHIP

The qualities that go to make a leader of a military organization are: willpower, intelligence, resourcefulness, health, and last, but not least, professional knowledge and training.

It is a mistaken idea that leaders are born and not made. It is true that a certain amount of personal magnetism may be of assistance in the making of a leader, but if an officer cultivates and develops the factors enumerated above, he will necessarily develop into a leader. Of prime importance is a study of psychology and its relation to discipline and morale.

Leadership may be either actual or directive. Actual in the lower grades of the commis-

First to Fight—"Democracy's Vanguard"—U.S. Marine Corps. U.S. Marine Corps recruitment poster designed by Sidney H. Riesenberg, 1917.
Library of Congress Prints and Photographs Division

sioned personnel of a military organization and directive in the higher commands. It is, however, just as important in the one case as the other and the same factors are applicable in each.

While the preparation for leadership must be left to the individual the Marine Corps could materially assist its officers by pointing out the road and by establishing and maintaining schools where officers could receive the best theoretical and practical training.

DISCIPLINE

Years ago, [British Rear Admiral Richard] Kempenfelt wrote: "The men who are the best disciplined, of whatever country they are, will always fight the best."

In some countries, the form of government naturally tends to promote discipline among all classes and the recruit, when called to the "colors," enters the Service already more or less inculcated with the habit of subordination. In other countries, however, where the method of living is more free, the recruit is not as susceptible to discipline and it is for this very reason that discipline in the military and naval organizations of such a nation assumes great importance.

It may be said that the [more] lax the rule, order, method of action, or living in a country the stricter should be the discipline in the military and naval organizations of such a country.

A study of the best method to be employed in obtaining excellent military discipline implies a study of the psychology of suggestion and its application to military life.

The recruit who has matured under certain free conditions of city or country life is suddenly placed in an entirely new atmosphere, and it is to overcome the perhaps bad impressions of such a sudden change of environment and to direct the mental attitude of the recruit along proper lines that psychology must be employed.

The study of this important subject by all commissioned officers of the Marine Corps should be made imperative, a proper course of study being outlined in general orders.

MORALE

The necessity for maintaining the "morale" of an organization at a high pitch, during both peace and war, is well recognized. This subject has been dealt with most thoroughly, in recent years, by students of psychology and in the present European war great attention is being devoted, on all sides, to this important factor.

It would therefore seem proper that special attention should be given by the Marine Corps to this subject, such, for example, as the appoint-

Sidney H. Riesenberg, *Flag Raising, Marines in the Caribbean*, 1913.
Art Collection, National Museum of the Marine Corps

U.S. Marine Corps – Service on Land and Sea. U.S. Marine Corps recruitment poster showing a Marine in dress uniform, marching along a dock, with ship, fort, and city skyline in the background by Sidney H. Riesenberg, 1917.
Library of Congress Prints and Photographs Division

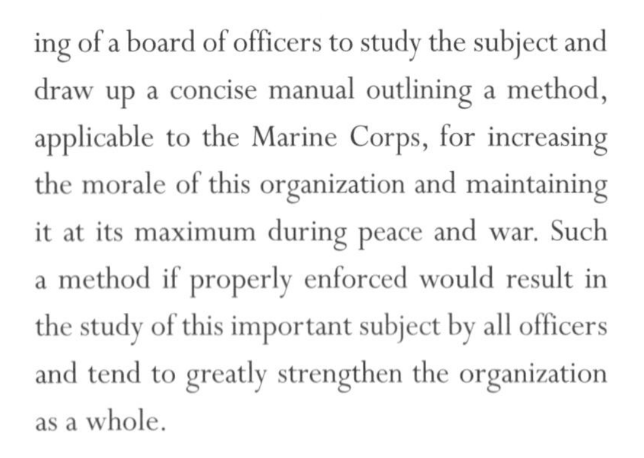

ing of a board of officers to study the subject and draw up a concise manual outlining a method, applicable to the Marine Corps, for increasing the morale of this organization and maintaining it at its maximum during peace and war. Such a method if properly enforced would result in the study of this important subject by all officers and tend to greatly strengthen the organization as a whole.

DOCTRINE

During the past few years, a number of articles that have become classics have been published on the subject of doctrine and its relation to war. The writer, therefore, feels a decided hesitancy in even touching on this subject, but he believes its importance to the Marine Corps to be so vital that he cannot refrain from a general discussion of it in the hope that the seed once sown will quickly germinate and develop into the strong branch of action, and that the day is at hand when the Marine Corps will be indoctrinated.

It is well understood by military men of the present time that the art of war has its theories and its principles; otherwise, it would not be an art. It follows that it also has the application of its principles or doctrine.

The common acceptation of the word doctrine makes it synonymous with principle. This is not true. A principle is a fundamental truth. A military principle is a fundamental truth arrived at by a study of the military history of wars and adapted to the circumstances and characteristics not only of the military organization but of the nation it represents. Napoleon [Bonaparte] aptly said: "The principles of war are those which have directed the great leaders and of which history has transmitted to us the main facts."

The word *doctrine*, as applied to military life, means a teaching that provides for a "mutual understanding" among the commissioned personnel of a military organization. In plain words "teamwork."

Military doctrine is born of military principle. It is the application of principle. A principle cannot be wrong; it is a fact. A doctrine, on the other hand, may be wrong. As it becomes ripened by experience or to suit new conditions, it is altered. It is thus, at first, tentative and gradually built up by a process of evolution.

The historical study from which we derive certain principles is nothing more or less than an estimate of the situation. The principles deduced represent our decision. Having once made a decision, it becomes necessary to put it into execution, in other words, to apply the principles. This is true military doctrine.

In the preparation of a doctrine the general mission of the organization must never be lost sight of. Let the doctrine be clear, concise, and founded on the accomplishment of the general mission in the shortest possible time. With doctrines covering sub-missions, confusion is certain to arise and we would have some officers indoctrinated for one situation and some for another—a grave error.

Such a work as the formulation of a doctrine, however, is not the task for one man but is rather a labor for a general staff, or lacking a general staff for a conference, a reflective body.

All the great powers of the world, except the United States, have instilled into their armies and navies doctrines of war [that] have inspired them with new life.

Without a doctrine, all the drill regulations, all the field service regulations, all the text books are as one writer puts it: "But dead bones and dry rust."

General [Hippolyte] Langlois, one of France's most astute generals and foremost military writers, has well said: "*Sans doctrine, les textes ne sont rien: a des textes sans doctrine, serait beaucoup preferable um.: doctrine sans textes, ce qui etait le cas a l'epoque napoleonienne.*"[2]

General [Aleksey] Kuropatkin, in his book on the Russian campaign in Manchuria, tells us: "Although the same drill books and manuals are used by the whole army, there is considerable variety in the way the tactical instruction is imparted, owing to the diverse views held by the District Commanders."

The first phase of the British campaign in South Africa resulted, as a clever British writer puts it, in "the unforeseen [*sic*] spectacle of a highly trained and well-disciplined regular army, whose armament and equipment were abreast of the requirements of modern war, checked at all points by the levies of two insignificant Republics whose forces were but loose gatherings of armed farmers."

During the period of Frederick the Great [King of Prussia], military forces were maintained in mass formations and maneuvered in combat by commands.

During the Napoleonic age, conditions changed, the rigidity of the mass formation was replaced by open and flexible formations, resulting in a consequent separation of units. This

[2] French translation: "Without doctrine, texts are nothing: texts without doctrine, would be much preferable to doctrine without texts, which was the case in the Napoleonic era."

gain in flexibility and ability to maneuver was obtained only by a corresponding loss of control or command. No longer could one man directly control the entire force. For example, Napoleon had to depend on the ability of his subordinates to interpret the meaning of his orders and instructions. But few of these had been trained in the same school of thought. There existed no common bond to assure a unity of mind and action. A link in the chain of command was missing; there was nothing to unite command and execution.

When that great German student of the art of war, [Helmuth von] Moltke, became chief of staff, he at once started to forge the missing link in the chain of command of the Prussian Army.

The successes of the Prussian campaign in Austria were soon followed by the victories of the Franco-Prussian War and clearly demonstrated the wisdom of Moltke's policy. The doctrineless armies of France lost the war, but thanks to their many able military students and writers, the lessons learned were clearly set forth; and at the present moment, the indoctrinated armies of France are holding at bay the indoctrinated German troops.

Flexibility of command spells "initiative." Initiative may be either reliable or unreliable. The introduction of doctrine means reliable initiative.

Moltke, the great exponent of doctrine, required of detachment commanders "a high degree of technical skill with minds trained to work in unison with that of the higher command, even when separated from Headquarters by a distance which made control impossible."

It was the inculcating of doctrine into the Prussian Army, which permitted the introduction of the "cult" of the offensive, which now permeates the German Army.

Even with the modern systems of communication [that] bind together the various units of an organization the need is as great, if not greater, for a unity of thought and action permitting of a reliable initiative.

The usual illustration for the necessity of a doctrine is that of a number of separate columns advancing on a broad front. Each column commander knows that, on making contact with the enemy, he can boldly take the offensive with the full assurance of the absolute support of the columns to his right and left and the knowledge that their interpretation of the various situations that may arise will be the same as his own.

Consider the well-worn simile of the football team. Let us take two teams, A and B. The first has been indoctrinated; the second has not. When a certain signal is given by the captain of A team, all the members of that team know that the ball is to be kicked, they know that the fullback will fall back, each member of the team on the line knows that he must hold his man at all cost (the strong defensive), the ends know that they must take a strong offensive, break through the opposing line and get down the field as the ball is snapped back.

On the other hand, B team has no doctrine. There exists no mutual understanding as to what is expected of each and every member of the team. The end knows that he should get down the field, but the man next to him does not know it and permits an opponent to block him. The line does not realize the necessity for putting up a strong defensive, and consequently, A team succeeds in breaking through and blocking the kick. On which team would you bet to win?

In this case, the units are in touch with each other. How much more difficult is the situation in the case of a military organization where the units, or some of them, are separated?

Leon Alaric Shafer, *Spirit of 1917 Marine Corps*, 1917.
Art Collection, U.S. Navy

Let us examine, for a moment, our *Field Service Regulations* (1914), the sacred book of every officer.

Under Articles I, II, III, IV, V, and VI, we find at the beginning of each article certain "general principles" to which in most cases many pages are devoted. As a matter of fact, a casual reading of these pages will show that principles, doctrine, instructions, regulations, and customs are all jumbled together in one almost intangible mass, which many officers no doubt take at their heading value—general principles.

Military principles and doctrine should form a creed for every officer, but when we obscure them by mixing them in with numerous regulations, instructions, customs of the Service and other data, they at once lose all force, if they do not become unrecognizable.

Why not cull out the principles and doctrine? Add to them what is deemed necessary, place all in clear and concise language, and make it form the military creed of our officers.

For example, in Article IV, under the heading "General Principles," we find the following: "The march is habitually at route order." This is certainly not a military principle; it is essentially a doctrine. There is a military principle of the conservation of energy. From this principle flows the doctrine: in campaigns the march is habitually at route order.

Other sentences in the above-mentioned article and under the same heading are: "When possible, ample notice is given so that preparations can be made without haste. Troops are informed of the length of halts so that they can take full advantage of the same. The men are kept under arms no longer than necessary, nor required to carry burdens when transportation is available. As a rule troops on the march pay no compliments; individual salutes, etc." All of this and much more in this paragraph consists of neither principles nor doctrine. It is purely administrative.

Again, the first sentence of Article IV reads: "A successful march, whether in peace or war, is one that places the troops at their destination at the proper moment and in the best possible condition." The first part of this doctrine, for doctrine it is, flows from the principle of the economy of forces and the second part from the principle of the conservation of energy.

Under Article VI, [*Field Service Regulations*,] we find under the heading "General Principles" no principles but definitions, administration, instructions, etc. The military principle covering all of these, but which is not stated in the text, is the principle of the conservation of energy. Turning to Article I we likewise find no principles.

The second paragraph of Article V under "Combat," placed in the text in the nature of a comment, reads as follows: "*Decisive results are obtained only by the offensive.* Aggressiveness wins battles. The purely passive defense is adopted only when the mission can be fully accomplished by this method of warfare. In all other cases, if a force be obliged by uncontrollable circumstances to adopt the defensive, it must be considered as a temporary expedient and a change to the offensive with all or part of the forces will be made as soon as conditions warrant such change." The underscoring is not in the text.

If we cut out of this paragraph all except the underscored words, we have a military principle, not stated as such in the text, from which naturally would flow the doctrine of the offensive except when the defensive is adopted as a temporary expedient. As a corollary, we would have, the defensive is a method of creating opportunity for offensive action. In the same article, under the heading "Combat Principles," we

Travel? Adventure? U.S. Marine Corps recruitment poster designed by James Montgomery Flagg, 1917. *Library of Congress Prints and Photographs Division*

find few if any military principles, [but] much doctrine and instructions. For example, "Avoid putting troops into action in driblets" is not a principle; it is pure doctrine. Again, "flank protection is the duty of the commanders of all flank units down to the lowest, whether specifically enjoined in orders or not." This is pure doctrine and cannot in any way be construed as a military principle.

In Article II, the service of security is covered by the military principle that a command protects itself from observation, annoyance, or surprise by an enemy. From this principle springs the doctrine that the "primary duty of an Advance Guard is to insure [*sic*] the safe and uninterrupted advance of the main body." The greater part of the information contained in the paragraphs in this article under the heading "General Principles" are definitions or instructions.

Turning now to Article III. This article deals with the subject of orders, and contained in the paragraphs under the heading "General Principles," we find definitions, information, instructions, but little doctrine and few military principles.

An examination of our *Drill Regulations* (1915) shows a similar condition to prevail. We find, for example, "Combat Principles" for the battalion, regiment, and brigade (pp. 209–18). A careful reading fails to disclose a single principle under these headings.

[For] a military organization to be efficient and powerful, [it] must be so indoctrinated as to acquire a uniformity of mind and action on fundamental military truths. Would not a commander in the field be reassured if he knew that an unsuccessful attack by the enemy would be a signal for a strong counterattack by all parts of the line attacked or that the offensive, once begun, would be carried on by all parts of the line with great vigor until order to cease? All the German military teaching is based on the cult of the offensive. Their teachings say: "It is not even necessary to delay looking for too many advices about the enemy; the time for research is being wasted from the operations; it allows the adversary to do as he pleases and to impose his plan on us when we should impose our plan on him." This is part of the doctrine with which every German officer is indoctrinated. The offensive, in spite of everything, has permeated their very blood and marrow. But to permit of the placing in the hands of subordinates so powerful a weapon as "initiative" the subordinates must one and all be carefully trained to a uniformity of thought and action. It has been well said: "Initiative is a double edged weapon, dangerous to trust in the hands of subordinates who are liable to misconceive the mind of the Chief and are unable to read a situation as he would read it."

We demand initiative of subordinates and yet fail to train them for an intelligent initiative. What then can we expect?

In our *Field Orders*, the first paragraph is the information paragraph. The second contains the "General Plan" and the third the details of the plan, etc. A subordinate officer of an indoctrinated force serving with a detached command receiving the order reads the information paragraph and "understands the train of thought to which the information paragraph has given rise. The information being so and so, naturally, the Commander wishes to do this, therefore, I must do that. Obedience at once becomes intelligent because the purpose of the superior is understood and unconsciously approved."

Colonel (now General) [Ferdinand] Foch in his conference lectures at L'École Supérieure de Guerre, puts it as follows: "An activity of the mind to comprehend the views of the Superior Commander and to enter into his views. An activity of the mind to find the material means of realizing them. An activity of the mind for realizing, in spite of the methods of the adversary, the conserving of freedom of action."

If an organization is doctrineless, a subordinate cannot arrive at an intelligent understanding of orders as now written in the Moltke style. For a doctrineless force, detailed orders are necessary with a consequent absence of initiative and poor results. Since we have gone halfway and adopted the modern system of writing orders, why should we not adopt the modern method of inculcating a doctrine? The one is dependent on the other.

Our *Drill Regulations* tell us that "in extended order the Company is the largest unit to execute movements by prescribed commands or means," and further, "in every disposition of the battalion for combat the orders of the [battalion commander] should give subordinates sufficient information of the enemy, of the position of supporting and neighboring troops, and of the object sought to enable them to conform intelligently to the General Plan."

How can they conform intelligently if they have no military doctrine, no interpretation of the military principles to act as a guide for them? It is as impossible as the command of the famous king that all clocks and watches in his kingdom should keep the same time. He established no method of regulating them and yet he ordered that they must all synchronize.

The mind of the subordinate must be "tuned" by the introduction of doctrine to work in harmony with the mind of the commander.

The Marine Corps has no doctrine and the lack of this important factor must necessarily greatly reduce the efficiency of the Corps. It is possible, some say probable, that the Marine Corps may be called on in the near future to face trained, seasoned, highly disciplined, and indoctrinated troops. Lacking a doctrine, no matter how good our organization, equipment, personnel, discipline and morale, we would unquestionably be badly handicapped, perhaps fatally. We have no creed to bind us together, to help us to understand one another, to guide us to assist one another, to concentrate all our effort; we are as helpless as a ship without a rudder.

The formulation of a doctrine rests with the Marine Corps. It does not require congressional action or outside advice. It would require but slight expense and little effort.

For the purpose of formulating a doctrine, it is suggested that a similar course be employed as to that suggested for determining on the general mission. Field officers of the Marine Corps, or as many as are available, should be assembled, under the direction of the Major General Commandant of the Corps, for a conference. The result of the work of such an experienced reflective body would be a tentative doctrine or creed for the Marine Corps to be preached by every commanding officer and taught to young officers on entry. It would thus soon permeate the very blood and marrow of the commissioned personnel.

Such a doctrine, or at least the results of the first conference, would only be tentative and might require changes in it as we became more experienced, but it would certainly be a start in the right direction and establish a bond of sympathy among the officers of the Corps.

Why should we not, in terse language, lay down certain military principles that we believe are applicable to the Marine Corps? Why should we not formulate a concise and clear doctrine to bind us together? Why should we not formulate our traditions and incorporate them in our doctrine? Why should we not have a cult of the offensive?

Such action would greatly increase the usefulness, efficiency, and prestige of the Marine Corps and tend to unite this organization into one organic whole.

Let us remember the words of General Langlois: "Without doctrine, text books amount to nothing; a doctrine without text books would be much better than text books without doctrine, as was the case in the Napoleonic age."

BARNETT LETTER TO OLIVER

by Major General Commandant George Barnett

HEADQUARTERS
UNITED STATES MARINE CORPS,
COMMANDANT'S OFFICE,
Washington, DC, 9 February 1917[1]

MY DEAR MR. OLIVER: In connection with your telephonic communication with me this date, requesting information as to how young men designated as second lieutenants, graduates of certain colleges, were appointed, I have to state as follows:

When the naval appropriation bill passed on August 29 last, I took the question of filling the vacancies (255 in number) up with the secretary of the Navy. This matter was brought up at a council meeting with the secretary and fully discussed. I proposed at this meeting that I be authorized to fill certain vacancies in the Marine Corps by the appointment of graduates of the military colleges designated by the president in general orders each year as "distinguished colleges." After a full discussion of this matter, the secretary and the whole council decided that, as only a very few graduates of the Naval Academy could be spared, it would be a good thing to fill a reasonable number of the vacancies by the appointment of graduates of these distinguished military colleges. The secretary of the Navy then authorized me to communicate with the presidents of these colleges and to designate not to exceed 60 of the graduates recommended by the presidents of the colleges. From many of the colleges, we received no recommendations whatever, having heard that a great many of their graduates had gone into the Army.

Each graduate authorized to appear for physical examination was required to present his graduating diploma together with a letter of recommendation from the president of the institution, and also numerous letters as to moral character and general standing in the community from which he came. Most of the applications came from the Virginia Military Institute,

[1] The original content came from MajGen Commandant George Barnett letter to Congressman William B. Oliver (D-AL) 9 February 1917, Congressional Record 54. Minor revisions were made to the text based on current standards for style, grammar, punctuation, and spelling.

Louis H. Gebhardt, *Major General Commandant George Barnett, 12th Commandant of the Marine Corps* (1914–20). *Art Collection, National Museum of the Marine Corps*

Lexington, Virginia; the Citadel, Charleston, South Carolina; some from St. John's College, Annapolis, Maryland: and some from Norwich University, Northfield, Vermont. I am appending herewith a complete list of the colleges from which responses were received and also the number of appointments made from the colleges from which recommendations were received.

Some of the young men who appeared for physical examination failed to pass the required test. As I think I stated to you over the telephone, quite a number of the institutions had no recommendations to make; in fact, the majority of the institutions are on this list. In the selection of these young men, no influence whatever was used by any human being. The only recommendations made were made by the presidents of the institutions referred to and the Army officers on duty at said institutions. As I stated to you, I would gladly have taken a great many more of the graduates than I was able to get, because, as stated above, I only secured 39 in toto from these institutions.

Before presenting this question to the Secretary of the Navy, I had heard so much of the good qualities of the Virginia Military Institute that I visited that institution last June and spent several days in going over their curriculum and witnessing drills of all kinds and talking with the superintendent and officers on duty there and with many of the cadets. In fact, this visit was the deciding factor [that] led me to make the proposition to the secretary of the Navy. In selecting any graduates from one of these institutions the state he came from was never considered. Since these young men were designated, we have held examinations all over the United States, at which any young man who made application or made known his desire to appear before the board was allowed to appear; and out of the total number examined (86 in all), only 29 successfully passed for entry into the Marine Corps from civil life. In this connection, I sent over 1,400 letters to young men all over the country who had in any manner requested information as to how he might get a commission in the Marine Corps. This 1,400 included the names of all young men recommended for appointment from any source.

Of the young men who have already been commissioned from these designated military schools, they have been ordered to duty at once in Haiti and Santo Domingo, and a late inspection of the Marine Corps posts in these countries developed the fact that these young men are doing unusually well and their commanding officers speak in the highest terms of them. Candidates from civil life, without such preliminary training at a military school, have to be sent to our school at Norfolk for 18 months before they can be assigned any military duty. Therefore, it may be seen that appointments from these designated colleges give far better returns to the government than would be possible without the military training they have received.

I wish to reiterate here what I stated above, that in the selection or in the attempt to get designations from the different colleges, every college in the United States designated as a "distinguished college" by the president was given no favoritism of any kind, and no influence of any kind by word or letter was ever used or presented by any individual, nor was the slightest attention paid to the section of the country from which these young men came, the only requisite being that they should be graduates of these well-known distinguished colleges; and I wish to unhesitatingly state that in my opinion it is the best possible means of procuring second lieutenants, excepting graduates of Annapolis and West Point.

In accordance with the above procedure, we have secured altogether 39 graduates of these institutions, but so far have been unable to obtain the authorized number, which, as I stated above, is 60.

Thanking you for your interest in this matter, and with kindest regards in this matter, I am,

Sincerely, yours,
GEORGE BARNETT
Major General, Commandant

Hon. WILLIAM B. OLIVER,
House of Representatives, Washington, DC

MILITARY COLLEGES DESIGNATED BY THE WAR DEPARTMENT IN ITS GENERAL ORDER OF 16 JUNE 1916 AS "DISTINGUISHED COLLEGES"

University of California, none
University of Illinois, none
Kansas State Agricultural College, 1
St. John's, Annapolis, Maryland, 3
University of Minnesota, none
University of Missouri, none
Cornell University, none
The Citadel, South Carolina, 20; 1 since killed in action
Agricultural and Mechanical College of Texas, none
University of Vermont and State Agricultural College, none
Virginia Military Institute, 12
Norwich University, Vermont, 3
University of Wisconsin, none

This Marine trio shows early enlisters in the first Women's Reserve during World War I. PFC Marry Kelly (left) of New Jersey was secretary to Col Alfred S. McLemore, who headed the Reserve. PFCs May O'Keefe (center) and Ruth Spike (right) of New York City, the youngest of 305 enlistees, served as messengers for MajGen George Barnett.
Marine Corps History Division

AUTHORIZATION TO ENROLL WOMEN MARINES

by Josephus Daniels, Secretary of the Navy

NAVY DEPARTMENT
WASHINGTON, DC
8 August 1918[1]

To: Major General Commandant, U.S. Marine Corps
Subj: Enrollment of women in the Marine Corps Reserve for clerical duty
Reference: Letter of Major General Commandant, USMC, dated 2 August 1918

1. Referring to [the] letter of the Major General Commandant, USMC, as per above reference and in particular to the statement contained in the second paragraph thereof, that it is thought that about 40 percent of the work at the Headquarters, U.S. Marine Corps, can be performed as well by women as by men; authority is granted to enroll women in the Marine Corps Reserve for clerical duty at Headquarters, U.S. Marine Corps, Washington, DC, and at other Marine Corps offices in the United States where their services may be utilized to replace men who may be qualified for active field service with the understanding that such enrollment shall be gradual.

Josephus Daniels

[1] The original content came from Josephus Daniels, secretary of the Navy, letter to MajGen Commandant George Barnett, "Authorization to Enroll Women Marines," 8 August 1918, Marine Corps History Division, Quantico, VA. Minor revisions were made to the text based on current standards for style, grammar, punctuation, and spelling.

THE GREAT WAR CRUCIBLE

by Brigadier General Edwin H. Simmons
Naval History, 2005

The First World War has been so overshadowed by the second that it seems largely forgotten. Arguably, however, it was the defining event of the twentieth century. Certainly, it was for the U.S. Marine Corps. Before the war, the Marines had been popularly regarded as a kind of colonial era infantry given over to exotic adventures in the Caribbean and Far East. Their role in World War I earned them recognition as a strategically important fighting force.[1]

When the fighting began in 1914, the prospects of the United States entering a European war appeared unlikely. America was a very isolationist country. Europe seemed far away—five days by fast steamer. Indeed, Woodrow Wilson won reelection as president in 1916 largely on the slogan "He Kept Us Out of War."

But events were pushing the United States into the conflict. In early 1917, the Germans resumed unrestricted submarine warfare, and suddenly American ships were being sunk. Then, the British intercepted a cable that indicated that the Germans were trying to spur Mexico into invading the United States. In exchange, Germany promised to help Mexico regain its lost territories in the U.S. Southwest. That notion was not as absurd as it may sound; U.S. forces, including a brigade of Marines, had landed at Veracruz in 1914, and in response to cross-border bandit raids, an Army punitive expedition had entered Mexico in March 1916 and was there until February 1917. Such German provocations prompted Wilson to reverse his stance. At his request, Congress declared war against Germany on 6 April 1917.

Almost immediately, military missions from Britain and France arrived to tell Wilson's government how American manpower should be used. The French suggested sending small U.S. units, perhaps up to regimental size, that could be melded into brigades with veteran French units. The British had an even simpler plan: send American youths to England, where they would be channeled through British regimental depots and used to replenish British battalions.

[1] The original content came from Edwin H. Simmons, "The Great War Crucible," *Naval History* 19, no. 6 (2005): 16–23. Minor revisions were made to the text based on current standards for style, grammar, punctuation, and spelling.

In this digital positive of the original photograph from early September 1918, MajGen John A. Lejeune (center) "plans" with his staff for the employment of the 2d Infantry Division at Saint-Mihiel. *George Eastman Museum, gift of Kodak Pathe, courtesy Charles Chusseau-Flaviens*

To the United States, both [concepts] were out of the question. Wilson wanted a powerful American army for political reasons—so the country would later have more influence in peace negotiations. U.S. Army Major General John J. Pershing, named commander of the American Expeditionary Forces [AEF], was adamant that U.S. troops be deployed under an American command for military reasons.

The mobilization of American manpower and industry for war was remarkably rapid, achieving in days or weeks what today would require months or years. The Army had to expand to 30 times its peacetime strength; it went from 130,000 men at the outbreak of the war to more than four million by the Armistice, 11 November 1918. [Approximately] two million of those soldiers went to France, and a million fought in the Meuse-Argonne, the last great battle of the war.

By comparison, the Marine Corps grew fivefold, from 14,000 to more than 75,000.[2] The less-dramatic expansion carried some advantages for the Marines; it left them with a higher percentage of trained officers and [noncommissioned officers] NCOs, many of whom were veterans of expeditionary service in China, the Philippines, Panama, Nicaragua, Haiti, and Santo Domingo.

In line with the Marine motto, "First to Fight," George Barnett, the Corps' Major General Commandant, insisted that a Marine regiment be in the first convoy to sail for France. Political strings were pulled, and Barnett got his way. A new regiment, the 5th Marines, was hurriedly activated—one battalion at the Philadelphia Navy Yard and two at the new Marine Corps base at Quantico, Virginia, which had been erected at breakneck speed near a fishing village on the Potomac River. Small companies of Marines were brought in from all over the United States and overseas, mainly the Caribbean, and filled to war strength with recruits. The 5th Marines, under command of Colonel Charles A. Doyen, sailed from New York on 14 June 1917 in the first convoy of troops bound for France.

Pershing did not quite know what to do with his Marines. Nominally, they were to be a regiment in the 1st Infantry Division, which was just being formed. At first, they were parceled out by companies as line-of-communications troops, that is, military police, guard units, and port companies. Not until the late summer of 1917 were they brought together as a regiment. By then, another regiment, the 6th Marines, was being formed at Quantico. It would be sent to France battalion by battalion during the fall and winter. Pershing decided that the two regiments

[2] Officially, Marine Corps strength on 6 April 1917 was 462 commissioned officers, 49 warrant officers, and 13,214 enlisted men. At the time of the Armistice, it totaled 63,714. A peak of 75,101 officers and men was reached on 11 December 1918. Maj Edwin N. McClellan, *The United States Marine Corps in the World War* (Washington, DC: Historical Branch, Headquarters Marine Corps 1920), 9, 13.

should be brought together as a brigade in the 2d Infantry Division then being formed. The 4th Brigade (Marines) was activated in October 1917, with Doyen, newly promoted to brigadier general, commanding.

The numbering of brigades and divisions in Pershing's AEF was a very orderly business. These were "square" divisions; that is [to say], each had two infantry brigades of two regiments. Each division also had a brigade of artillery, a regiment of engineers, and many other supporting troops. The square American divisions were very large—28,000 men at full strength, two or three times the size of the average war-worn British, French, or German divisions. General Pershing had two good reasons for their size: first, it gave them staying power; and second, he did not have enough field-grade and general officers to staff a multitude of smaller divisions. The size would prove both an advantage and a disadvantage.

In the 1st Infantry Division, the two infantry brigades were designated as the 1st and 2d Brigades of Infantry. In the 2d Infantry Division, the two brigades were the 3d Brigade of infantry and the 4th Brigade of Marines. The latter was quickly shortened in everyday use to 4th Brigade. There were three kinds of divisions in the AEF: Regular Army, National Guard, and National Army divisions. But that did not mean there were many "Regulars" in the Regular Army divisions nor even a preponderance of National Guardsmen in the National Guard divisions. Increasingly all three types of divisions were filled with draftees. By contrast, the Marine Corps continued to recruit volunteers.[3]

In addition to the two infantry regiments, the 4th Marine Brigade included the 6th Machine Gun Battalion. Eventually, each infantry battalion would have a machine gun company armed with French Hotchkiss heavy machine guns, reliable but clumsy weapons. Previously, the Marines had Lewis light machine guns. The British used these guns throughout the war and liked them, but for mysterious reasons, the Lewis guns were taken away from the Marines. In their place, the leathernecks were issued Chauchat [light machine guns] (or Sho-shos), rather odd-looking and temperamental French automatic rifles. Not until after the Armistice would the Marines be rearmed with superb Browning M1917 water-cooled machine guns and well-regarded Browning M1918 automatic rifles. It was said that Pershing had not wanted these weapons issued prematurely for fear the Germans would copy them.[4]

General Pershing still was intent on building an American army, which he planned to have ready to use in 1919 as the instrument that would win the war. Events caused him to change his mind, but only slightly. The 1917 Russian Revolution led to the collapse of the Russian Army, which in turn enabled the Germans to transfer a large number of seasoned divisions from the eastern front to the western in early 1918. In the calculus of war, this gave the Germans the advantage of having 200 divisions in France—a

[3] Another division that would attract a great deal of early attention was the 42d Infantry Division or "Rainbow Division," nominally a National Guard division, but with men and units from all over the United States. The Rainbow Division made the reputation of Gen Douglas MacArthur. He modestly claimed that it was his idea to form the division. As a colonel, he was the division's chief of staff. Then, with a promotion to brigadier general, the youngest in the U.S. Army, he commanded the 84th Brigade. (The two brigades of infantry in the 42d Division, in accordance with systematic numbering, were the 83d and 84th.) In the closing days of the war, MacArthur briefly commanded the division.

[4] The 5th Brigade, under command of BrigGen Eli K. Cole, arrived in France in September 1918. Assigned to port duties at Brest, it would see no combat. In April 1919, command passed to the colorful BrigGen Smedley D. Butler. In addition to the 4th and 5th Brigades, the Corps also sent a "1st Marine Aviation Force" of four light bomber squadrons to France, which will be the subject of an article in a forthcoming issue of *Naval History*.

With "tin hats" and stripped for action down to light marching order packs, leathernecks of the 55th Company, 5th Marines, form up in a village street for the march to the front. A French *poilu* (WWI infantryman), who has seen it all many times before, watches insouciantly with hands in pockets.
Still Picture Records Section (Record Group 128), National Archives and Records Administration

full 20 more than the Allies. With the divisions came the powerful team of Field Marshal Paul von Hindenburg and General Erich Ludendorff to direct the German effort in the west. They calculated that Germany could win the war—or at least a favorable peace—by launching an offensive against the French and British before the Americans could arrive in sufficient numbers to make a difference.

By early 1918, the 4th Brigade had come together in a training area near Bourmont in eastern France. It numbered almost 10,000 men, the same size as many of the French and British divisions, but it was not yet combat-ready. Pershing was insistent on training in long marches and open warfare; the French were equally insistent that training concentrate on trench warfare. By March 1918, however, the 2d Infantry Division, under command of [Army] Major General Omar Bundy, was ready for some "on the-job" training.

The division deployed to a quiet sector of trenches near Verdun, where the American battalions were paired off with French battalions. In theory, the process was to be gradual. As soon as the Americans were considered combat-ready, the French battalions were to depart and American regiments—and later American brigades—would take over. But it did not turn out quite that way.

During the third week of March, the Germans, in the first of their five 1918 offensives, came close to driving a decisive wedge between the French and British armies near Amiens. In desperation, the British and French agreed to a joint western front command. General Ferdinand Foch became the overall commander, but with limited powers. Pershing agreed to the

In *The 5th Marines at Champagne, France, 1918*, Col John W. Thomason represented his firsthand experience as a captain in the bitter fighting for Blanc Mont.
Art Collection, National Museum of the Marine Corps

temporary assignment of American divisions and regiments to the British and French armies.

The French thinned out their lines near Verdun, and briefly the Marines held a sector vacated by a French division. In early May, Brigadier General Doyen was sent home because of illness—he would be dead in five months—and Pershing gave command of the 4th Brigade to an Army officer, James G. Harbord, who had been a close friend since the two men served together in the U.S. 10th Cavalry. As chief of staff of the AEF, Harbord had jumped from major to brigadier general in a year.

Within a couple of weeks, the brigade was withdrawn from the front and moved with the rest of the 2d Division to an assembly area near Paris. Most of these long-distance moves were carried out using the French military railway system, which was much ridiculed by the Marines for its dinky boxcars, the "40-and-8s" that could carry 40 men or eight horses. The division had very few trucks of its own for motor marches. The French furnished trucks (*camions*). Many of the drivers were "Annamites."[5] A later generation of Marines would know them as Vietnamese.

By this time the German onslaught seemed to have halted, but on 27 May, the Germans jolted the Allies with their third big 1918 offensive, breaking through the French lines on the Cham-

[5] Term comes from the mountain range across Vietnam and Laos.

Capt Harvey Dunn, one of eight artists commissioned by the Army in 1917, captured this high point of the German advance on Paris, as the Germans came through wheat fields from Lucy-le-Bocage toward Mares Farm on 4 June 1918.
Art Collection, National Museum of the Marine Corps

pagne front. This sector had been quiet since a failed French attack the previous year that had been so disastrous much of the French Army mutinied. After 1917, the French Army, now under General Philippe Pétain, was reliable only for defense. Even then, the French had authorized a so-called "flexible defense"—essentially permitting their commanders to give up ground at their discretion.

For the Germans, their success on the Champagne front was unexpected. They drove toward the crossing of the Marne River at Château-Thierry, which, in turn, opened a route to Paris, only 40 miles away. On Memorial Day, 30 May, the 2d Division prepared to move to the front. It arrived west of Château-Thierry on 1 June, posting itself astride the Paris-Metz road. The Germans, meanwhile, poured through a four-kilometer gap torn in the French lines to their north. From 1 to 5 June, the Marine brigade fought a defensive battle. On 6 June, the Marines attacked. Their principal objective was a small wooded area about a kilometer and a half wide and three kilometers long called the Bois de Belleau, which was the hunting preserve of the comte de Belleau. The battle for Belleau Wood lasted almost until the end of June. Nearly 90 years later, the forest remains an eerie place, seemingly filled with ghosts.

The battle was not well fought. It was a confused crisscrossing of battalions and companies stumbling blindly through gas-choked woods and suffering horrendous losses from German machines guns and field artillery. The Marines lost almost half their men, but they beat the best the Germans had to offer.

After Belleau Wood, the battered Marine brigade had only two weeks to prepare for the next major battle. During the brief interlude, Pershing sacked Major General Bundy, the 2d Division commander, and gave command of the division along with a second star to James Harbord. Colonel Wendell Neville, commander of the 5th Marines, got the Marine brigade and a star.

The Germans launched their fifth and final offensive—the last throw of the dice—on Bastille Day, 14 July 1918. Anticipating the attack, General Foch had already prepared a counteroffensive. Attacking southwest of Soissons, the U.S. 1st and 2d Divisions and the French 1st Moroccan Division would be the spearhead. For the Marine brigade, it was a two-day battle. On 18 July, the 5th Marines, led by Lieutenant Colonel Logan Feland, attacked, coming out of the Forêt de Retz. The next day, the 6th Marines, commanded by Lieutenant Colonel Harry Lee, passed through the 5th [Brigade] and continued the assault. It was a very violent battle, more violent than Belleau Wood had been. The casualties were half those of Belleau Wood, but they were incurred in just two days of fighting. By comparison, however, it was a well-fought attack, heavily supported by French, as well as American, artillery; by French tanks; and by French aircraft. French cavalry waited to gallop through any gap in the German lines in the best Napoleonic style. Such a gap did not open, but after Soissons, the Germans never again mounted an offensive.

By the summer of 1918, the Germans were short of infantry, but they had enormous numbers of machine guns and quick-firing field guns. Moreover, half of the shells the German artillery was firing contained deadly gas, mostly mustard gas. Attacks had to be carefully planned with very fixed schedules. Most artillery fire was prearranged from map data; very little was controlled by forward observers. Unfortunately, these carefully planned attacks almost always broke down into kind of a shapeless melee.

"Devil Dog" is a motivational nickname for a U.S. Marine. According to lore, it is based on the alleged use of "teufel hunde" by German soldiers to describe Marines fighting in WWI. German historians dispute this claim, as the correct translation of "devil dog" would be "Höllenhunde."
Art Collection, National Museum of the Marine Corps

After Soissons, the 2d Division went into a rest area near Nancy to rebuild and refit. Marine Brigadier General John A. Lejeune had arrived in France at the end of June. He paid a call on Pershing, telling him he was authorized to propose the build-up of the Marine brigade to a

S. J. Wolff's portrait catches the indifference, intelligence, and war weariness of an unidentified Marine of the 4th Brigade.
Art Collection, National Museum of the Marine Corps

division. Another brigade and an artillery regiment were in training at Quantico, the command of which Lejeune had just left. Pershing said he would accept another Marine brigade, but he did not want the artillery regiment. As for deploying a Marine division, he reported to the secretary of war: "While Marines are splendid troops, their use as a separate division is inadvisable." In fairness to Pershing, he knew that forming a Marine division would require pulling the Marine brigade out of the 2d Division, where it was functioning very well. Moreover, a Marine division could not be combat-ready until 1919.

Lejeune was initially assigned temporary command of the 32d Division's 64th Brigade. Cross-Service assignments were not unusual. Harbord, an Army brigadier, had commanded the Marine brigade, and numerous Marine field-grade officers had been assigned to the Army for duty; at least two commanded Army regiments. A number of new Army second lieutenants—90-day wonders—were detailed to duty with the Marines.[6] Quantico also had been turning out second lieutenants, but by 1918, the brigade preferred to commission its own lieutenants from its sergeants, many of whom were then sent to Army schools in France.

Lejeune, after a short stint with the 64th Brigade, was transferred to the 2d Division. He briefly led the Marine brigade, and then Harbord left to take over the troubled Services of Supply.[7] With a promotion to major general, Lejeune became the commander of the 2d Infantry Division in late July, the first Marine to command a division.

Assistant secretary of the Navy Franklin D. Roosevelt visited the brigade at Nancy. Back in January, General Pershing had ordered the leathernecks out of the green Marine uniforms they had worn to France and into Army olive-drab uniforms. As a mark of distinction, Roosevelt now authorized enlisted Marines to wear Marine Corps collar emblems, [which were] until then, an officer's privilege. The emblems took the form of round disks with embossed eagle, globe, and anchor. To further distinguish themselves, some Marines also had eagle, globe, and anchor devices affixed to the front of their British-style helmets.

During the first week in August, the 2d Division moved to the Marbache sector, a quiet 12-mile stretch of front south of the southern tip of what was called the Saint-Mihiel salient, a deep

[6] The term *90-day wonders* comes from the reference for how long it takes lower ranking officers to be turned into lieutenants at Fort Benning, GA.

[7] Harbord had become a great favorite of the Marines. A portrait of him hangs in the ballroom of the Army-Navy Club in Washington, DC, paid for by subscription by his Marine officers.

The Last Night of the War, by Frederick Yohn in 1920, portrays the Marines' crossing of the Meuse River, a senseless and costly attack ordered by the high command in spite of the fact that the Armistice was imminent. *Art Collection, U.S. Navy*

penetration the Germans had held since 1915. The Marines had a rather pleasant time there, but something big was coming; Pershing was ready to take command of an American army in the field. He had organized his best divisions into two corps, and their objective would be the reduction of the Saint-Mihiel salient. The attack began on 12 September. Lejeune's 2d Division went into action the next day, and by the 15th [of September], the Marine brigade had met its objectives. It was an easy victory for Pershing's army, helped considerably by the fact that the Germans had already begun to withdraw from the sector.

All along the western front, the Germans were falling back to a line of prepared positions that the Allies called the Hindenburg Line. Foch, now a full-fledged marshal, ordered a general offensive in which the AEF was given the Meuse-Argonne sector. Americans were coming to France at the rate of a quarter-million men a month. In the haste to build up rifle strength, the new divisions were arriving without their artillery or combat support units. With his most experienced troops still at Saint-Mihiel, Pershing had to begin the Meuse-Argonne offensive with half-trained divisions and soon found that commanding both the AEF and a giant field army was too much for his headquarters. The field army was divided into the First, Second, and eventually Third Armies. In effect, Pershing had become an army group commander.

The 2d Division, meanwhile, was detached for service with the French Fourth Army. By the end of September, the French had been stopped near Somme-Py in the Champagne sector. The key terrain there was Blanc Mont—the "White Mountain" [was] a low ridgeline held by the Germans since 1914. The Fourth Army's commander, General Henri Gouraud, wanted to break up the 2d Division into brigades or regiments to support his weakened French divisions—or so he said. Or perhaps he tricked Lejeune into saying the 2d Division could take Blanc Mont if it remained intact. The 2d [Division's] attack began on 3 October. The Marine brigade made a frontal assault against the ominous ridgeline; the 3d Infantry Brigade came up on the right flank. It was costly, but it worked. By 6 October, the 2d Division had taken the ridge and the Marines had moved on to the village of Saint-Étienne.

After Blanc Mont, the 2d Division returned to the U.S. First Army and was assigned to the V Corps, commanded by Major General Charles P. Summerall. Lejeune got along well with most of his Army peers and seniors; he was a graduate of the Army War College, and many of the ranking generals were his classmates and friends. Summerall, however, was the exception.

The 2d Division was given a narrow front, just two kilometers wide, and the mission of driving a wedge into the German lines. By now, the division knew its business well. It attacked on 1 November, with the Marine brigade out front. The offensive went like clockwork, and the Germans retired behind the Meuse River.

Elsewhere, the German Army was in full retreat, but in front of the Americans it exhibited a streak of stubbornness. And then Lejeune was ordered to make a night crossing of the Meuse. He protested; everyone knew that an armistice was imminent. Summerall nevertheless ordered that the attack be carried out on the night of 10 November. Footbridges were thrown across the Meuse, and the crossing was made under heavy German fire. The Armistice came at 1100 the next day. Some Marines were still fighting at 1400, and a patient German officer had to tell them that the war was over. The Marines blamed the costly attack on Summerall, but it was not really his fault. Pershing had ordered his field armies to capture the best possible defensive positions in case the Germans continued to fight after the scheduled Armistice. And Pershing, in turn, had received his orders from Foch.

So the war ended, and the Marine brigade marched into Germany as part of the Army of Occupation.[8] During the war, the strength of the Corps had grown to just over 75,000, about 32,000 of whom served in France. Casualties there, nearly all of them in the 4th Marine Brigade, totaled 11,366. Of these, 2,459 were killed or missing in action.[9] Only 25 Marines were taken prisoner. The Corps had made its mark.

[8] John A. Lejeune would succeed Barnett as Major General Commandant in 1920. He was enormously proud of the Marine brigade and the 2d Infantry Division and their record in France. But he and some other thinkers saw that the future of the Corps did not lie in being simply a reinforcement for the Army. They foresaw that the next war would be against Japan in the Pacific and that there would be an amphibious role for the Marines. Charles Summerall became chief of staff of the Army in 1926 and after retirement became president of the Citadel. In a curious parallel, Lejeune, after his retirement, became superintendent of Virginia Military Institute.

[9] Officially, Marine Corps deaths were 1,465 killed in action, 991 died of wounds, 27 died from accidents, 269 died of disease, and 12 died of other causes, for a total of 2,764. McClellan, *The United States Marines in the World War*, 65.

THROUGH THE WHEAT TO THE BEACHES BEYOND

The Lasting Impact of the Battle for Belleau Wood

by General Charles C. Krulak
Marine Corps Gazette, 1998

"It was a hell of a mess . . ."

~General Gerald C. Thomas,
Assistant Commandant of the Marine Corps,
reflecting on his experience as a sergeant during the fight for Belleau Wood, 6–26 June 1918

On the 31st of May 1998, I delivered the Memorial Day address at the American Cemetery at Belleau Wood, France. It has become a tradition for the Commandant to visit this historic battlefield on Memorial Day to join with Marines from all over Europe, veterans groups, and the French people to pay tribute to the Marines who sacrificed their lives in the epic battle that raged for over 20 days in June of 1918. Of all the traditions associated with the commandancy, this is one of my favorites. It certainly causes me to think deeply about the legacy of the Corps and, equally important, our preparations for the future.[1]

The battle for Belleau Wood, and the exploits of the 4th Brigade during the First World War, have fascinated me since I was a child. In my formative years, I met and was influenced by Marines such as Clifton [B.] Cates, Lemuel [C.] Shepherd, [Gerald C.] Thomas, and my godfather [Holland] M. ("Howling Mad") Smith. Their reputation as leaders, innovators, and tacticians is legendary. In the 1920s and 1930s, they played a pivotal role in transforming the Corps from what it had become during the First World War—a second land army—into the world's finest amphibious power projection force. In the 1940s and 1950s, these men planned and led amphibious assaults on Guadalcanal, Tarawa, Guam, Iwo Jima, Okinawa, Inchon, and many others. Over the course of their careers, they tenaciously strove to ensure that the Corps would be the nation's force in readiness—the air-ground striking force that was most ready

[1] The original content came from Charles C. Krulak, "Through the Wheat to the Beaches Beyond: The Lasting Impact of the Battle for Belleau Wood," *Marine Corps Gazette* 82, no. 7 (1998): 12–17. Minor revisions were made to the text based on current standards for style, grammar, punctuation, and spelling.

United States Marines in France during World War I.
Official U.S. Marine Corps photo (Record Group 128 G), Still Picture Records Section, National Archives and Records Administration

when the nation was least ready. All of them had one thing in common. They all participated in the battle for Belleau Wood. Throughout their careers, in every decision they made, their experience in the assault and conquest of Belleau Wood factored very heavily. In fact, I believe that the fight for Belleau Wood was the birthplace of the modern day Marine Corps. Let me explain why.

The First World War had been raging for four years by the time the Marines got into the fight. The face of warfare had changed dramatically over those years. In weaponry alone, the rate of technological advance was staggering. Innovations such as large caliber, high velocity artillery, machine guns, poison gas, and aircraft had exponentially increased the tempo and lethality of the battlefield. Even though the Corps began to prepare in earnest for combat in Europe in 1917, we were too late—we had not kept up with the technological and tactical advances unfolding in the World War. As a result, in June of 1918, the Corps found itself on a futuristic battlefield it had not prepared for, one that it did not anticipate, and the Marines who fought there paid the price in blood. Those who survived never forgot, and to a man they vowed never again—never again. The story of the battle for Belleau Wood is well known to all Marines. But to prepare for my Memorial Day speech, I researched the oral histories of the Marines who fought there who eventually went on to become the future leaders of the Corps. I read the after action reports of the division, and the regimental and battalion commanders. Then I reread three books, [John W.] Thomason's *Fix Bayonets*, [Elton E.] Mackin's *Suddenly We Didn't Want to Die*, and [Robert B.] Asprey's *At Belleau*

Barry Faulkner painted this large decorative map of Belleau Wood from actual air maps and documents gathered by the Marine Corps for their official records. It was designed to hang above the fireplace in the Memorial Room to Capt Phillips Brooks Robinson, designed by Murphy and Dana, Architects, at Quantico, VA.
Official U.S. Marine Corps image

Wood. In the process, I gained a new appreciation of the importance of this battle in the transformation of the Corps in the 1920s and 1930s. I gained additional perspective on why the veterans of this battle fought so tenaciously for organic air and artillery for the Corps. It also became very clear to me why they could see the incredible potential in amphibious assault when all the self-proclaimed "experts" considered it futile in light of the 1915 debacle at Gallipoli.

Sketch of typical uniforms and equipment of Marines on the western front by Col Donald L. Dickson.
Art Collection, National Museum of the Marine Corps

This year, after the Memorial Day speech, the Sergeant Major of the Marine Corps and I walked the battlefield and retraced the battle as fought by the 4th Brigade. Today, the battlefield looks much as it did on the 6th of June 1918—the day the Marines initiated their attack. It was easy to see how the Germans were so successful in turning Belleau Wood into a natural fortress. The forest is surrounded on all sides by wide-open, and relatively flat, wheat fields. The foliage in the forest is incredibly thick, making it difficult to see more than 50 meters in any direction. Glaciers deposited huge slabs of rock in such a way that they form superb natural pillboxes. The remains of the old trenches and fighting positions give credence to the veterans' accounts of how the Germans took full advantage of what Mother Nature had given them. The after action reports describe how the Germans had positioned observation balloons on the ridge to the north of the town of Belleau, registered their artillery on the approaches to the forest, and emplaced machine gun and trench mortar positions so that every square inch of the wheat fields and the forest was covered by murderous interlocking fires.

Sergeant Major Lee and I started off in the wheat fields where the 4th Brigade would lose over 1,000 men on the first day of the attack. In fact, more Marines were lost on the 6th of June in 1918 than in the previous 142 years of Marine Corps history. The senior commanders wanted to achieve an element of surprise in the attack, so they only gave the forest a short preliminary artillery barrage, followed by a rolling barrage to support the attack. Neither proved adequate. After the initial barrage, the Germans quickly remanned their machine gun positions and waited for the Marines. As soon as the assault began, it became readily apparent that some of the units had not made it to their assigned positions in time for the attack. Those that did immediately came under a withering barrage of artillery, mortar, and machine gun fire. Huge gaps opened up in the attack formations as the German gunners mowed the Marines down. One veteran described it, "as if a huge scythe had been swept across the field at boot-top height." The attack faltered, and many Marines went to ground seeking cover. Luckily, combat veterans from the Corps' numerous small wars, such as Gunnery Sergeant [Daniel J.] Daly, stood tall and rallied the young Marines to rise up and press on with the attack. Throughout their 800-yard assault, the Marines were raked by machine gun fire and high explosive artillery. The wheat field still bears witness to the carnage that raged there 80 years ago. As Sergeant Major Lee and I walked that field, I found a freshly plowed sector. I picked up a handful of soil and found several pieces of shrapnel. I picked up another handful and found even more. After all those years, long after the flesh and blood disappeared, the soil tells the story of the price the 4th Brigade paid crossing that now-hallowed ground.

Once they got into the forest, the Marines' problems intensified. The foliage was so thick that units became disoriented. Some of the officers' land navigation skills broke down and entire units collided. Units started to report that they had reached objectives that, in fact, they were nowhere close to. Artillery strikes were called in on the wrong coordinates. To top it off, the Marines had to rely on runners to pass the word because the communication wires were continually broken by enemy artillery. It was a command and control nightmare. As a result, the fighting degraded to small unit actions. But, it was in these small unit actions that the 4th Brigade distinguished itself. In fact, the Marines' initiative, tenacity, and endurance surprised the German defenders. Those three factors proved critical in the 4th Brigade's ultimate victory.

The high intensity and duration of the combat taxed the logistics system beyond its abilities to cope. Compounding the problem was the German local air superiority. To sustain the Marines in Belleau Wood, the resupply effort had to cross exposed wheat fields. Every time the Marines attempted to move supplies across those fields, they came under observation of the German balloons to the north of Belleau. Shortly thereafter, they would be engaged by intense artillery and machine gun fire. The Marine commanders asked the French for air support to shoot down the balloons, but the French could not spare enough sorties to get the job done. As one commander reported, "A number of Ger-

Frank E. Schoonover, *Marines at Belleau Wood*, 1919.
Art Collection, National Museum of the Marine Corps

man planes over this morning, and they have been busy all day. It is almost impossible to make a move in this area without coming under the eye of a balloon observer. Our aviation is either passive or non-existent." Throughout the battle, the balloons continued to wreak havoc on the 4th Brigade. The Marines in the forest were reduced to scavenging food, water, weapons, and ammunition off dead Germans and fellow Marines to continue the fight. Wounded Marines had little hope of evacuation. In fact, many of the Marines who died in Belleau Wood fought until they succumbed from their second or third wound.

Sergeant Gerald Thomas was right—the fight for Belleau Wood was a mess. Yet the Marines carried the day. The cost was staggering. Of the 8,000 Marines who participated in the battle, over 4,700 were casualties: 1,035 were killed. Unfortunately, many who survived the fight for Belleau Wood would lose their lives in the following four months of combat.

When the 4th Brigade returned home after the war, the nation demobilized, and, as is our nation's tradition, we shifted money and resources away from the military. While the money may have been in short supply, ideas certainly were not. In many ways, the 1920s and 1930s were some of the most productive years for innovative military thought in American history. The Marine Corps led the way.

In the 1920s, Army and Navy planners looked to the Pacific and saw a growing threat from Japan. In response, they developed a plan by which the United States would fight its way across the Pacific to defeat the air, land, and sea forces of the empire of Japan. The plan hinged on access to key islands for refueling, resupply, and air and sea control of sea lanes. Most of the planners believed that Japan would attempt to deny us access to these islands by invading them and then turning them into defensive fortresses. War Plan Orange encouraged visionaries such as Lejeune and [Earl H.] Ellis, both veterans of land combat in France, to foresee a new mission for the Marine Corps—amphibious assault. They knew that if war came in the Pacific, the Marine Corps would be called upon to take those islands from the Japanese defenders. They also realized how difficult it would be to create a Marine Corps capable of doing it. To attack a fortified island in the Pacific, one surrounded by coral reefs, posed seemingly insurmountable problems. But, the Corps' young officers, many of whom were veterans of the battle for Belleau Wood, believed it could be done. After all, to them Belleau Wood was much like an island surrounded by wheat instead of water. They felt that if they could rectify the problems associated with the attack on that forest, they could build the amphibious force General Lejeune envisioned.

TRANSFORMING THE CORPS—AMPHIBIOUS ASSAULT

They started at Newport and Quantico—the school houses. In 1920, Major H. M. Smith, who many consider the father of amphibious warfare in the Corps, was sent to the Naval War College, where he challenged the status quo, relentlessly advocating the employment of Marines in an amphibious strike role. At Quantico, the Corps began to restructure the Field Officers Course to stress amphibious operations. In 1924, the curriculum featured only two hours of classes on amphibious operations. By 1927, that time had grown to 100 hours, and by the end of 1939, more than 155 hours. Students such as Smith, Cates, Shepherd, Thomas, [Graves B.] Erskine, [Keller E.] Rockey, and [Roy S.] Geiger studied amphibious operation ranging from Alexander [the Great] at Tyre to [Sir Ian] Hamilton at Gallipoli. They wargamed the amphibious assaults they believed necessary for War Plan Orange to succeed. These wargames were used to model the fleet landing exercises that experimented with new amphibious assault doctrine, tactics, and equipment. During the course of 20 years, they not only discovered where the problem areas were and what to do about them, but also uncovered new opportunities resident in projecting Marine combat power from the sea.

The students looked at amphibious assault force's exposed transit from ship-to-shore as they did the wheat fields surrounding Belleau Wood. They knew that they had to make the transit as quickly as possible, transported in landing craft that offered protection from enemy fires. Additionally, they needed a landing craft that would not get hung up on the coral reefs associated with the Pacific Islands, especially Guam, which was a critical objective in War Plan Orange. These requirements, and 20 years of experimentation and study, led to the development and procurement of the Higgins boat and the amphibious tractor, two innovations that proved critical to victory during the Second World War.

The veterans of the fight for Belleau Wood all remembered the poor state of logistics resupply during the battle. German air and artillery made it almost impossible to resupply the Marines inside the forest. If the enemy had the opportunity to destroy the logistics resupply during an amphibious operations ship-to-shore phase—the Marines could be thrown back into the sea. They knew that they needed to build up combat power ashore as quickly as possible. Through wargaming and experimentation they discovered that they had to load the ships in such as way that they could be unloaded quickly and, more importantly, in the sequence needed by the landing force. They drafted requirements for ships that could proceed directly to the beachhead and then unload heavy vehicles and supplies. They experimented with several types of landing craft to transport materials ashore. They looked at tracked armored vehicles that could take the materials in the heat of battle directly from the ships to objectives well inland. They believed that an opposed amphibious assault on a Pacific island would consume supplies at a very high rate, and they were correct. Luckily they developed a robust sea-based logistics system tailor-made to support amphibious operations. In so doing they revolutionized the art of war.

The Marines knew that they would need massive preinvasion fire support from the Navy. They knew what is was like to storm a well-defended fortress without it. As such, they strove to convince the Navy to rethink how they would use naval gunfire to support an amphibious assault. They brought Navy officers, such as Lieutenant Walter [C. W.] Ansel, onto their team to help. In the course of their experimentation and wargaming, they discovered that naval gunfire could not address all of their fire support needs. Thus, the Marines looked to organic aviation to provide the overhead observation, fire support, and protection they needed during the preparation and amphibious assault phases. Once phased ashore, they thought they could use a combination of air, naval gunfire, and organic artillery to provide the firepower superiority they needed to consolidate the objective.

Reading the oral histories of the veterans of the battle, I found each of them talked about the effect the German observation balloons had on the fight. To a man, they were furious that they remained in place, unmolested by allied air. This is why these infantry officers proved to be such vocal and passionate supporters of Marine aviation in the interwar years. If the Corps had its own aircraft as part of its amphibious force, Marines would never again be put in that position. They believed the Marine commander on the scene should be able to establish the priorities for air support, not some detached headquarters. Amphibious landings and the subsequent dynamic land campaigns that followed, did not lend themselves to preplanned and scripted air support plans. The most critical phase of the landing operation was the initial assault and breakout from the beachhead. With no artillery ashore, the Marines needed their aircraft to provide highly responsive fire support. This requirement led to the development of the Marine air ground, combined arms philosophy.

From the First World War on, the Marine Corps would fight as an integrated air-ground team. Throughout the interwar years they experimented with Marine aviation in support of expeditionary operations. Wherever the grunts went, the air went too. When the team was assembled to crate the *Tentative Manual for Landing Operations* [1935], the "bible" of amphibious assault doctrine, the room was filled with ground and air Marines. They developed it as a team—a combined-arms team.

On 19 February 1945, V Corps, a Marine amphibious assault force three divisions strong, landed on the black sand of the Japanese island fortress known as Iwo Jima. The landing and the 36-day battle that followed proved to be one of the most significant feats in the annals of military history. It was the epitome of amphibious excellence. Fittingly, V Corps was commanded by a Marine who probably did more than anyone else to create the Corps' amphibious assault

Built of brick and masonry and often octagonal in shape, shelters for hunting parties were a common feature in many French forests. The remains of this hunting lodge lie in the northwest corner of Belleau Wood on a hillside behind the cemetery chapel, as seen here in *Bois de Belleau*, by Jean F. Boucher, ca. 1918.
Art Collection, National Museum of the Marine Corps

capability, Lieutenant General H. M. Smith. His three divisions were commanded by Clifton Cates, Graves Erskine, and Keller Rockey, all of whom fought as junior officers in the battle for Belleau Wood.

THE LEGACY OF BELLEAU WOOD

Perhaps the most enduring impact that flowed from the battle for Belleau Wood was an attitudinal one—our institutional commitment to change. Belleau Wood, in many ways, constituted a strategic inflection point for the Marine Corps. In the business world, a strategic inflection point occurs when your competition develops a new product or your market changes so that what you produced in the past is no longer desired. At Belleau Wood, the Marine Corps discovered that warfare had changed, and we had failed to adapt to those changes. The 4th Brigade paid the price in blood. Those who survived never forgot that lesson, and they vowed that the Corps would never again be caught unprepared. They became the innovators, risk takers, and visionaries who championed amphibious assault in the 1920s, close air support in the 1930s, and vertical envelopment in the 1950s. They were the architects that built the force-in-readiness that we are the proud stewards of today.

As I walked through Belleau Wood and the wheat fields that surround it, I tried to distill what lessons the current generation of Marines could glean from the veterans of Belleau Wood. In my mind the most important one is that we can never rest upon our laurels. Yes, we are the world's finest naval air-ground, combined-arms fighting force. Without doubt, the Marine Corps performed brilliantly in Desert Storm. In many ways, we are still basking in the warm glow of that victory, much like the Marine veterans of the First World War were tempted to do in the 1920s. But, those veterans knew they must change. They knew that they needed to accept the risks associated with developing new concepts for the ever-changing battlefield. They weathered the failures and the setbacks. They sparred with the naysayers. But, they never rested upon their laurels. As a result, instead of being the victim of a strategic inflection point as they were at Belleau Wood, they caused one in the Second World War.

That is exactly what today's Corps must do. That is exactly why we are pushing ahead, breaking new ground, developing our new warfighting concept, *Operational Maneuver from the Sea* (OMFTS, 1996), a concept tailor-made for the twenty-first century battlefield. OMFTS promises to once again revolutionize the art of war. Like amphibious assault, it will require the efforts of each and every Marine to discover and maximize the opportunities resident within this new concept, while sealing any exploitable seams. It is up to us, the current generation of Marines, to make OMFTS a reality.

Among the fog-shrouded trees at Belleau Wood stands a sentinel—a monolith—a statue designed and erected by the Marines of the 4th Brigade. It features a Marine rifleman, the shirt ripped from his back, rifle in hand, advancing, his face pointing toward the tree line that was the 4th Brigade's objective. You cannot see his face—why? Because he is not looking back toward the past—his face is pointing away—toward the future. That face and that simple statue contains a message for all of us—and for all those who will follow us. That Marine shows us the penalty that he and 4,700 of his fellow Marines paid because the Corps failed to recognize and adapt to change. That Marine knows the only way to avoid that fate is to advance toward the future—to steal a march on change—to be the cause of strategic inflection points, not the other way around. That Marine is showing the rest of us our course for the future.

Felix de Weldon's Belleau Wood memorial plaque dedicated on 18 November 1955.
American Battle Monuments Commission

CLIFTON BLEDSOE CATES BIOGRAPHY

"I have only two out of my company and 20 out of some other company. We need support, but it is almost suicide to try to get it here as we are swept by machine gun fire and a constant barrage is on us. I have no one on my left and only a few on my right. I will hold."

~First Lieutenant Clifton B. Cates
96th Company, Soissons, 19 July 1918

General Clifton Bledsoe Cates was born in 1893 and reported for active duty from the Marine Corps Reserves on 13 June 1917. As a lieutenant, he served in the 6th Marines; fighting in Verdun, at Bouresches and Belleau Wood, at Soissons, in the Saint-Mihiel offensive and in the Meuse-Argonne offensive. He was awarded the Navy Cross, Army Distinguished Service Cross, and an Oak Leaf Cluster in lieu of a second Distinguished Service Cross for heroism in the Bouresches and Belleau Wood fighting, in which he was both gassed and wounded. He earned the Silver Star Medal at Soissons, where he was wounded a second time, and an Oak Leaf Cluster in lieu of a second Silver Star Medal in the Blanc Mont fighting of the Meuse-Argonne offensive.

Interviewed in 1973, he described one of his earliest engagements, at Bouresches during the Battle of Belleau Wood:

> *We were deployed across this wheat field and taking very heavy fire—my platoon was. We received word that Captain Duncan had been killed, the company commander. So with that I yelled to this Lieutenant Robertson, I said, "Come on, Robertson, let's go." And with that we jumped up and swarmed across that wheat field towards Bouresches. About two-thirds of the way I caught a machine gun round flush on my helmet. It put a great big dent in my helmet and knocked me unconscious. So Robertson with the remainder of my platoon entered the west part of Bouresches. Evidently I must have been out for five or ten minutes. When I came to, I remember trying to put my helmet on and the doggone thing wouldn't go on. There was a great big dent in it as big as your fist.*
>
> *The machine guns were hitting all around and it looked like hail. My first thought was to run to the rear. I hate to admit it, but that was it. Then I looked over to the right of the ravine and I saw four Marines in this ravine. So I went staggering over there—I fell two or three times, so they told me—and ran in and got these four Marines, and then about that time I saw Lieutenant Robertson who, with the remain-*

Clifton B. Cates in World War I. The inscription on the photograph reads: "Taken at Verdun Apr 1918."
Collection of Clifton B. Cates/COLL3157, Archives Branch, Marine Corps History Division

Bjorn Egeli, *Clifton B. Cates, 19th Commandant of the Marine Corps* (1948–51).
Art Collection, National Museum of the Marine Corps

der of my platoon, was leaving the western end of the town. By that time, we were right on the edge of the center of town. I yelled at him and I blew my whistle and he came over and he said, "All right, you take your platoon in and clean out the town and I'll get reinforcements." Which I thought was a hell of a thing.

Well, anyway we did. We went on in and after getting into the town, we took heavy fire going down the streets. In fact, one clipped my helmet again and another hit me in the shoulder. We cleaned out most of the town, but by that time I had, I think it was, twenty-one men left.[1]

[1] Gen Clifton B. Cates, intvw with Mr. Benis M. Frank, Historical Division, HQMC, 28 March 1973, Oral History Collection, Simmons Center for Marine Corps History, Quantico, VA, 18–19.

Following the Great War, he remained in the Marine Corps advancing steadily, attending schools and writing doctrine in between assignments with the 4th and 6th Marines. In May 1942, he took command of the 1st Marines, commanding the regiment throughout the Guadalcanal campaign. He later commanded the 4th Marine Division in the Marianas operation, the Tinian campaign, and the seizure of Iwo Jima. In 1948, General Cates became Commandant of the Marine Corps, holding that position throughout most of the Korean War.

A true three-war Marine, General Cates's career illustrates the Corps' transformation from a purely colonial infantry force into the large, professional Fleet Marine Force of the latter half of the twentieth century.

Sgt Tom Lovell, *Tarawa*, 1943.
Art Collection, National Museum of the Marine Corps

CHAPTER TWO

The Lessons of World War II and Korea

by Paul Westermeyer

The performance of the Corps in the First World War had established its *bona fides* as a modern, professional fighting organization. It was no longer in search of a mission; the officer corps had embraced the idea of Marines as the Navy's landing force that was intended to seize and defend advanced naval bases. Marine zeal for this mission, which promised the Corps a specific role in the strategic calculus of America's defense, coincided with the growing realization of the American military that Japanese aggression in the Pacific was the most likely future threat to the interests of the United States.

Directed and supported by legendary Commandant Major General John A. Lejeune, Lieutenant Colonel Earl H. "Pete" Ellis produced the first doctrinal look at amphibious warfare and how it would fit into a Pacific campaign. Other Marines continued Ellis's work, notably future Commandant General Thomas Holcomb and General Holland M. Smith, long considered the "father" of modern U.S. amphibious warfare. These Marines took the lessons in modern warfare they learned fighting a positional war of attrition in the mud of France and applied them to land forces supporting a naval campaign in the Pacific. Reducing the Saint-Mihiel salient, for example, seems to have little in common with securing a coral island such as Midway from powerful enemy land, sea, and air forces. However, Marines recognized the underlying fundamentals of supply, training, and organization that modern warfare required for success in either field.

In *Coral and Brass*, General Smith said, "If the Battle of Waterloo was won on the playing fields of Eaton, the Japanese bases in the Pacific were captured on the beaches of the Caribbean."[1] The Corps did not merely create a doctrine for fighting its expected amphibious war in the Pacific; it experimented and trained for two decades. In the process, the Corps developed specialized equipment and techniques for every aspect of amphibious assaults and landings.

Simultaneously, the Corps continued its traditional missions as colonial infantry, with service in Latin America and China. But even those Marines traditionally associated with colonial warfare, such as Major General Smedley Butler, advanced amphibious capabilities. Butler encouraged the development of the Christie amphibious tank for the 1924 Fleet exercises and commanded a combined arms Marine brigade in China in the late 1920s, which was a useful experience for Marines working with armor and aircraft while on expeditionary duty in a harsh environment. As the Marines' colonial infantry role slowly faded, the Corps produced the remarkable *Small Wars Manual* in 1940, capturing the lessons of such conflicts for later Marines.

When World War II finally arrived, the Marines Corps was ready, putting its plans for defending advanced naval bases to the test early at Wake Island and Midway. In August 1942, the Corps put the entire concept to the test, seizing and then defending Guadalcanal from Japanese counterattacks. Throughout the remainder of the war, the Marines earned accolades as they refined and employed their amphibious warfare doctrine supporting a highly successful naval campaign that won the war against Japan.

Though the broad outlines of the war followed predictions first made by Ellis and other prewar thinkers, technological changes altered geographic realities of operations in the Pacific, particularly increased steaming ranges and the development of underway replenishment by the U.S. Navy, which reduced the need for advanced naval bases. But the dramatically increased value of airpower required airfields, and the Corps' techniques worked as well to seize and defend islands for airfields as they did naval bases.

The dramatic opening of the Atomic Age put the Corps' future in doubt, especially for those who believed its usefulness was limited to dramatic amphibious assaults against heavily fortified beaches. The Korean War would soon put the Corps' concepts to the test again. A Marine brigade was in Korea by August 1950, supported by a Marine air group and illustrating the power and flexibility of the Marine air-ground team. The 1st Marine Division would demonstrate the continued relevance of amphibious warfare at Inchon, and the indomitable spirit of the Corps at Chosin.

A helicopter troop landing during the Korean War pointed the way toward the Corps' increased relevance as a strategically mobile naval infantry force, capable of responding to a wide variety of situations as needed by the nation at short notice. Working closely with the Navy, the Corps remained ready to provide power projection ashore for the Fleet.

[1] Gen Holland M. Smith and Percy Finch, *Coral and Brass* (New York: Scribner, 1948), 19.

VALUE OF AVIATION TO THE MARINE CORPS

by Major Alfred A. Cunningham
Marine Corps Gazette, September 1920

In common with every new weapon introduced to the military Service, Marine Corps aviation has traveled a rocky and uphill road.[2] Its small size has tended to make the jolts more frequent and severe. Nothing short of the firm conviction that it would ultimately become of great service to the Corps sustained the enthusiasm of the small number of officers who have worked to make it a success. The past year has seen the completion of the first of the stages through which our aviation must pass. Prior to this, we had practically no official status or recognition. While we sent 182 officers and 1,030 men to the front in France, and they made a splendid record under severe conditions, we had no aerodromes at home, no shops, or other facilities; in fact, nothing permanent [existed] and could very readily have been disbanded entirely. When it was realized that the Marine Corps' permanent strength of 17,000 was entirely inadequate and that a larger permanent strength must be requested, the figure decided upon was approximately one-fifth the authorized strength of the Navy, or about 26,380. It was desired to utilize this number for ground duties; therefore, Congress was asked to authorize an additional 1,020 men for aviation duty, making the total 27,400. This gave us permanently our aviation personnel. The next task was to secure well-equipped home stations for our personnel, and it required the surmounting of many discouraging obstacles before the Navy Department, which handles the expenditure of all aviation funds, approved the construction of flying fields at Quantico, Parris Island [South Carolina], and San Diego. With this much accomplished and our men and pilots well trained, we feel that the time has arrived when we can demonstrate our usefulness to the Corps, which I am confident will be great.

One of the greatest handicaps that Marine Corps Aviation must now overcome is a combination of doubt as to usefulness, lack of sympathy, and a feeling on the part of some line officers that aviators and aviation enlisted men

[2] The original article came from Alfred A. Cunningham, "Value of Aviation to the Marine Corps," *Marine Corps Gazette* (September 1920). Minor revisions were made to the text based on current standards for style, grammar, punctuation, and spelling.

are not real Marines. We look upon the first two criticisms complacently, knowing that we can abundantly prove our usefulness even to the most skeptical, and that when we have done so, we will receive the sympathy and hearty support of all Marine officers. The last criticism we resent vehemently as an injustice, so far as it applies to loyalty, supreme pride in the Corps, and a desire to do what is assigned to us as quickly and as well as it can be done. Conditions arising from the necessity of organizing and training in a short time an aviation section, with practically nothing to start with and the nature of the duty, which does not allow the older officers to keep their juniors continually under their observation and guidance as is allowed in ground work, may have prevented the instillation in the younger pilots of all the qualities necessary in a Marine officer to the same degree as is done in infantry work. We have realized this difficulty and have made an earnest effort to overcome it and believe, with some few exceptions, that we have been successful. Now, since the rush of organizing for war service is over, this difficulty will be easily and simply overcome and the task of aviation officers made much more simple by taking into aviation only those young officers who have had enough service with infantry troops to be thoroughly indoctrinated with Marine Corps discipline and spirit.

It is fully realized that the only excuse for aviation in any Service is its usefulness in assisting the troops on the ground to successfully carry out their operations. Having in mind their experience with aviation activities in France, a great many Marine officers have expressed themselves as being unfriendly to aviation and as doubting its full value. I am confident that this must have been caused by some local condition, as the French, British, and Belgian troops in the sector in which the First Marine Aviation Force and the British squadrons operated were enthusiastically "full out" for aviation. In our own aviation section we intend, before asking a vote of confidence from the remainder of the Corps, to demonstrate to their complete satisfaction that we can contribute in a surprising degree to the success of all their operations, save many hours of weary, fruitless "hiking" and materially shorten each campaign. Previous to now, we have had no opportunity to do this. During the war, we were unfortunately not allowed to serve with the 4th Brigade, but were placed in a sector containing only British, French, and Belgian troops. Since the war, all our effort has been required

On 22 May 1912, 1stLt Alfred A. Cunningham reported to the Naval Academy at Annapolis, MD, for duty in connection with aviation. As a captain, Cunningham was ordered by MajGen Commandant George Barnett to form a "Marine Corps Aviation Company" in early 1917 in anticipation of the entry of the United States into World War I.
U.S. Naval Institute Photo Archives 9875656

to secure flying fields and the construction of buildings and hangars on them. We would have been hopelessly handicapped without these facilities. Now, since they are nearing completion, we are looking forward with enthusiasm to our real work of cooperating helpfully with the remainder of the Corps. All we ask is a spirit of cooperation and encouragement, and that judgment be reserved until the proper time.

Judging from the unfamiliarity of the average Marine officer with what has been accomplished by Marine aviation, we have failed woefully to advertise. A short résumé of what has been accomplished will perhaps be of interest.

In May 1912, when the [author] was detailed for aviation, the Marine Corps took very little interest in the subject. In those days, it was looked upon more as a crazy sport than as anything useful, and when I look back on the old original Wright [Model A] 35-horsepower planes I flew, where one sat on a board projecting out into atmosphere, I am inclined to agree with that view. About eight months later, another Marine officer was assigned to aviation, and during the next year, we accumulated six Marine enlisted men. There was very little increase in personnel until the World War began. On 6 April 1917, Marine aviation amounted to four officers and 30 men, all part of the complement of the Naval Air Station Pensacola, Florida. From this time, we began to work energetically for expansion. Our ambition was to organize a first-class aviation force to operate with the Marine forces we hoped would be sent to the front. During the next few months we secured a flying field at Philadelphia, organized a full squadron of land planes, and began intensive training, so that we would be ready to go to France with the other Marine Corps forces. In order to have the latest aviation information, the commanding officer of this squadron was sent to France to serve with the French aviation forces for three months. This officer made every possible effort, both with the War Department in Washington [DC] and the American Expeditionary Forces authorities in France, to secure authority for our Marine aviation squadron to serve with the Marine brigade in France. No success whatever attended these efforts. [U.S.] Army aviation authorities stated candidly that, if the squadron ever got to France, it would be used to furnish personnel to run one of their training fields, but that this was as near the front as it would ever get. Confronted with this discouraging outlook, the squadron commander set about to find some other way of getting his squadron into the fight. The only aviation operations abroad planned by the Navy at that time were antisubmarine patrols in flying boats. After visiting the Navy flying station at Dunkirk, France, and talking with officers of the British destroyer patrol, it was realized that Marine aviation's opportunity to get into the fight lay right here. The situation was as follows: submarines were causing enormous losses to shipping; their main operating bases and repair shops were at Ostend, Zeebrugge, and Bruges [Belgium], all within easy reach by plane from Dunkirk; and the water for 10–15 miles off these bases is so shallow that a submarine cannot safely negotiate it submerged. If these waters could be patrolled continuously during daylight with planes carrying heavy bombs, submarines attempting to enter these bases could be destroyed. Destroyers were prevented from patrolling these shallows efficiently in daylight by the heavy shore batteries, but could, under the cover of darkness and with mines, close the channels at night. This was evidently such an effective plan that inquiries were made as to why it was not put into effect. These inquiries developed that the Germans re-

In April 1918, the First Marine Aviation Force trained on Curtis Jennies to accompany British bombers flying an antisubmarine patrol over the North Sea and English Channel. World War I-era aircraft were not particularly sleek or fast, but what they lacked in performance was compensated for by individual daring in the execution of aerial maneuvers.
Defense Department photo 518425

alized the danger of such a plan and energetically suppressed any attempts of the British Navy to patrol these waters with seaplanes, sending out their best land pursuit planes to shoot them down. An inquiry as to why the British did not patrol this area with bombing planes protected by fighting land planes developed the fact that they were so hard pressed on the front in Flanders and northern France that they could not spare the planes for this work.

Why could not the Marine Corps man the necessary number of planes to allow this operation to be carried out? Jubilant at having discovered a prospective field of usefulness for Marine Corps aviation, our squadron commander hurried home and placed the whole scheme before the Major General Commandant, had a hearing before the General Board and the Secretary [of the Navy], and as a result orders were issued soon afterward to organize four Marine land squadrons as quickly as possible and secure from the Army the necessary planes to carry out the operation. It may well be imagined that, with the prospect of getting into some real thick fighting, all hands turned to with a rush, and by May 1918, we had our planes and four of the best-trained fighting squadrons that ever went to war. A short time before going overseas, a British ace and all-round aviation expert was ordered to spend a week with these squadrons to give them their finishing touches. After three days, he stated that they were the most thoroughly trained squadrons he had seen away from the front, and that he could offer no suggestions for improvement—that they were then ready to go over the front lines.

John Englehardt, *AEF France – First Marine Aviation Force*, ca. 1918.
Art Collection, National Museum of the Marine Corps

Before the Marine squadron arrived in France, the Navy decided to make the main objective the destruction of the bases at Ostend, Zeebrugge, and Bruges, and to increase the number of land squadrons by manning the additional squadrons with Navy personnel and assigning a naval officer to command the whole operation. It was somewhat of a disappointment that the status of this operation, which was originated and organized by the Marine Corps as a Marine operation, should have been changed. But with the prospect of getting into the fight, nothing could discourage the squadrons.

The Northern Bombing Group, which was the title given the combined Navy and Marine Corps land plane bombing operation in Belgium and northern France, although supposedly operating under the British, was in reality almost an independent body. It was composed of four Marine squadrons of 18 [Airco] DH4 biplanes each, known as the Day Wing, and was to have had four Navy squadrons of six Caproni night bombing planes each, known as the Night Wing. Only one Navy squadron was organized and it got into difficulties and sent, prior to the Armistice, only one plane over the front on one raid. Although handicapped on account of the inability of the naval bases at Pauillac, France, and Eastleigh, England, to furnish us our planes, spare parts, and tools, the four Marine squadrons accomplished a great deal. The results of one of our raids, verified after the enemy had evacuated Belgium, showed that we totally destroyed a troop train, killing about 60 officers and 300 men. The Marine aviators also introduced an innovation at the front. A French regiment was isolated during an offensive near Stadenburg [Germany], and it was decided to feed them by plane. Sacks of food were bundled into planes and they flew low over the isolated regiment and made good deliveries of much-needed subsistence. This necessarily had to be done at a low altitude and under a heavy fire from every weapon the enemy could bring to bear. It is believed to have been the first instance of its kind. This organization participated in the Ypres-Lys offensive and the first and second Belgian offensives.

The following is a table of what was accomplished over the front lines. The objectives of some of these raids were 75 miles from our aerodromes, nearly all of the distance over German territory:

Number of raids with French and British	43
Number of independent raids	14
Pounds of bombs dropped	52,000
Number of food-dropping raids	5
Pounds of food dropped	2,600
Number of enemy planes shot down	12
Pilots and observers cited for decorations (two for the Medal of Honor)	25

The recruitment poster *Aviation Fly with the U.S. Marines*, designed by Howard Chandler Christy for the U.S. Marine Corps in 1919, depicts a giant-size eagle in flight through a cloudy sky with a smiling Marine seated on his back. Below them is a biplane in flight with Marine insignia and a white plane in the background with a red, white, and blue tail.

National Air and Space Museum Poster Collection

Two Marine aviators earned the Medal of Honor flying a de Haviland DH-4 during their very first bombing raid on Thielt, Belgium, depicted here in *Raid on Thielt, 14 October 1918,* by James Butcher in 1985.
Art Collection, National Museum of the Marine Corps, gift of British Aero Space

De Havilland DH-4B, one of five stationed at Santo Domingo with a Lewis .30-caliber machine gun on the scarf mount, ca. 1919.
Marine Corps History Division

In the meantime, other activities were being worked out by Marine Aviation. An organization of 12 officers and 133 men was organized and sent to the Naval Base at Punta Delgada, Azores, where they carried on an antisubmarine patrol with seaplanes and flying boats until the Armistice. A temporary flying field was secured at Miami, Florida, where approximately 282 pilots and 2,180 aviation mechanics were completely trained, including advanced and acrobatic flying, gunnery, bombing, photography, and radio. A Marine aviation unit of six officers and 46 men was organized and attached to the Naval Air Station Miami and performed practically all the long overseas patrols for that station.

In March 1919, a squadron of six land planes and six flying boats was organized and attached for duty to the 1st Brigade in Haiti, and in February 1919, a flight of six land planes was organized and attached to the 2d Brigade in Santo Domingo. These organizations have been seriously but unavoidably handicapped by a lack of suitable planes and not enough personnel to properly carry on the work. These handicaps will be removed in the near future. However, both brigade commanders have requested that the number of planes be increased, and very complimentary reports as to the value of the aviators' work have been received. They patrol regularly the whole island and have saved many long, hot, and fruitless "hikes." They have located bands of *cacos*, [or guerrilla fighters,] dispersed them with machine gun fire, and performed many useful services that will be explained later.

Naturally, our first and most important peace-time duty was to secure permanent well-equipped flying fields as close as possible to large Marine Corps posts, so that we could by actual demonstration prove our usefulness. The difficulty of accomplishing this was greater than all our previous endeavors. We received abundant proof that, whether the government wastes money or not—as is claimed by the public—it certainly does not waste it on the Marine Corps. It was finally accomplished, however, and we now have nearing completion well-equipped stations at Quantico and Parris Island, and the establishment of a similar station at San Diego is approved and work on it will begin when the ground at the Marine base is in condition.

LtCol John J. Capolino, *Capt BF Johnson and Bandits. Art Collection, National Museum of the Marine Corps*

The question regarding aviation that is of most interest to the Marine Corps is: of what practical use is it to us? We see the planes flying around and they seem to be enjoying themselves, but how will they help us perform our mission? It is confidently believed that this question can be answered to your satisfaction. This article will mention some of the ways aviation will be helpful to the Marine Corps. These suggestions will not attempt to cover every use aviation can be put to or to mention anything which is not practical with the present development of planes and the art of flying them. No one can more than prophesy what the future development of aviation will allow it to do.

It is my opinion that a great part of the evident lack of belief in aviation shown by officers serving with ground troops is caused by the entirely unnecessary amount of flying that is done with no specific object in view, except the practice the pilot gets in handling his plane. This naturally creates an impression that the only use the planes have is to give their pilots practice in handling them. This impression should and will be removed. There are many important military problems which must be worked out by aviation and so many interesting opportunities to work in cooperation with troops on the ground that flights should rarely be made in future except with some useful military purpose in view.

The following paragraphs give some of the duties we believe we can perform satisfactorily, provided always that suitable equipment is furnished. They require no equipment impossible to secure with the present state of development.

Every officer who served at the front in the World War was given a rather impressive demonstration of the damage and demoralizing effect of bombs dropped from the air, and was perhaps extremely annoyed by being shot up by a machine gun in an airplane that it seemed impossible to hit from the ground. It will be remembered that troops in this war were, for the most part, well protected in trenches and dugouts from aerial attack as well as from attacks on the ground, and that both bombing and gunnery from airplanes will be much more effective against guerrillas and troops with less permanent protection. During the late war, proper advantage was not taken of the possibilities of radio and radio-telephonic communication between planes and between ground troops and planes.

Let us assume that the commanding general of a Marine expeditionary or advanced base force with his troops on board transports is approaching a port at which he is supposed to land in the face of enemy opposition. Would it be of value to him if one or more of his Marine aviators left his ship a hundred or more miles off shore, flew over the port, photographed the harbor, and returned in time to have the finished photographs in the hands of all subordinate commanders before land was sighted? This would allow the commander to plan his operation, not with inaccurate maps, but with actual photographs showing every detail of any effective plan of resistance. Pilots would hardly be available at an enemy port. The photographs of the harbor in practically all tropical waters show clearly the channels, buoys, reefs, sandbars, and minefields, if any exist, allowing the ship to be navigated into the harbor without a local pilot.

If the commanding general desired to prevent the removal from any locality of enemy stores, railway equipment, and locomotives, would it not be of service to him if the aviator left the ship before the enemy was aware of its presence and destroyed the railway tracks or bridges and made the highways impassable by bombing? During the actual landing, the planes could, with machine gun fire and small fragmentation bombs, so demoralize resistance as to make the task of landing much easier and safer.

After having landed, the following are a

few of the ways the planes can be useful to the troops:

They can locate quickly bodies of the enemy and communicate instantly their approximate strength, location, disposition, and actions. The enemy can be watched and any movement instantly reported. In this connection, there has been developed a portable radio and radio-telephony ground set that is so small and easily set up that one can be carried by two or three men or on the back of a mule, horse, or donkey. In future operations, every unit that has one of these—and every unit should have one—will be in instant communication with the planes and through them with any other station.

Photographs of enemy defenses, proposed battle terrain, or any other object or area of reasonable size within a radius of 50 miles can be taken, developed, and the desired number of prints delivered to the troops in time to use them in the plan of attack or defense. I have personally seen photographs distributed to the various organizations 45 minutes after the plane that took them had landed.

Planes continuously in communication with headquarters can patrol wide areas daily or hourly, which duty would require large bodies of troops and much fatigue to accomplish otherwise.

By bombing and machine gunnery, the enemy can be harassed and prevented from making orderly dispositions.

Enemy troops and population well in rear of the line of resistance can be kept in a demoralized condition, and enemy ammunition and supply depots and other military objects destroyed.

Any railways, bridges, and roads within a radius of 100 miles can be quickly made impassable.

Rapid communication can be furnished between detached bodies of our troops in difficult country, and officers can be quickly transported anywhere on urgent missions.

In thick and rough country, the planes can keep headquarters informed at all times of the disposition, progress, and needs of our troops.

In the event the enemy has planes, we can protect our troops from observation and annoyance and prevent the enemy from securing benefit from his planes.

For the field artillery, the following are some of the ways in which we can be helpful:

Difficult and temporary targets can be located quickly, accurately described and changes in targets promptly reported.

The bursts of our shell can be accurately spotted and corrections for the next shot instantly reported.

Targets invisible from the ground can be kept under accurate fire by corrections given by the planes.

Photographs of targets can be furnished showing progressively the results of artillery fire.

The location of hidden artillery batteries causing damage can be discovered and reported.

At night, designated areas can be kept lighted by parachute flares, etc.

Through its speed and remarkable visibility, and by the use of its radio and radiotelephone, together with visual signals that must be developed, the airplane will coop-

LtCol Albert Michael Leahy's *Managua Tri-Motors* depicts a Marine Atlantic-Fokker TA-2 tri-motor transport aircraft arriving at Managua, Nicaragua, in December 1927 to support the 5th Brigade to quell a civil war there—the first sustained tactical airlift operation by any American military service branch. The TA-2 could transport a load in two hours that would have taken a mule train three weeks to deliver.
Art Collection, National Museum of the Marine Corps

> *erate with the signal and communication troops so as to greatly increase their effectiveness.*

For advanced base work:

> *In addition to the duties mentioned above that aviation will perform—and nearly all these will enter into advanced base work as well—the planes will cooperate in the following ways:*
>
> *Offshore patrols to prevent surprise raids by enemy light forces.*
>
> *Antisubmarine patrols.*
>
> *Spotting for shore batteries in attacks by enemy ships.*
>
> *Communication between the base and our vessels offshore.*
>
> *Photographing, bombing, and torpedoing enemy craft and bases within reach.*
>
> *On account of the aviator's ability in most localities to pick up and chart enemy mine fields, airplanes should furnish valuable assistance in countermining and mine sweeping.*

A large part of the work performed by the Marine Corps is to combat guerrilla and bandit warfare, usually in tropical countries where roads are few and ground communications almost nil. We must not overlook the valuable assistance aviation can render in this kind of fighting or fail to realize its many helpful possibilities in the occupation of such territories whether fighting is in progress or not. The enemy encountered under these conditions are usually unstable and cannot withstand punishment. They are nearly always superstitious and easily stampeded or cowed by methods of warfare with which they are unfamiliar. They base their hope for success on their ability to make raids and get away before the necessary number of our troops arrive. When an attempt to round them up is made, their knowledge of the country and their ability to travel light and fast allow them to lead our troops an exhausting chase for some time before they are dispersed if that is accomplished. The work of the Marine Corps aviators in Haiti and Santo Domingo has abundantly shown the possibilities in this class of operations. Difficult country can be patrolled so completely and frequently that it is impossible for bands to form without being discovered. To cover an area as thoroughly and frequently as can be done by airplanes would require a prohibitive number of troops and a weary amount of "hiking." The planes in Haiti have already proved that they can, without assistance from the ground, disperse and almost destroy bands of *cacos* with gunnery and small bombs. When these *insurrectos* [rebels] realize that they cannot congregate without being attacked within a very short time thereafter by our planes, their enthusiasm quickly disappears and the unfamiliar form of attack from the air greatly assists in their discouragement. If the planes could perform no other service for our expeditionary troops than to make unnecessary the long marches formerly required in searching for *cacos* they would be worth their keep, but a little imagination will suggest to any experienced Marine officer numerous duties the planes, on account of their special abilities, can perform for them.

It is believed that enough has been said to show those who are students of Marine Corps operations that an intelligent development of aviation and an encouraging spirit of cooperation between it and our troops can only result in enabling the Corps to perform its function much more quickly and efficiently. Marine officers very properly "like to be shown," and nothing is more desired by Marine Corps Aviation than a chance to work out with our troops the problems suggested above, as they feel assured that such an opportunity can result only in mutual respect and confidence.

Before closing this article, I would like to mention something that might interest prospective pilots. Above all, aviation is a young man's game. It requires a young heart, nerves, lungs, eyes, and reflexes. It has been said that, after a man reaches a certain age, he has too much sense to do what an aviator is required to do. There are exceptions, of course, and older men have been good fliers, but I believe they are exceptions, and my eight years' experience and observation has shown me that, provided they have the necessary amount of judgment, the younger the pilots are the better. I believe it is good policy to set the maximum age for applicants for pilot's duties at around 25 years. I am also led to believe that the average term of usefulness for a pilot flying regularly is not more than five years. At the end of that time, they know the work thoroughly, but those who are still alive have lost the "pep" and enthusiasm that is essential.

The established policy regarding pilots is that they will not be ordered to aviation duty until they have had enough experience with troops to have become thoroughly qualified Marine officers. The ordinary length of the detail will be five years, after which they will return to duty with troops.

Aviation is probably the most highly technical branch of the military Service. It differs from other arms in the unusually fast development of its equipment, planes, motors, etc. The administrative and technical part of it is really a profession that requires long experience and constant study to fit one to properly make decisions, which decisions must necessarily be correct, as the life of the pilot, even in peace times, depends upon their soundness. For self-evident reasons, it is necessary for any aviation organization to have enough old experienced aviation officers to run the technical part of it on sound principles proved by experience and to prevent the enthusiasm and inexperience of younger pilots from causing harm. This necessity for officers of long experience is recognized and unquestioned, and for this reason, a very small number of pilots who show special aptitude will be continued in aviation duty indefinitely to furnish the number of expert and experienced officers required.

The men in aviation are enlisted especially for aviation duty, and are sent through the regular recruit course at Parris Island or Mare Island [California], after which they are given a thorough education in gasoline motors, as shop machinists, and in practical and theoretical airplane repair and upkeep. Our main trouble with enlisted men has been that, after giving them an excellent education, they discover that men performing identically the same duties in the Army and Navy draw much more pay than they. As a result, they become dissatisfied and do not reenlist. It is hoped and believed that this will be remedied shortly and their pay put on a par with men doing similar work in the Army and Navy.

For fear that, by mentioning in this article, the skeptical feeling regarding aviation that is supposed to exist among some officers, I have given an erroneous impression, I would like to state that I believe the number of officers who hold this attitude constitutes a small minority of the officers of the Corps. The subject is only mentioned here because the whole article is an effort to show Marine Corps officers that, with encouragement and cooperation, we can be of real service to them, and to show commanding officers what parts of their problems they can use aviation to perform. Naturally, the ones we wish most to convert are those who at present do not fully believe in us.

ADVANCED BASE OPERATIONS IN MICRONESIA, FMFRP 12-46

by Major Earl H. Ellis

Major Earl H. Ellis was a brilliant Marine officer whose superb skills as a planner helped forge the modern Marine Corps and its Fleet Marine Force.[3] Though Major General Commandant John A. Lejeune was the guiding light, he trusted Ellis with the job of translating concepts into concrete plans.

Ellis's brilliance in planning was only one aspect of a complex personality. He served the Marine Corps with a single-mindedness that left no time for marriage and that damaged his health and ultimately cost him his life. Ellis frequently worked on assignments around the clock, without sleep, until physically and mentally exhausted. This proclivity, plus alcoholism, put him in the hospital with mental breakdowns on more than one occasion. (In an era of promotions by seniority, these problems did not end his career as they would today.) His death in the Japanese-controlled Caroline Islands probably resulted from the effects of excessive drinking.

Ellis, who was not a college graduate, enlisted in the U.S. Marine Corps in 1900. His talents earned him a commission just more than a year later. His service prior to World War I involved the assignments typical for officers of the era. He became a first lieutenant in 1903, a captain in 1908, and a major in 1916. During World War I, Ellis received a temporary promotion to lieutenant colonel, but reverted to his permanent rank after the war. During the war, he served as a principal staff officer to Lejeune when the latter commanded the 4th Brigade and then the 2d Division in France. The close relationship, which actually began in the Philippines in 1908, continued after Major General Lejeune became Commandant of the Marine Corps.

Service at the Naval War College as a student and faculty member (1911–13) was a turning point in Ellis's career. This tour of duty came during the period when the Naval War College was a major participant in the development of the Navy's war plans. One of these was War Plan

[3] The original Fleet Marine Force Reference Publication (FMFRP) came from *Advanced Base Operations in Micronesia*, FMFRP-46 (Washington, DC: Headquarters Marine Corps, 1992), 77–82. Minor revisions were made to the text based on current standards for style, grammar, punctuation, and spelling.

Earl Hancock "Pete" Ellis was a U.S. Marine Corps Intelligence Officer and author of *Advanced Base Operations in Micronesia*, which became the basis for the American campaign of amphibious assault that defeated the Japanese in World War II.
Art Collection, National Museum of the Marine Corps

Orange, which grew out of the need to defend the recently acquired Philippines and from the perception that Japan was the most likely enemy in a future war in the Pacific. War Plan Orange served as the basis for a groundbreaking paper by Ellis on the theoretical basis for doctrine covering the defense of advanced bases.

After World War I, the naval Services again turned their attention to War Plan Orange and the problems of a naval campaign against Japan. Study was necessary because the strategic situation had changed radically. Japan, which was on the Allied side in World War I, had captured islands previously occupied by Germany. These islands provided Japan with bases suitable for launching attacks on the Philippines and other American possessions in the Pacific. The altered strategic situation meant that a war with Japan would have to include amphibious assaults for capturing island bases for subsequent fleet actions. This reality served as the basis for the Marines' development of amphibious warfare doctrine and for the formation in 1933 of the Fleet Marine Force.

The first tangible step by the Marines came from Ellis. Working virtually around the clock during early 1921, he developed Operation Plan 712, *Advanced Base Operations in Micronesia*. This plan stood the test of time; 20 years later, during World War II, the actual American campaign for Micronesia diverged from Ellis's plan only in areas affected by technological advances.

The effort to develop Operation Plan 712 was not without cost to Ellis. He suffered several reoccurrences of his earlier mental and physical problems. He was under medical supervision and treatment for much of the time he worked on the plan.

At the completion of his plan, Major Ellis received permission to travel to Japanese-held Micronesia in the guise of a civilian. En route, he had to be hospitalized in Japan for alcohol poisoning and nephritis, a disease of the kidneys. He managed to get to Micronesia but was under close Japanese surveillance. In May 1923, the Japanese authorities announced that Ellis had died on the Micronesia island of Korror [now in Palau]. While there was some speculation that the Japanese had killed him, the most likely cause of death was alcohol poisoning and nephritis.

Major Earl Ellis's writings and plans made him a major architect in the development of the modern Marine Corp. Ellis Hall, the home of

The Marines Have Landed, a Marine Corps recruitment poster designed by James Montgomery Flagg in 1942, shows three Marines landing on a beach with ships in the background.
Art Collection, National Museum of the Marine Corps

the Marine Corps' Command and Staff College at Quantico, Virginia, pays tribute to his memory and contributions.

SUMMARY FROM OPERATION PLAN 712, *ADVANCED BASE OPERATIONS IN MICRONESIA*

Strategy

Governing Factors:

A. A main fleet action will decide the war in the Pacific.

B. Our fleet, on taking the offensive, will be at least 25 percent superior to that of the enemy.

C. The enemy will hold his main fleet within his defensive line and endeavor, during preliminary operations, with his lesser craft (old gun ships and torpedo, mine, and bomb craft) and land forces to "wear down" our fleet to an extent where he believes he may reasonably risk a main fleet action.

D. Fleet fighting units, being comparatively irreplaceable in war, must be husbanded for action against enemy fleet units.

E. Operations preliminary to a fleet action must be carried out by (as far as possible) the minimum naval forces and those of least fleet value in fleet action.

F. Marine forces of reduction, occupation and defense must be of such strength and composition (so far as maybe compatible with the conditions under which they must operate) as to require the least possible naval support.

G. An offensive projection into the enemy's strategic front must be made in a series of well-defined and rapid moves (sea objectives) in order to afford the battle fleet the greatest protection for the greatest portion of the time. (Long-drawn-out operations, with the fleet and its base subject to close attack by the enemy light forces, are to be avoided).

H. A sea objective must be more or less isolated and include an area that can be reduced practically simultaneously.

I. The sea objective should include an anchorage suitable for the fleet, so situated as to facilitate offensive operations against further sea objectives.

J. Subobjectives in any sea objective will be as follows, in order of importance:

1. Reduction of a base for the fleet;
2. Reduction of enemy bases;
3. Reduction of any anchorages that may be used as enemy emergency bases;
4. Reduction of other areas.

Tactics

Governing Factors:

A. The enemy will use land forces freely and by a universal shore resistance in strategic areas gain time and create opportunities for his "wearing down" operations.

B. The enemy defense of land areas will consist, in general, of a mobile land defense and a mine defense, thus enforcing extensive landing and sweeping operations for their reduction.

C. The enemy will have ample time in which to prepare his defense.

D. The main points of enemy resistance

U.S. Marines go ashore from Navy motor-sailers in the prewar era before the advent of Andrew Higgins's landing craft.
Defense Department photo (U.S. Navy) 58920

will be his own bases and those of greatest value to us.

E. In the reduction of any island position (island or group) the immediate mission may be any of the following, depending upon the particular strategic situation:
   1. Reduce land masses necessary to control anchorages and landing fields, thus preventing their use by the enemy;
   2. Reduce land masses necessary to control anchorages, landing fields and the passages thereto, thus permitting of their use by our forces;
   3. Reduce entire island or group, for our unrestricted use and entire denial to the enemy for any purpose whatsoever, including observation.

F. Depending upon weather and sea conditions and enemy resistance, the procedure in the execution of the mission may be:

Marines landing under fire at Santo Domingo City. Copy of illustration by Dickson, ca. 1916.
Records of the U.S. Marine Corps (Record Group 127), National Archives and Records Administration

1. Land direct on objective from open sea;
2. Land on land masses controlling a reef passage, thus securing entrance for effecting a landing on objective from reef bound waters.

G. In landing on any land mass the immediate mission must be to secure and consolidate a "boat head."

H. The choice of boat heads must depend upon the ease and rapidity with which it can be obtained and its position relative to the objectives.

I. Owing to the restricted area of land masses, the jungle terrain that generally obtains, and the paucity of existing communications, the enemy's main line of resistance and the bulk of his resistance will be practically on the seacoast in all cases.

J. As a decision is to be reached by a very short advance inland, the enemy defense will consist of a closely linked and intricate obstacle and strongpoint system in the back beach jungle.

K. The greatest effort of our troops must be put forth at the time of landing.

L. The forces and weapons provided should be those best suited to beach and jungle combat's close, rapid fighting.

M. In order to effect a concentration on the enemy, operations must be carried out with surprise and rapidity.

N. In the defense of bases, the primary object of the defense forces will be to prevent the enemy from damaging property within a certain area (anchorages, port facilities, etc.), not necessarily to destroy enemy craft. The defense required is only that necessary to render an enemy attack so dangerous as to be unreasonable, taking into consideration the conditions under which the enemy is operating.

O. The base fixed defense must concentrate on good observation, quick communication, and rapid, accurate gunfire: the best fixed defense against all types of sea and air forces.

Materiel

Governing Factors:

A. Owing to the restricted ship space available, only articles of the widest use should be included. Articles of special, limited use have no place, provided the necessary service can otherwise be obtained approximately.

B. Delicate or complicated materiel that cannot reasonably be depended upon to withstand rough handling, exposure to the elements and service by ordinarily trained personnel, should not be included.

C. Whenever the addition of special materiel is effected, its weight, timesaving qualities, and reliability in service should always be considered.

D. All materiel should be of sizes and dimensions favoring rapid transportation between ship and shore and quick installation on the latter.

E. Owing to the lack of local resources in the theatre of operations, practically all materiel absolutely necessary for the installation and maintenance of the military forces on shore must be carried on transports.

F. Owing to the fact that the bulk of troops will be located near the seacoast, that ample small boat transport will be available and that land communications are few and poor, not more than 50 percent of the land transport usually required need be furnished.

G. In order to simplify training and supply, the materiel provided should be standard Army, Navy, or Marine Corps.

H. The materiel considered for the advanced base force must be that which we now have or may reasonably be expected to have at the outbreak of hostilities.

Personnel

Governing Factors:

A. The greatest fighting (and losses) will occur in the ship-shore belt, and troops suitable and trained for combat in that area must be provided.

B. In sea operations, where vulnerable, floating troop centers are necessary, specialist organizations (employed only as such) must be cut to the limit. Wherever it is possible, troops must be given specialized training for emergencies without withdrawing them from the necessary fighting organizations.

C. Owing to the conditions obtaining in the theatre of operations, the following specialized training among fighting personnel is particularly necessary:

1. Field Engineering: dock, road and shelter construction; obstacle and trench work; pioneer work; transportation of heavy materiel.
2. Communications: all types for linking up isolated and dispersed forces.
3. Water Transportation: motor, sail, or car.

Organization

Governing Factors:

A. The number of transports must be cut to the efficient minimum in order to reduce as far as possible the activity of fleet fighting units for their protection.

B. The loss of one-third of any particular floating force should not prevent the complete functioning of the remainder of the force in the performance of its normal task.

C. Personnel should not be subjected to such conditions on board as would tend to prevent their putting forth their highest effort at the moment of landing.

D. There must be no wastage in the employment of transports or of troops.

E. No shifting of troops or materiel between ships on blue water is practicable.

F. Task forces must be formed before leaving base port and must be embarked as such.

G. Personnel and materiel best adapted to perform the normal tasks must be provided.

H. A task unit (in its necessary elements) should not be split up between transports, but an economical use of space obtained by the subtraction or addition of infantry units.

I. All training in the performance of tasks must be carried out prior to leaving home ports.

THE U.S. MARINE CORPS, AMPHIBIOUS CAPABILITIES, AND PREPARATIONS FOR WAR WITH JAPAN

by David J. Ulbrich, PhD

MCU Journal, Spring 2015

The U.S. Marine Corps' amphibious mission had its genesis at the dawn of the twentieth century.[4] Following the Spanish-American War in 1898, American strategists worried about the possibility of war between the United States and Japan because both nations vied for influence in East Asia and the western Pacific. Acquiring the Philippines, Guam, and Wake islands after winning the Spanish-American War gave the United States a presence in East Asia and the Pacific Ocean. China especially represented an important commercial resource for the United States. Americans wanted expanded markets in China and hoped to maintain an "Open-Door" trade policy with that nation's large population. These commercial interests required sufficient forces to protect them. A few years later in 1905, victory in the Russo-Japanese War turned Japan into the dominant nation in the region. Due to severe deficiencies in natural resources, the Japanese leaders coveted the raw materials and agricultural production of the Asian mainland. Any southward or westward expansion would inevitably bring this rising power into conflict with America's strategic and commercial interests in that region.

As early as 1900, the senior admirals in the U.S. Navy argued that the new strategic situation required American power to be projected across

[4] This article grew out of a presentation titled "Marine Corps Doctrine and the War with Japan" at the 2013 Chief of Army History Conference in Canberra, NSW. A longer version with this same title was published in the conference's proceedings Peter Dennis, ed., *Armies and Maritime Strategy* (Newport, NSW: Big Sky Publishing, 2014). This article is reprinted by permission of the Army History Unit. Portions of this article have also been drawn from Ulbrich's award-winning "Clarifying the Origins and Strategic Mission of the U.S. Marine Corps Defense Battalion, 1898–1941," *War and Society* 17, no. 2 (October 1999): 81–109; "Document of Note: The Long-Lost *Tentative Manual for Defense of Advanced Bases* (1936)," *Journal of Military History* 71, no. 3 (October 2007): 889–901; the award-winning *Preparing for Victory: Thomas Holcomb and the Making of the Modern Marine Corps, 1936–1943* (Annapolis: Naval Institute Press, 2011); and as coauthor with Matthew S. Muehlbauer, *Ways of War: American Military History from the Colonial Era to the Twenty-First Century* (London and New York: Routledge, 2014). Ulbrich gratefully acknowledges assistance from the Marine Corps Heritage Foundation and the Marine Corps University's History Division, Reference Branch, and Archives Branch. Jack Shulimson, *The Marine Corps' Search for a Mission, 1880–1898* (Lawrence: University Press of Kansas, 1993), 168–210; and Jack Shulimson, "The Influence of the Spanish-American War on the U.S. Marine Corps," in *Theodore Roosevelt, the U.S. Navy, and the Spanish-American War*, ed. Edward J. Marolda (New York: Palgrave, 2001), 81–93.

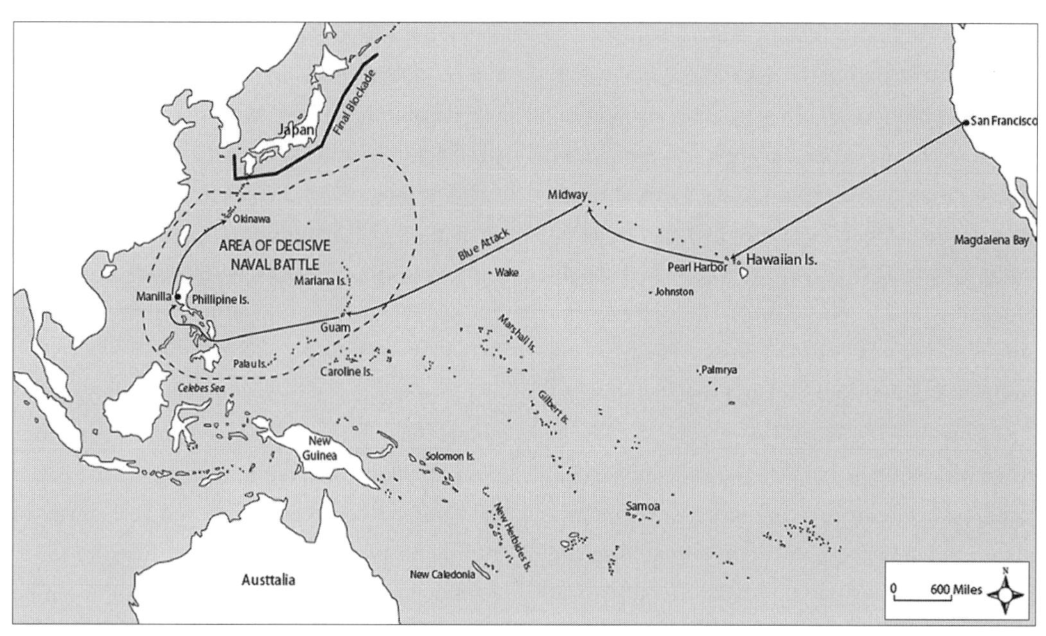

Map of War Plan Orange in the 1920s–30s.
Map courtesy of U.S. Naval Institute

the vast Pacific Ocean. American strategists prepared a number of scenarios with potential allies and enemies designated by colors. The U.S. Navy's planners focused their attention on the Pacific Ocean and on Japan, otherwise known by the color designation "Orange" in American war plans. This potential threat gave the Marine Corps two new roles: amphibious assault and island defense. Marines would no longer expect to subsist in nineteenth-century duties as shipboard police, legation guards, and constabulary troops. Doing so would only relegate the Corps to insignificance and eventual extinction.[5]

As the plan to defeat Japan, War Plan Orange spanned the next several decades until 1938. All its variations shared several tenets. American strategists expected that the Japanese would launch a preemptive strike, likely without a formal declaration of war. That attack would presumably be directed against American bases on the Philippines and Guam. Following the initial Japanese onslaught, the U.S. Fleet would sortie from Hawaii and sail across the Pacific. During this offensive campaign, the Marines would seize and hold "temporary advanced bases in cooperation with the Fleet and . . . defend such bases until relieved by the Army."[6] These roles constituted a new dual mission for the Marine Corps. The newly captured bases would subsequently function as coaling stations, safe anchorages, repair facilities, supply depots, and eventually aircraft bases. The U.S. Fleet would either relieve besieged American forces in the Philippines or liberate the archipelago if it already had fallen. As the U.S. Fleet menaced

[5] For the seminal works, see David C. Evans and Mark R. Peattie, *Kaigun: Strategy, Tactics, and Technology in the Imperial Japanese Navy, 1887–1941* (Annapolis: Naval Institute Press, 1997); and Edward S. Miller, *War Plan Orange: The U.S. Strategy to Defeat Japan, 1897–1945* (Annapolis: Naval Institute Press, 1991).

[6] Gen Holland M. Smith, *The Development of Amphibious Tactics in the U.S. Navy* (Washington, DC: History and Museums Division, Headquarters Marine Corps, 1992), 22.

the Japanese home islands, American planners hoped that the Imperial Japanese Navy would contest the American offensive. This ensuing naval battle, as was unquestioningly assumed in every iteration of War Plan Orange, would result in a decisive American victory. If the Japanese chose not to fight, then the U.S. Fleet would blockade their home islands. Regardless, the American victory would consign Japan to the status of a diminished, isolated regional power.[7]

This article traces the progression from America's strategic plans to the doctrine formulation phase, to the force structure development phase, and to the equipment procurement phase in the decades leading up to the Second World War. During the planning process behind War Plan Orange and subsequent plans, missions were dispensed downward from Navy to the Marine Corps. Once strategic priorities were set for offensive or defensive portions of the Marines' dual mission, the Corps' own planners worked to fulfill those needs.[8] The doctrines for advanced base defense and amphibious assault formed the pivot point for the Marine Corps to match operational, force structure, and material capabilities to the Navy's strategic needs in the Pacific Ocean. It should also be noted that the Marines embraced amphibious capabilities as a means of institutional survival during the resource-poor interwar years. Lastly, this article also highlights a few of the personalities that helped drive this process.

Among the personalities was Thomas Holcomb, whose career spanned more than four decades from his commissioning as an officer in 1900 to completing a seven-year term of command of the U.S. Marine Corps in 1943. As much as any other Marine, Holcomb can be considered a touchstone because he influenced so heavily the strategic, doctrinal, technological, and organizational evolution of the Corps' amphibious capabilities. Even so, other Marine officers like Earl Ellis, Holland Smith, Robert H. Dunlap, and John Lejeune also played critical roles in preparing the Corps for this amphibious mission. This article will highlight their contributions. Ultimately, the efforts by Holcomb and the others, particularly in the 1920s and 1930s, would bear much fruit in the Second World War. Their habits of mind, as well as their actual ideas about amphibious warfare, likewise provide examples that can be applied in the twenty-first century.

ESTABLISHING THE CORPS' PLACE IN AMERICAN STRATEGY, 1900–33

The U.S. Marine Corps made positive strides in developing its amphibious capabilities from 1900 to 1915. Marines specializing in this new type of warfare could attend their own Advanced Base School, where they studied operational is-

[7] James O. Richardson, *On the Treadmill to Pearl Harbor: The Memoirs of Admiral James O. Richardson as Told by George C. Dyer* (Washington, DC: Naval History Division, Department of the Navy, 1973), 256–68; George W. Baer, *One Hundred Years of Sea Power: The U.S. Navy, 1890–1990* (Stanford, CA: Stanford University Press, 1994), 44, 51–53, 90–92, 119–28; Steven T. Ross, ed., *American War Plans, 1890–1939* (Boulder, CO: Lynne Rienner, 2002), 7–9, 49, 80, 137, 167–74; and Miller, *War Plan Orange*, 202–3, 226.

[8] S. L. Howard, "The Marine Corps in War Plans" lecture, 3 May 1929, Box 7, Strategic Plans War Plans Division (SPWPD), Series I, Record Group 38 Records of the Office of the Chief of Naval Operations (RG 38), National Archives and Records Administration, College Park, MD (NACP); Memo for the Officer in Charge, War Plans Section, Headquarters Marine Corps, 4 May 1936, Box 22, Division of Plans and Policies War Plans Section General Correspondence 1926–1942 (DPPWPGC 1926–42), Record Group 127 General Records of the U.S. Marine Corps (RG 127), National Archives, Washington, DC (NADC); Donald F. Bittner, "Taking the Right Fork in the Road: The Transition of the U.S. Marine Corps from an 'Expeditionary' to an 'Amphibious' Corps, 1918–1941," in *Battles Near and Far: A Century of Overseas Deployment–The Chief of Army Military History Conference 2004*, ed. Peter Dennis and Jeffrey Grey (Canberra, Australia: Army History Unit, 2005), 116–40; and Miller, *War Plan Orange*, 181, 197–99, 226.

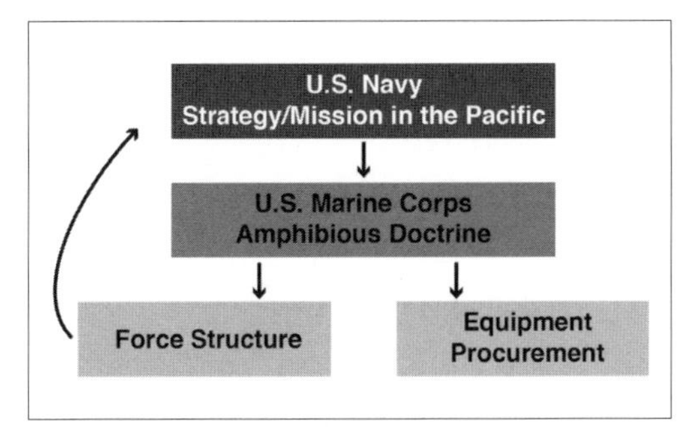

Process to dispense an amphibious mission downward to the Marine Corps. Marines then formulated the doctrine, created the force structure, and procured equipment necessary to fulfill the amphibious mission.
Chart courtesy of MCU Press

sues important to any base defense, such as artillery placement, communications, logistics, and staff organization. Academic study and practical experience coalesced in 1914 with a simulated assault on Culebra, a small island near Puerto Rico in the Caribbean. Warships from the U.S. Atlantic Fleet attacked 1,700 Marines defending the island. The Marine Advanced Base Brigade succeeded beyond expectations by quickly fortifying the island, harassing the Navy warships, and repulsing amphibious assaults.[9]

In 1915, the Marine Corps' Assistant Commandant Colonel John A. Lejeune created an ad hoc war plans committee comprised of himself and three promising Marine captains assigned at Headquarters Marine Corps: Ralph S. Keyser, Earl "Pete" Ellis, and Thomas Holcomb. Of these, Ellis emerged as the premier amphibious assault theorist until his untimely death in 1923. However, Holcomb and another rising officer, Holland M. Smith, provided the Corps with the continuity of purpose and the baseline of knowledge from 1915 through 1943 as they rose through ranks.

Among other issues, Lejeune's war plans committee set to work examining the Navy's evolving strategic needs and determining how the Marine Corps could best fulfill them. By 1916, however, it was not the specter of a war with Japan or the possibility of amphibious operations in the Pacific that absorbed the Marines' energies. Instead, it was the bloody conflict raging in Europe. Lejeune and his war plans committee worked diligently to determine how new weapons technology and battlefield tactics might affect their Service's combat capabilities. Mobilizing and fighting the First World War in France demanded the Corps' entire attention. Although Marines acquired the nickname "Teufel Hunden" (Devil Dogs in English) and gained invaluable combat experience, the Great War did little to help the Corps as an amphibious assault or base defense force. Indeed, Marines worried that their Service might be seen as a second American land army that could be disbanded during postwar demobilization.[10]

Indeed, similar soul-searching is occurring as the present-day Marine Corps attempts to return to its roots as the United States' premier amphibious force, after having served as a second land army in Iraq and Afghanistan for more than a decade. There is a generation of midcareer Marines who have limited knowledge or experience of amphibious operations.

Returning to the postwar anxieties in the 1920s, the Marine Corps did not disband after the conflict's end. Instead, recently promoted Major General John A. Lejeune helped solidi-

[9] Graham A. Cosmas and Jack Shulimson, "The Culebra Maneuver and the Formation of the U.S. Marine Corps' Advanced Base Force, 1913–1914," in *Changing Interpretations and New Sources in Naval History: Papers from the Third United States Naval Academy History Symposium*, ed. Robert William Love Jr. (New York: Garland Press, 1980), 293, 299–306.

[10] See David J. Ulbrich, *Preparing for Victory: Thomas Holcomb and the Making of the Modern Marine Corps, 1935–1943* (Annapolis: Naval Institute Press, 2011), 14–27.

fy its place in American naval strategy when he became Commandant in 1920. At his behest, then-Major Pete Ellis authored two definitive reports on amphibious operations that very next year. His *Navy Bases: Their Location, Resources, and Security* (1921) and *Advanced Base Operations in Micronesia* served as primers on how advanced bases could support Fleet operations. Two decades before the American entrance into the Second World War, Ellis predicted with uncanny accuracy the base defense and amphibious assault operations that characterized that conflict in the Pacific.[11]

Because Japan was "the only purely Pacific world power," Ellis saw it as the only principal threat to the United States. His report, *Navy Bases*, anticipated that Japan would take the offensive and try to capture outlying American island bases. These bases would then form a strategic defense-in-depth.[12] Ellis's other report, *Advanced Base Operations*, stood as a companion work to *Navy Bases*. It outlined a strategy for seizing and defending various Pacific islands, including the Marianas, Marshalls, and Carolines, which the Japanese already controlled. Imagining a potential campaign in the Pacific, Ellis outlined targets for amphibious assaults and anticipated certain sea battles. He suggested that Marines receive simultaneous training for the offensive and defensive components of their mission. Knowledge of how to defend an island against an enemy amphibious assault could only improve the attackers' abilities to make a successful assault in the future, and vice versa.[13]

Both of Ellis's seminal reports cast the Marine Corps in roles mandated by War Plan Orange. Later in 1926, the inter-Service report, *Joint Action of the Army and Navy*, similarly called for training, supply, and maintenance of Marine units for the following priorities: "For land operations in support of the fleet for the initial seizure and defense of advanced bases and for such limited auxiliary land operations as are essential to the prosecution of the Naval campaign."[14] A dual mission was now the Marine Corps' strategic *raison d'être*, as well as the ongoing key to survival during an era of restricted resources. In this way, military necessity blended with institutional pragmatism.

Ellis was hardly alone in his advocacy of an amphibious focus for the Corps in the interwar years. Other ardent supporters included Lejeune, Holcomb, James C. Breckinridge, John H. Russell Jr., Ben H. Fuller, Robert H. Dunlap, and Holland M. Smith. Naval officers such as Rear Admiral Clarence Stewart Williams, in his role as head of the Navy's War Plan Division, also recognized the Corps' potential as an amphibious force in the early 1920s.[15]

Although the Corps' new mission was clearly distilled, two obstacles remained. First was the continued emphasis and commitment to "Banana Wars" in Latin America. A power clique among Marine officers remained dedicated to

[11] *Advanced Base Operations in Micronesia*; Earl H. Ellis, *Navy Bases: Their Location, Resources, and Security* (Washington, DC: Government Printing Office, 1992, 1921); LtCol Frank O. Hough, Maj Verle E. Ludwig, and Henry I. Shaw Jr., *History of U.S. Marines Corps Operations in World War II*, vol. I, *Pearl Harbor to Guadalcanal* (Washington: Historical Branch, G-3 Division, Headquarters Marine Corps, 1958), 8–10, 459–61, hereafter *Pearl Harbor to Guadalcanal*; and Allan R. Millett, *Semper Fidelis: The History of the United States Marine Corps*, rev. ed. (New York: Macmillan, 1991). For the best biography of Ellis, see Dirk Anthony Ballendorf and Merrill L. Bartlett, *Pete Ellis: An Amphibious Warfare Prophet, 1880–1923* (Annapolis: Naval Institute Press, 1997).

[12] Ellis, *Navy Bases*, 3–6, 10–23, 30, 48.

[13] Ellis, *Advanced Base Operations*, 39–50.

[14] *Joint Army and Navy Basic War Plan–ORANGE*, 6 October 1920, quoted in Frank J. Infusino, "U.S. Marines and War Planning, 1940–1941" (masters thesis, San Diego State University, 1974), 145.

[15] See relevant chapters in Allan R. Millett and Jack Shulimson, eds., *Commandant of the Marine Corps* (Annapolis: Naval Institute Press, 2004); and Leo J. Daugherty III, *Pioneers in Amphibious Warfare, 1898–1945: Profiles of Fourteen American Military Strategists* (Jefferson, NC: McFarland, 2009).

Tom Lea, *First Wave: Going In, Peleliu,* 1944.
Life Collection of Art WWII, U.S. Army Center of Military History, courtesy of the Tom Lea Institute

constabulary security as the Corps' primary role. Lejeune, Russell, Holcomb, and others needed to overcome this internal resistance against amphibious development. External to the Corps, obtaining the resources and writing the doctrine to fulfill that mandate became Lejeune's primary goals in the final years as Commandant. Reductions in budgets and personnel, however, persisted throughout the 1920s, despite his best efforts. The Corps was not alone in experiencing these years of famine. The U.S. Army and Navy also saw declining budgets.[16]

Meanwhile, Lejeune decided to maximize the resources and expertise within the Corps. He put a premium on military education for his Marines and founded the Marine Corps Schools in the 1920s. The next Commandant, Major General Wendell C. Neville, followed in Lejeune's footsteps. He ensured that Marines would receive advanced training in all aspects of warfare at schools in the Marine Corps, the Navy, the Army, and at the prestigious institutions like the École *Supérieure de Guerre* in Paris, France. These opportunities afforded Marine officers to consider warmaking in systematic ways, as well as to interact with peers from other Services.[17]

One Marine who took advantage of advanced military education was then-Colonel Thomas Holcomb. As a highly decorated veteran of Belleau Wood in 1918 and a member of Lejeune's war plans committee two years earlier, Holcomb applied past experiences to his studies at the Naval War College from 1930 to 1931. He exemplified the type of professional development advocated by Neville and Lejeune. Holcomb's year at the Army War College from June 1931 to June 1932 proved to be still more fer-

[16] See Ulbrich, *Preparing for Victory*, 38–42; and Keith B. Bickel, *Mars Learning: The Marine Corps Development of Small Wars Doctrine, 1915–1940* (Boulder, CO: Westview, 2001), 205–8, 211–13.

[17] W. C. Neville, "The Marine Corps," *Proceedings of the United States Naval Institute* 55, no. 10 (October 1929): 863–66; and Donald F. Bittner, "Foreign Military Officer Training in Reverse: U.S. Marine Corps Officers in the French Professional Military Education System in the Interwar Years," *Journal of Military History* 57, no. 3 (July 1993), 481–510.

Albin Henning, untitled.
Art Collection, National Museum of the Marine Corps

tile time in his development as a senior officer. He worked with other students to formulate plans for attacking enemy nations and defeating enemy forces. Some scenarios were fabricated, while others were realistic. In one course project, Holcomb played the role of naval commander of an American force conducting an amphibious assault on Halifax, Nova Scotia. This assignment reinforced his conviction that planning down to the minutest details was necessary for a successful landing operation. Another career officer, the Army's Major George S. Patton Jr., also worked on this group project with Holcomb. These academic exercises doubtlessly helped Patton during his amphibious operations a decade later.[18]

Additionally, while working independently at the Army War College in 1932, Holcomb wrote a special report titled *The Marine Corps' Mission in National Defense, and Its Organization for a Major Emergency*. He asked an important question about the Corps: What should be the most suitable organization for a major emergency? His lengthy answer outlined the principles of seizing and defending advanced bases, and he discussed all aspects of training and supplying Marine units. Amphibious operations represented the Corps' future role in the nation's war plans. No longer did Holcomb see the Ma-

[18] "Report of Committee No 6. Subject: Plans and Orders for the Seizure of Halifax," 29 March 1932, File Number 386-6, and "Analytical Studies, Synopsis of Report, Committee No. 5," 2 March 1932, File Number 388-5, Army War College (AWC), U.S. Army Heritage and Education Center, Carlisle Barracks, PA (AHEC); and Daugherty, *Pioneers of Amphibious Warfare*, 359–99.

Men of the 2d Battalion, 2d Marines, debark from the USS *Zeilin* (APA 3) into medium landing craft on D-Day in Col Richard Gibney's *Down the Nets, Tarawa*. Amid confusion in the preassault darkness, hundreds of Marines had to cross-deck from landing boats into tracked landing vehicles, climbing across the gunwales of the two pitching craft.
Art Collection, National Museum of the Marine Corps

rine Corps as a constabulary force fighting small wars, or "other minor operations" as he called them.[19]

Although Holcomb's report drew on existing ideas and documents, its significance as an original endeavor should not be discounted. In an appendix, he also anticipated the creation of the Fleet Marine Force that next year in 1933, the end of the Corps' constabulary duties in Central America in 1934, the creation of a triangular Marine division-size unit, and lastly the publication of doctrinal manuals on amphibious assault operations in 1934 and base defense operations in 1936. The degree to which Holcomb's report circulated beyond the confines of the Army War College is not clear. This report, however, did constitute a blueprint for the Corps' future that Marines could follow and that he himself did follow later in the 1930s and during the war years.[20]

After graduating from the Army War College in 1932, Holcomb's critical academic study and practical experiences prepared him for his next duty station in the Navy Department, where he served at the Navy's War Plans Division and offered advice on amphibious operations and strategic planning relating to War Plan Orange. In this position, Holcomb advocated what military historian Edward S. Miller calls a "cautionary" strategy.[21] The U.S. Navy would strike at the Japanese forces across the Pacific using island bases seized and held by Marine Corps units as stepping stones, rather than seeking a single climactic battle between Japanese and American fleets as a primary goal. Japan's acquisition of the Micronesia Islands in the Pacific from the Germans after the First World War necessitated this more realistic and cautious strategic mind-set.[22]

Thomas Holcomb has been overshadowed by John Lejeune, Earl Ellis, and Holland Smith, all of who became household names in Marine

[19] Thomas Holcomb, "The Marine Corps' Mission in National Defense, and Its Organization for a Major Emergency," 30 January 1932, File 387-30, AWC, AHEC, 1–4.

[20] Holcomb, "The Marine Corps' Mission in National Defense, and Its Organization for a Major Emergency," 13.

[21] Edward S. Miller, *War Plan Orange: The U.S. Strategy to Defeat Japan, 1897–1945* (Annapolis: Naval Institute Press, 1991).

[22] Miller, *War Plan Orange*, 36, 181, 183, 329, 377–78.

Col Richard Gibney's *Kamikazi Attack, Okinawa*, 1 April 1945, is a tribute to the Marines and sailors at the guns on board offshore ships.
Art Collection, National Museum of the Marine Corps

Corps lore and history. These giants certainly played integral roles in the 1910s–20s in the case of Lejeune and Ellis, and from the 1920s through the Second World War in Smith's case. However, none of them transcended the years between the 1910s and 1940s as Holcomb did, nor did they exert so much influence as a student, educator, strategic planner, staff officer, and Commandant during these years.

CODIFYING DOCTRINE, CREATING FORCE STRUCTURE, AND PROCURING EQUIPMENT, 1933–38

Despite the best efforts of President Franklin D. Roosevelt and his prodefense allies in Congress, the U.S. military's funding slipped to low levels. The Marines also felt this crunch in which the Corps' annual expenditures ran between $15 and $25 million from 1935 to 1939 (in 1930s dollars). To put this in perspective, these figures amounted to between 3 and 4 percent of the U.S. Navy's annual expenditures. Nevertheless, the decade of famine also saw the flourishing of force structure improvements, doctrinal developments, and technological adaptations that easily surpassed any decade before or since in the history of the Corps, if not the entirety of American military history. In 1933, for example, the creation of the Fleet Marine Force (FMF) gave the Corps a platform, albeit modest in size, to

Col Donna J. Neary's *Fourth Marine Division Landing on Iwo Jima* depicts elements of the 4th Marine Division's eighth wave landing on Yellow Beach 1 into intensive artillery and small arms fire.
Art Collection, National Museum of the Marine Corps

support amphibious assault and base defense units.[23]

With an amphibious force structure on paper, the Marine Corps needed to codify the amphibious doctrines to be employed by the FMF in future conflicts. Much work had already been underway at the Marine Corps Schools in the mid-1920s when Brigadier General Robert H. Dunlap was the schools' commandant. His ideas and efforts, as well as the ideas outlined by Ellis, formed the foundations for the *Tentative Manual for Landing Operations* (1934) and the *Tentative Manual for Defense of Advanced Bases* (1936) produced by the Marine Corps Schools faculty and students.[24] The two "tentative" surveys looked to the future, while a separate doctrinal survey titled the *Small Wars Manual* (1935 and 1940) enumerated past lessons from Marine deployments as constabulary units in Latin America. Taken together, these three manuals constitute what

[23] MGC to Chief of Naval Operations (CNO), "Expeditionary Force," 17 August 1933, File 1975-10, PDGC 1933–38, Box 135, RG 127, NADC; William J. Van Ryzin, intvw with Benis M. Frank and Graham A. Cosmas, 1975, transcript, Marine Corps University Archives (MCUA), 74–76; Millett, *Semper Fidelis*, 319, 330–37; and Jeter A. Isely and Philip A. Crowl, *The U.S. Marines and Amphibious War: Its Theory and Its Practice in the Pacific* (Princeton, NJ: Princeton University Press, 1951), 74–75.

[24] *Tentative Manual for Landing Operations*, 1934, History Amphibious File (HAF) 39, MCUA; and *Tentative Manual for Defense of Advanced Bases*, 1936, War Plans and Related Material 1931–1944, Box 7, Entry 246, RG 127, NADC; Isely and Crowl, *The U.S. Marines and Amphibious War,* 36–44; LtCol Kenneth J. Clifford, *Progress and Purpose: A Developmental History of the U.S. Marine Corps, 1900–1970* (Washington, DC: History and Museums Division, Headquarters Marine Corps, 1973), 139–43; Allan R. Millett, "Assault from the Sea: The Development of Amphibious Warfare between the Wars—The American, British, and Japanese Experiences," in *Military Innovations in the Interwar Period*, ed. Williamson Murray and Allan R. Millett (Cambridge: Cambridge University Press, 1996), 74–75; and Daugherty, *Pioneer of Amphibious Warfare*, 194–212.

Marines wade ashore on Tinian from landing barges, which could not make the beach. The amphibious tractors in the assault wave came all the way onto the beach and then crossed the island.
Official Marine Corps photo 88088, Marine Corps History Division

Marine Corps Chief Historian Charles D. Melson has called the "holy trinity" of Marine Corps doctrine.[25]

Classes at the Marine Corps Schools were suspended from November 1933 to May 1934, so that faculty and students could compile the *Tentative Manual for Landing Operations*. They completed their work in June 1934. Not only did this resulting document outline lessons learned from past amphibious operations, but it also anticipated challenges in future operations. Despite the British amphibious fiasco at Gallipoli during the First World War, for example, American Marines postulated that careful planning, adequate training, and proper equipment could overcome the tactical advantages enjoyed by an enemy defending a shoreline. This document created a rational framework that would facilitate American amphibious assault operations in the Second World War. This process of systematic analysis regarding practical lessons of the past likewise demonstrated the institutional adaptability that has been the hallmark of the Marine Corps.[26]

Nevertheless, this landing operation manual made no detailed examination of the complexities of advanced base defense, the other half of the Corps' new dual mission. Two years later in 1936, the *Tentative Manual for Defense of Advanced Bases* filled that void by providing a doctrinal foundation for advanced base defense that

[25] Charles D. Melson, intvw with author, July 2003, cited in Ulbrich, *Preparing for Victory*, 36.

[26] *Tentative Manual for Landing Operations*, paragraphs 1.1, 1.2, 1.5, 1.8, 1.22, 3.120; James C. Breckinridge to John H. Russell, 6 November 1934, Holcomb Papers, Box 11, MCUA; Isely and Crowl, *The U.S. Marines and Amphibious Warfare*, 5, 36–44; Hough, Ludwig, and Shaw, *Pearl Harbor to Guadalcanal*, 14–22; Bittner, "Taking the Right Fork in the Road," 124–25; Gunther E. Rothenberg, "From Gallipoli to Guadalcanal: The Development of U.S. Marine Corps Amphibious Assault Doctrine, 1915–1942," in *Assault from the Sea: Essays on the History of Amphibious Warfare*, ed. Merrill L. Bartlett (Annapolis: Naval Institute Press, 1983), 177–82.

Sgt John Fabian, *LST's Off Tinian.*
Art Collection, National Museum of the Marine Corps

had been so intrinsically tied to the Corps' roles since 1898.

In the meantime, Thomas Holcomb received his first star and became commandant of the Marine Corps Schools in February 1935. During the next 22 months of his tenure, the schools made various revisions to the *Tentative Manual for Landing Operations* that would subsequently be folded into the U.S. Navy's *Landing Operations Doctrine: United States Navy, 1938,* Fleet Training Publication 167 (FTP 167) in 1938. Holcomb also supervised the completion of manuals on base defense and small wars. Because of his previous work on war plans and his military studies, Holcomb brought especially significant knowledge about amphibious warfare to the writing of the Corps' new base defense manual. Just as he routinely conducted spot inspections in classrooms and on parade grounds at Quantico, it is reasonable to infer that he sat in on discussions about artillery placement, unit deployments, or other topics, as well as read drafts of the manual.[27]

Although no documents cite Holcomb by name, his tacit influence can be seen in the following lines from the *Tentative Manual for Defense of Advanced Bases*: "Defense of advanced bases will involve the combined employment of land, air, and sea forces. Depending on the nature of the hostile attacks against a base, one arm or service may play the major role, but in the event of a general landing attack, the land forces will constitute the basic element of the defense. In any case, the ultimate success of the defense will depend upon the closest cooperation and coordination between the naval defense forces, the shore defense forces, and the aviation forces."[28] This quote highlighted the need to utilize coordinated combined air, naval, and ground forces to mount a successful defense that was reminiscent of the report penned in 1932 by Holcomb at the Army War College. In summary, the Marines looked up from the operational and tactical levels to the U.S. Navy's strategic needs and then formulated operational and tactical doctrines to fulfill those needs.

In 1936, Thomas Holcomb was promoted to Commandant of the Marine Corps. He jumped over several more senior Marine generals for several reasons. He maintained a friendship with President Roosevelt dating back to the First World War. Holcomb also fit a particular political profile inside the Corps that placed him in the ascendant clique. He favored the new dual mission of amphibious assault and base defense over the outmoded mission of constabulary security in small wars. Indeed, Holcomb's interest in amphibious doctrine and strategic planning dated back 20 years to his membership on Lejeune's ad hoc war plans committee in 1916.

[27] Clifford, *Progress and Purpose*, 45–48, 58–59, 139–42; Bittner, "Taking the Right Fork in the Road," 125–26; and Ulbrich, "The Long-Lost *Tentative Manual for Defense of Advanced Bases* (1936)," 889–901.

[28] *Tentative Manual for Defense of Advanced Bases*, preface, no pagination.

Harry Jackson, *Dawn Beachhead.*
Art Collection, National Museum of the Marine Corps

This made Holcomb an ideal candidate for the sitting Commandant, John H. Russell, who was one of the most fervent amphibious warfare advocates in the Corps.

Holcomb's career track provides other concrete justifications for his promotion. In the first 36 exemplary years of his career, he climbed steadily through the commissioned ranks, gained valuable experience in the First World War, distinguished himself in the military's education system, demonstrated administrative skills in performing staff duties, supervised significant doctrinal developments at the Marine Corps Schools, and maintained cordial contacts with civilian and naval officials alike. He enjoyed high levels of prestige as a "China Hand" and one of the "Old Breed" of the First World War.[29] Holcomb benefited from such high-ranking patrons as Lejeune and Russell, both of whom helped him into many key postings. Holcomb was the right person, in the right place, at the right time to become Marine Corps Commandant in 1936, just as he always seemed to be the right person for a given post throughout his career.

With the Fleet Marine Force established and amphibious doctrines codified, the next stage of readying the Corps for amphibious operations

[29] The term *China Hand* refers to those with expert knowledge of the Chinese culture, people, and language, particularly soldiers, journalists, and diplomats before, during, and after World War II. The term *Old Breed* refers to Marines with 5–10 years of service.

Cpl Richard Gibney, *Tarawa Landing*, sketch.
Art Collection, National Museum of the Marine Corps

entailed conducting several Fleet Landing Exercises between 1934 and 1941. When Holcomb became Commandant, he continued these efforts despite facing severe budget constraints. Known as FLEXs, these simulated amphibious assaults and base defenses gave the Marine Corps and Navy several opportunities to experiment with doctrine, troubleshoot problems, and field test equipment. The Navy performed several types of long-range shore bombardments, including counterbattery and interdiction fire. The Marines tested existing weapons and vehicles that they might employ in an actual amphibious assault, and they established a defensive position against possible counterattacks from land or sea. In so doing, the Marines discovered deficiencies in the Navy's landing craft. Only with great difficulty could Navy whaleboats or motor launches transport troops from ships through the surf to the beach. These craft offered little protection to their occupants, moved too slowly, lacked seaworthiness in rough surf, and failed to traverse coral reefs. The Marines also found such weaknesses as combat loading, which would need careful consideration to ensure that transport vessels might be packed so that equipment could be off-loaded more efficiently. It became abundantly clear that existing Navy warships, although absolutely necessary as weapons platforms, were not ideal for moving men or equipment. It took several years before the Corps found suitable landing craft and the money to pay for them in part because the Navy would not fund these efforts. Eventually, however, the Marines identified two ideal civilian designs for landing craft: Andrew Jackson Higgins' "Eureka" boat and Daniel Roebling's "Alligator" amphibian tractor. Both could be adapted to military use, and both surpassed anything in the Navy or Marine Corps' existing inventory.[30]

Cpl Richard Gibney, *Run for the Beach, Saipan.*
Art Collection, National Museum of the Marine Corps

Meanwhile, tensions in East Asia grew more acute. The year 1937 represented a watershed because Japanese forces invaded China. By year's end, Beijing, Shanghai, and Nanjing fell to Japanese control. This did not, however, bring Japan victory in this Sino-Japanese conflict in 1938 or thereafter. Instead, the fighting dragged on with no end in sight. In Europe, Nazi Germany steadily expanded its territory by annexing Austria and occupying the Sudetenland in 1938. The fluid situations in East Asia and Europe reduced the utility of War Plan Orange. The new set of threats dictated that the United States prepare for several scenarios.[31]

The Japanese, for their part, also planned for a possible war with the United States. Military historians Mark R. Peattie and David C. Evans argue that the Japanese had long followed a "wait-and-react" strategy. The Japanese anticipated three phases for naval operations during the conflict: "first, searching operations designed to seek out and annihilate the lesser American naval forces . . . in the western Pacific; second, attritional operations against a westward-moving American main battle force coming to assist in the relief or reconquest of American territories there; and third, a decisive encounter in which the American force would be crushed and the Americans forced to negotiate."[32] It was a given that the Japanese would capture American-held advanced island bases in the western Pacific. The Japanese expected to use their own bases

[30] B. W. Galley, "A History of the U.S. Fleet Landing Exercises," 3 July 1939, HAF 73, MCUA; Thomas Holcomb to Harold Stark, 26 May 1941, Box 50, SPWPD, Series III, NACP; Smith, *The Development of Amphibious Tactics*, 25–38; Isely and Crowl, *The U.S. Marines and Amphibious War*, 45–58; and Millett, *Semper Fidelis*, 339–40.

[31] Mark R. Peattie, *Ishiwara Kanji and the Japan's Confrontation with the West* (Princeton, NJ: Princeton University Press, 1975), 295–308; Saburo Hayashi with Alvin D. Coox, *Kōgun: The Japanese Army in the Pacific War* (Quantico, VA: Marine Corps Association, 1951, 1959), 9; Akira Iriye, *The Origins of the Second World War in Asia and the Pacific* (London: Longman Press, 1987), 41–51; Michael A. Barnhart, *Japan Prepares for Total War: The Search for Economic Security, 1919–1941* (Ithaca, NY: Cornell University Press, 1987), 18–20, 84–90, 116, 131; D. Clayton James, "American and Japanese Strategies in the Pacific War," in *Makers of Modern Strategy from Machiavelli to the Nuclear Age*, ed. Peter Paret (Princeton, NJ: Princeton University Press, 1986), 710, 717; Ross, *American War Plans*, 177–83; and "Joint Army and Navy Basic War Plan–Orange (1938)," Joint Board No 325, Serial 618, p. 1, Microfilm 1421, Reel 10, NACP.

[32] Peattie and Evans, *Kaigun*, 464. For an excellent study of Japanese amphibious capabilities, see the chapter titled "The Development of Imperial Japanese Army Amphibious Warfare Doctrine" in Edward J. Drea, *In the Service of the Emperor: Essays on the Imperial Japanese Army* (Lincoln: University of Nebraska Press, 1998), 14–25.

Marines wading ashore on D-Day at Bougainville, as seen from a beached LCVP.
Official U.S. Marine Corps photo 54384, Marine Corps History Division

in the Marshalls, Marianas, and other Micronesian islands in offensive and defensive operations. Construction of airfields began on these islands as early as 1934 and accelerated military building programs thereafter. The Japanese plan to defeat the U.S. Fleet mirrored the American Orange Plan. It seems that each side was playing into the other's hands. Japan's wait-and-react strategy remained intact until 1940, when such priorities as natural resources and such realities as American naval expansion caused the Japanese to shift toward an offensive mind-set.[33]

In the United States, the outmoded War Plan Orange did not affect the Marine Corps, which continued to play an important role in the last iteration as well as in subsequent war plans. Because the Corps' contributions were tactical and operational rather than strategic, the Marines kept their focus squarely on defending friendly bases or attacking enemy-held bases. They adapted to the evolving situations in 1938 and thereafter.[34]

Two important measures bore witness in 1938 to the U. S. Navy's acceptance of the Marine Corps as its amphibious assault and base defense force. First, the Navy adopted the Fleet Training Publication 167 (FTP-167) as its blueprint for amphibious operations. Commandant Holcomb had ordered a committee to modify

[33] Peattie and Evans, *Kaigun*, 465–73; Ross, *American War Plans*, 168–69; Specter, *Eagle Against the Sun*, 43–45; and James, "American and Japanese Strategies in the Pacific War," 705–7.

[34] Commander-in-Chief, U.S. Fleet (CINCUS) to CNO, 27 July 1937, Marine Corps Budget Estimate (MCBE) FY 1936–43, Box 1, Entry 248, RG 127, NADC; and Gerald C. Thomas to Alexander A. Vandegrift, 9 August 1945, HAF 204, MCUA.

As supporting naval and air units pave the way with high explosives, Marine-laden assault craft form the first wave and move in for the attack on Peleliu in the Palau Islands. The leathernecks hacked out a mile and one-half long beachhead, and after bitter fighting, began the advance on the Japanese airfield.
Official U.S. Marine Corps photo 94875, Marine Corps History Division, courtesy of Sgt William A. McBride

the Corps' own *Tentative Manual for Landing Operations* of 1934 according to the Navy's needs. The resulting revision added broad strategic and naval perspectives to the Marines' tactical and operational focuses.[35]

Second, U.S. Secretary of the Navy Claude A. Swanson appointed Admiral Arthur J. Hepburn to head a board of naval officers to assess the strategic roles of bases on Guam, Wake, Midway, and other islands in light of Japanese threats in the Pacific. In December 1938, the so-called "Hepburn Board" prioritized the advanced bases in the Pacific, according to strategic needs dictated by a given base's possible benefits for aircraft, submarines, and surface warships in a war with Japan. The board argued that Guam should become a "Major Advanced Fleet Base" for operations in support of American forces on the Philippines and in the western Pacific. Wake and Midway Islands should become patrol plane bases for reconnaissance or supply bases for defensive and offensive actions. The Hepburn Board members believed that construction should be started as quickly as possible on those islands. Apart from recommendations regarding the bases proper, the board's final report instructed the Marine Corps to organize "defense detachments" to hold those island bases against possible Japanese attacks in the opening stages of a conflict. This decision drew on ideas outlined in the Marine Corps School's *Tentative*

[35] Alexander A. Vandegrift and Robert B. Asprey, *Once a Marine: The Memoirs of A. A. Vandegrift, United States Marine Corps* (New York: Norton, 1964), 93, 118; Smith and Finch, *Coral and Brass*, 60–62; and Millett, *Assault from the Sea*, 76–77.

Manual for Defense of Advanced Bases of 1936.[36]

Other important steps toward operational readiness occurred in 1938. American entrepreneurialism provided the technological means for effective ship-to-shore transportation during an amphibious operation. The American military possessed no landing craft capable of providing speed, durability, and seaworthiness during this transit. Furthermore, any craft needed to be able to land on a beach and extract itself from that beach with relative ease. Ironically, the commercial designs of Roebling's Alligator amphibian tractor and Higgins's Eureka boat provided vessels to meet performance specifications. Both found enthusiastic supporters among Marine officers. Nevertheless, subsistence-level budgets restricted the Marines from supporting the two boat builders. To their great credit, Higgins and Roebling spent their own money to modify their civilian designs to fit the amphibious assault applications.[37]

The fast-rising tide of Nazi Germany in Western Europe and Militarist Japanese in East Asia made War Plan Orange obsolete by 1939. American strategists reacted by formulating the more realistic Rainbow Plans with five versions addressing several possible wartime circumstances that might confront the United States. The versions ranged from Rainbow Plan 1, which entailed a unilateral American defense of the Western Hemisphere and no involvement with conflicts in Europe or East Asia; to Rainbow Plan 5, which envisioned combined American, British, and French offensives to vanquish Germany as quickly as possible. The United States, meanwhile, would remain on the strategic defensive in the Pacific against Japan. Once Germany was defeated, all available American and Allied forces would be redirected to crush Japan. As a result of these new scenarios, the U.S. Army reoriented its strategic emphasis towards defense of the Western Hemisphere and war in Europe and away from Japan and the Pacific Ocean. East Asia held little or no interest among most Army planners, except for those who agreed with General Douglas MacArthur's delusional belief in the defensive viability of the Philippines in a war with Japan.[38]

All the Rainbow Plans expected the Corps to play active operational roles in the Pacific. It mattered little what the Navy did at the strategic level. If the U.S. Fleet launched an offensive campaign against the Japanese, then the Marines would capture enemy bases in support of the fleet and defend them against possible counterattack. Or, if the U.S. Fleet stood on the defen-

[36] "Report of the Board to Investigate and Report upon the Need, for Purposes of National Defense, for the Establishment of Additional SubMarine, Destroyers, Mine, and Naval Air Bases on the Coasts of the United States, its Territories and Possessions," 1 December 1938, Strategic Plans Division War Plans Division (SPDWPD), Series III, Misc. Subject File, Box 50, RG 38, NACP, 1–6, 62–70, 87–89, hereafter Hepburn Board Report; Miller, *War Plan Orange*, 241–43, 250–53; Gregory J. W. Urwin, *Facing Fearful Odds: The Siege of Wake Island* (Lincoln: University of Nebraska Press, 1997), 48–52; and Ulbrich, "Clarifying the Origins and Strategic Mission of the U.S. Marine Corps Defense Battalion," 81–107.

[37] Unreferenced quotation in Austin R. Brunelli, intvw with Norman J. Anderson, 1984, transcript, MCUA, 25; William Upshur to Holcomb, 26 February 1939, Holcomb Papers, Box 6, MCUA; Victor H. Krulak, *First to Fight: An Inside View of the U.S. Marine Corps* (Annapolis: Naval Institute Press, 1984), 88–92, 100–2; Timothy Moy, *War Machines: Transforming Technologies in the U.S. Military, 1920–1940* (College Station: Texas A&M University Press, 2001), 117–18, 150–57; and Jerry E. Strahan, *Andrew Jackson Higgins and the Boats that Won World War II* (Baton Rouge: Louisiana State University Press, 1994), 24–39.

[38] "Joint Army and Navy Basic War Plans, Rainbow Nos. 1, 2, 3, 4, and 5," 9 April 1940, JB 325, Serial 642, M1421, Reel 11, NACP; Alexander Kiralfy, "Japanese Naval Strategy," in *Makers of Modern Strategy: Military Thought from Machiavelli to Hitler*, ed. Edward Meade Earle (Princeton, NJ: Princeton University Press, 1941), 457–61, 480–84; Henry G. Gole, *The Road to Rainbow: Army Planning for Global War, 1934–1940* (Annapolis: Naval Institute Press, 2003), 108–9, 177–81; and Brian McAllister Linn, *Guardians of Empire: The U.S. Army and the Pacific, 1902–1940* (Chapel Hill: University of North Carolina Press, 1997), 177–82, 244–46; James, "American and Japanese Strategies," 708–11; Ross, *American War Plans*, 164–78; and Miller, *War Plan Orange*, 83–4, 214–29, 324.

sive, then the Marines would also be called upon to hold American bases and recapture any bases taken by the Japanese.[39]

As American strategies shifted to meet new threats, the Marines honed their amphibious assault techniques and improved their landing craft in additional FLEXs in 1939. The force structure for the other half of the Corps' dual mission also began to take shape during that summer with the unveiling the Marine Corps' "defense battalion."[40] As envisioned on paper, this 1,000-man unit boasted an impressive array of weapons: 12 Navy 5-inch artillery pieces for coastal defense, 12 3-inch antiaircraft artillery guns for air defense, 48 .50-caliber machine guns for either antiaircraft or beach defense, and 48 .30-caliber machine guns for beach defense. All units would also receive high-intensity searchlights and radar systems. Some defense battalions might even receive larger 7-inch artillery pieces. The proportion of Marines per heavy weapon far exceeded the Corps' typical light infantry unit. Indeed, the defense battalion's firepower rivaled that of an U.S. Navy light cruiser.[41]

Once ensconced on a fortified island, defense battalions provided the American naval or aviation forces with self-sufficient bases of operations. Nevertheless, the Marines did depend on the Navy for logistical support and eventually relief during a campaign. They could not hold out indefinitely against determined enemy assaults.

The defense battalions became part of the FMF and complemented the amphibious assault units therein. The defense battalions represented the reincarnation of the Marine Corps Advanced Base Force of the early twentieth century, as well as the realization of the *Tentative Manual for Defense of Advanced Bases* in 1936. The defense battalions thus fit strategic and doctrinal molds perfectly.[42]

Despite the fact that global war appeared ever more likely, the United States' armed forces remained ill-prepared for any conflict. Isolationism maintained its hold on an American public who did not wish to get entangled in the conflicts in Europe or Asia. Instead, they turned their attention to feeding their families during the last years of the Great Depression.

During the summer of 1939, the Navy conducted a detailed self-assessment to answer the question, "Are We Ready?" A negative answer came out in the final report. Both seaborne Services, according to Chief of Naval Operation Admiral Harold R. Stark, suffered from numerous and "critical deficiencies" in manpower and equipment. Of relevance to the Corps was "the lack of Pacific bases west of Hawaii." Stark further cited the inability of the Navy and the Marine Corps to seize any island bases or protect those bases once they had been captured. The CNO saw it as his major task to alleviate these deficiencies, and he spent the next 30 months in office trying to do so. Rarely did the Marine Corps enjoy a better advocate than Admiral Stark, who began deploying Marines to island bases in the Pacific. He subsequently asked

[39] "The Idea of the Fleet Marine Force," *Marine Corps Gazette* 23, no. 6 (June 1939): 61; Miller, *War Plan Orange*, 227; and Ulbrich, "Clarifying the Origins and Strategic Mission of the U.S. Marine Corps Defense Battalion," 93.

[40] Hepburn Board Report, 1–6, 62–70, 87–89; and CNO to MGC, 16 February 1939, Holcomb Papers, Box 6, MCUA, 1–2.

[41] Holcomb to Commanding General of Fleet Marine Force (FMF), 28 March 1939; Robert D. Heinl, "Defense Battalions," 15 August 1939; and unsigned memorandum, "Material Requirements for four Defense Battalions," 15 August 1939, all in DP-PWPSGC 1926–1942, Box 4, RG 127, NACP.

[42] Annual Report of the MGC to the Secretary of the Navy (SecNav) for FY 1940, 27 August 1940, MCUA, 24, 38–40; Col Robert Debs Heinl Jr., *Soldiers of the Sea: The United States Marine Corps, 1775–1962* (Baltimore, MD: Nautical & Aviation, 1991, 1962), 306–7; Urwin, *Facing Fearful Odds*, 192; and Maj Charles D. Melson, *Condition Red: Marine Defense Battalions in World War II* (Washington, DC: History and Museums Division, Headquarters Marine Corps, 1996), 2–5.

Marines go down cargo nets into landing craft, August 1944.
Defense Department photo (Marine Corps) 94712, Marine Corps History Division, courtesy of Bailey

the Corps to organize four fully manned and equipped defense battalions. This task, however, caused severe strains in the thinly stretched and underfunded Marines.[43]

SHIFTING AMERICAN STRATEGIES, CONSISTENT MARINE MISSIONS, 1938–41

When German forces rolled over the Polish border on 1 September 1939, the governments of France and Britain promptly declared war on Germany. That same month, President Roosevelt reacted by declaring a "limited national emergency" with two goals in mind: "safeguarding" American neutrality and "strengthening our national defense within the limits of peacetime authorizations."[44] War in Europe likewise affected American strategic planning and caused a rapid succession from War Plan Rainbow 2 with its focus on Japan, to War Plan 3 with its focus on Germany, and finally to War Plan 4. This last change occurred when France surrendered to Germany in June 1940. The strategic situation degenerated to a point that the United States stood only with beleaguered Great Britain against the Axis powers. War Plan Rainbow 4 reduced the United States to defending the Western Hemisphere against potential Axis incursions. American forces in the Pacific would set up a defensive parameter from the Panama Canal Zone to Hawaii to Alaska.

No more was there question of whether the United States would enter the Second World War. The new seminal questions concerned how much and how fast the nation could mobilize and prepare itself for conflict. President Roosevelt adopted a short-of-war strategy.[45]

The Marine Corps exercised little influence over the changes in strategic planning process, so the Marines focused on fielding a force adequate to meet those expectations of fighting on one and maybe even two oceans. Making mat-

[43] Ernest J. King, *The U.S. Navy at War, 1941–1945: Official Reports to the Secretary of the Navy* (Washington, DC: Department of the Navy, 1946), 37; Baer, *One Hundred Years of Sea Power*, 152–53; and Millett, *Semper Fidelis*, 342–43.

[44] F. Roosevelt, "The Five Hundred and Seventy-Seventh Press Conference (Excerpts)," 8 September 1939, in *The Public Papers and Addresses of Franklin D. Roosevelt, 1939*, vol. 8, *War and Neutrality* (New York: Macmillan, 1941), 483–84.

[45] "Joint Army and Navy Basic War Plans, Rainbow Nos. 1, 2, 3, 4, and 5"; Stetson Conn, "Changing Concepts of National Defense in the United States, 1937–1947," *Military Affairs* 28, no. 2 (Spring 1964): 1–4; Miller, *War Plan Orange*, 260–61, 270; James, "American and Japanese Strategies," 705–6, 710–11; Calvin L. Christman, "Franklin D. Roosevelt and the Craft of Strategic Assessment," in *Calculations: Net Assessment and the Coming of World War II*, ed. Williamson Murray and Allan R. Millett (New York: Free Press, 1992), 243–45; Linn, *Guardians of Empire*, 180–82; Peattie and Evans, *Kaigun*, 464–67; and Ronald H. Spector, *Eagle Against the Sun: The American War with Japan* (New York: Free Press, 1985), 63–65.

ters worse, the Corps could not hope to mobilize quickly enough to keep up with any of the Rainbow Plans' timetables.[46] The Marines did their best to augment their amphibious assault and base defense capabilities between the outbreak of war in Europe in 1939 and the end of 1940. Marine units participated in FLEX 6 in January to March 1940. The simulated attacks showed the greatest improvements and achieved the highest level of realism to date, though limitations and deficiencies in equipment and manpower still plagued the Americans. Doctrine intersected with practice as the Marines recognized the following principles as essential to successful assaults: naval gunfire and aviation close air support could be combined with Marine forces to effect an amphibious assault; logistical capabilities could be expanded to supply those troops on shore; and specially trained and equipped defense battalions could secure islands against counterattack by enemy forces. The Eureka boats and Alligator tractors proved themselves as superior to all competitors. Their respective designers, Higgins and Roebling, finally received large contracts for the Eureka and Alligator, and would become officially known as the Landing Craft, Vehicle, Personal (LCVP) and the Landing Vehicle Tracked (LVT-1).[47] Even so, funds took a long time to get disbursed to contractors, and the manufacturers procured new materials at an interminably slow pace. This sluggishness vexed senior Marine leaders like Holcomb and Holland Smith.[48]

The final months of 1940 brought into clear view the fact that the United States could expect only Britain to be an ally. In the Pacific, token resistance by British and Dutch forces could not hope to halt the determined Japanese expansion. Not even Rainbow Plan 5 accounted for the complexity or flexibility of the new circumstances.[49]

Consequently, the United States adopted a "Germany First" strategy. In so doing, the Navy's Chief of Operations Admiral Harold R. Stark conceded to what the Army's strategic planners wanted when he formulated Plan Dog. In this newest scheme, the war in Europe would be dominated by the Army, leaving the Navy in a subordinate role. The seaborne Services would play a larger, albeit defensive, role in the Pacific against Japan. Plan Dog formed the nucleus for America's wartime strategy.[50]

Although the Marines remained observers of the process surrounding Plan Dog and successive war plans, this did not mean that the Corps

[46] Memo for MGC, 16 June 1940, DPPWPSGC 1921–43, Box 34, RG 127, NACP; RAdm Julius A. Furer, *Administration of the Navy Department in World War II* (Washington, DC: Navy Department, 1959), 34–35, 587; and Gordon W. Prange, *At Dawn We Slept: The Untold Story of Pearl Harbor* (New York: McGraw-Hill, 1981), 38–40.

[47] VAdm George C. Dyer, *Amphibians Came to Conquer: The Story of Admiral Richmond Kelly Turner* (Washington, DC: Navy Department, 1972), 206–8; Robert D. Heinl Jr., "The U.S. Marine Corps: Author of Modern Amphibious Warfare," in *Assault from the Sea*, 189; Memo for Director of Plans and Policies, 3 July 1940, MCBE FY 1936–43, Entry 248, Box 1, RG 127, NADC; Annual Report of the MGC to the SecNav FY 1940, 15 August 1939, MCUA, 61-63; Krulak, *First to Fight*, 93–95, 101–4; Smith, *Development of Amphibious Tactics*, 29–33; Strahan, *Andrew Jackson Higgins*, 42–50; and Moy, *War Machines*, 159–60.

[48] J. B. Earle to CNO, 8 September 1939, Stark to SecWar, 24 October 1939, Director of Division of Plans and Policies to MGC, 12 October 1939, and H. B. Sayler to Holcomb, 3 October 1940, all in DPPWPSGC 1926–1942, Box 4, RG 127, NACP; and Holcomb to Robert L. Denig, 5 November 1939, Holcomb Papers, Box 6, MCUA.

[49] Conn, "Changing Concepts of National Defense," 5–6; and Jonathan G. Utley, "Franklin Roosevelt and Naval Strategy, 1933–1941," in *FDR and the U.S. Navy*, ed. Edward J. Marolda (New York: St. Martin's Press, 1998), 53–57.

[50] Harold R. Stark, memo to SecNav, 12 November 1940, in Stark, summary notes, Box 142, MCOHC, MCUA, hereafter Stark memorandum. Various drafts of the Stark memorandum can be found in Ross, *American War Plans*, 225–30; Mark M. Lowenthal, "The Stark Memorandum and the American National Security Process, 1940," in *Changing Interpretations and New Sources in Naval History*, 358–59; B. Mitchell Simpson III, *Admiral Harold R. Stark: Architect of Victory, 1939–1945* (Columbia: University of South Carolina Press, 1989), 70–75; Gole, *Road to Rainbow*, 102–21; and Linn, *Guardians of Empire*, 177–83.

Senior American leaders observing a joint Marine Corps-Army amphibious exercise at New River, NC, in July 1941. From left: MajGen Holland M. Smith; MajGen Commandant Thomas Holcomb; Secretary of the Navy Franklin Knox (looking through binoculars); and then-Col Teddy Roosevelt Jr., U.S. Army 1st Infantry Division. It is worth noting that Roosevelt's division was the only major unit in the U.S. Army with amphibious experience before the outbreak of war later that December.
Marine Corps History Division

was ignored as irrelevant. Stark and the Navy concentrated on strategic and national goals, which only concerned the Corps in terms of mobilization timetables and resource allocation, but mattered very little to it in terms of its dual missions. Both base defense and amphibious assault fit into operational requirements of Plan Dog, because they concerned the prosecution of the war. With help from the Marines, the U.S. Fleet would hold the defensive perimeter from Alaska to Hawaii to Central America against Japanese incursions. American forces were also expected to preserve the logistical lifeline through Australia to British-held Malaysia. Stark hoped that advanced bases on Wake, Midway, and other islands could be maintained as American for future operations. Japanese-held island bases would have to be assaulted and defended in turn. Any American islands taken by the Japanese would need to be recaptured by American forces. In sum, the Navy would conduct limited operations utilizing its air, surface, and amphibious forces to maintain the strategic status quo in the Pacific. Once Germany was eliminated as an enemy, the United States could turn its full weight against Japan. Herein lay the significance of Plan Dog and its successive plans for the Corps: Marines could expect to play active roles in both base defense and amphibious assault, whether in operations supporting defensive or offensive operations.[51]

Because naval campaigns outlined in the

[51] Stark Memorandum; Baer, *One Hundred Years of Sea Power*, 154–57; and Hough, Ludwig, and Shaw, *Pearl Harbor to Guadalcanal*, 64.

war plans would require larger amphibious assault units, the Corps received authorization to create more viable, larger division-size units of approximately 18,000 Marines capable of seizing enemy-held islands. The creation of two paper divisions in the FMF occurred in early February 1941. Later in July that summer, elements of the U.S. Army's 1st Infantry Division, the 1st Marine Division, and the U.S. Atlantic Fleet made simulated amphibious landings in the Caribbean and at New River, North Carolina. The new force structures and exercises followed the doctrinal principles laid down in the *Tentative Manual for Landing Operations* from 1934 and the FTP-167 from 1938. Although these exercises suffered some setbacks, the participating Marines, soldiers, and sailors learned what NOT to do.[52] This Marine Corps' emphasis on amphibious warfare took on another element as well—institutional survival.

FROM PREWAR DOCTRINE TO WARTIME APPLICATION

The last few months of peace in late 1941 passed very quickly. The Marines struggled to ready themselves on far-flung Pacific islands as well as mobilize back in the United States. Commandant Holcomb's efforts to meet expectations resembled robbing Peter to pay Paul as he ordered units with full complements to be split apart to create cadres for two separate units. The U.S. Navy and Army's senior leaders experienced similar problems in matching resources to needs.[53]

While American strategic planners anticipated Japanese attacks on the Philippines, Guam, or Wake, the idea of a massive air attack against the main U.S. Navy and Army bases at Pearl Harbor seemed too far-fetched to be plausible. Sadly, underestimating the skill and audacity of the Japanese had dire consequences on the Sunday morning of 7 December 1941. On that infamous day, the Japanese caught the Americans unawares and launched preemptive strike that destroyed the U.S. Fleet's battleship component and laid waste to the ground-based aircraft on Oahu in Hawaii.[54]

In the hours, days, and months thereafter, the Japanese launched attacks against Wake, Guam, the Philippines, and Midway. Those were consistent with the anticipated Japanese actions. Elements of a defense battalion on Wake Island proved its mettle for more than a fortnight before succumbing to overwhelming Japanese force in late December. The few Marines on Guam surrendered without a fight in December. The Philippines fell five months later after American and Filipino forces fought desperate holding actions, as waiting for the relief force envisioned in War Plan Orange would take nearly three years to arrive.

Although attacked, Midway was not secured by the Japanese. It would later be the scene of a decisive naval battle in 1942. Indeed, Marines in two defense battalions held Midway against Japanese aerial attacks. Their antiaircraft fire downed 10 Japanese planes during their aerial assault, which did not destroy the ground defenses on Midway in anticipation of an amphib-

[52] "Training of Units of the FMF," n.d. [ca. February 1941], GBSF, GB 425, Box 135, RG 80, NACP; H. Smith to CNO via MGC, 10 September 1941, Holland M. Smith to King, 14 November 1941, and Deputy Chief of Staff of the U.S. Army to CNO, 10 October 1941, all in Holcomb Papers, Box 27, MCUA; Holcomb to Marston, 22 November 1941, Holcomb Papers, Box 4, MCUA; Smith, *Development of Amphibious Tactics*, 36–38; Isely and Crowl, *The U.S. Marines and Amphibious War*, 63–65; and Millett, *Semper Fidelis*, 348–49.

[53] See Ulbrich, *Preparing for Victory*, 92–102.

[54] The best single volume survey of the Pacific War remains Spector, *Eagle Against the Sun*. See also relevant chapter in Millett, *Semper Fidelis*.

Marines aboard a Navy transport study a relief model of Tarawa Atoll a few days prior to the famed 76-hour battle by the 2d Marine Division for the former Japanese stronghold. Many of these relief models were made by the Relief Mapping Section of the 2d Division, so each Marine would be familiar with detailed terrain features of the island. The same system was used for the Saipan operation.
Defense Department photo (Marine Corps) 101807, Marine Corps History Division, courtesy of Sgt Porter

ious assault in the coming days. It is also worth noting that a defense battalion opposed daily Japanese aerial bombing raids and frequent Japanese Navy bombardments on Guadalcanal from August 1942 to February 1942. The Midway and Guadalcanal Marines' tactics and unit structure followed the doctrines laid down in the *Tentative Manual for Defense of Advanced Bases* (1936).[55]

During the War in the Pacific, the doctrines in the *Tentative Manual for Landing Operations* were successfully applied in the island-hopping and leapfrogging campaigns, though not without halting progress and severe casualties. At Guadalcanal, the 1st Marines made an unopposed landing on 7 August 1942. The real challenge came not in defending their tenuous beachhead and all-important airfield against Japanese air, land, and sea incursions, but only in the Navy's maintaining the supply lines to the American units on the island. Although suffering severe losses in men, aircraft, and ships, the U.S. Navy succeeded in this logistical mission and also destroyed the Japanese supply system.[56]

More than a year after the amphibious operation on Guadalcanal, the long-anticipated drive across the Central Pacific began in Novem-

[55] The more thorough examination of Wake Island is Urwin, *Facing Fearful Odds*. For an overview of defense battalions in the Pacific War, see Melson, *Condition Red*.

[56] Aptly titled is Richard B. Frank, *Guadalcanal: The Definitive Account of the Landmark Battle* (New York: Random House, 1990). See also David J. Ulbrich, "Thomas Holcomb, Alexander Vandegrift

ber 1943. The Marines' bloody assault against Tarawa was one example of how, even with the most sound doctrines, the fog and friction of war can conspire to bring about near defeat. The Marines and their Navy counterparts used a feedback loop that created a learning curve. The Americans adapted doctrines, equipment, and force structure to overcome the Japanese corresponding evolution of tactics in their defensive efforts on the likes of Peleliu, Saipan, Iwo Jima, and Okinawa.[57]

The value of the Marine Corps' doctrines extended beyond the Central Pacific into the Southwest Pacific and European theaters of operations, where the U.S. Army and Navy conducted several large-scale amphibious assaults.[58] The principles outlined the *Tentative Manual for Landing Operations* found their way into the Navy's FTP-167 (1938) and subsequently on to the War Department and Army in *Landing Operations on Hostile Shores*, FM 31-5 (1941). This document's preface stated that it "is based to large extent on the Landing Operations Doctrine, U.S. Navy, 1938. The arrangement of subject matter is similar to the Navy publication and many illustrations are taken from it." The Army's Chief of Staff General George C. Marshall's name appeared on the signature block "by order the Secretary of War."[59] Perusals of the tables of contents of the 1941 FM 31-5 and later revisions as well as wartime revisions of FTP-167 reveal that the U.S. Army and Navy continued to borrow and adapt the Marines' doctrines.

CONCLUSIONS

The operational and tactical applications of amphibious assault and base defense in the Pacific and European theaters remained a means to a strategic end as determined by the senior Allied leaders. Although untested in the 1920s and 1930s, the Marine Corps amphibious doctrines laid out in the tentative manuals, in ideas presented by the likes of Pete Ellis, Holland Smith, John Lejeune, and Thomas Holcomb, in simulated amphibious assaults, and in equipment procurement, proved to be remarkably forward-looking in fulfilling strategic needs in the Pacific and Europe. They took their doctrine, force structure, and equipment procurement cues from the American strategic plans and missions. The late military historian Russell F. Weigley saw great value in this process: "Simply by defining the specific problems into which amphibious operations divided themselves, the Marine Corps made it evident that the problems most likely were not insoluble; and the Corps went on to delineate many of the solutions."[60]

Such problem-solving efforts are needed as much in the twenty-first century as they were in the Second World War. In 2013, the new term is Anti-Access/Anti-Denial (A2/D2).[61] The new operational challenges to successful assaults can be seen in accurate long-rang rockets, advanced underwater obstacles, fast jet aircraft, and even tactical nuclear weapons. Overcoming these requires the amphibious assault forces to have plans

and Reforms in Amphibious Command Relations in 1942," *War and Society* 28, no. 1 (May 2009): 113–47.

[57] Even after more than 60 years since publication, the seminal work on amphibious operations in the Pacific War remains Isely and Crowl, *The U.S. Marines and Amphibious Warfare*. For the latest study, see Sharon Tois Lacy, *Pacific Blitzkrieg: World War II in the Central Pacific* (Denton: University of North Texas Press, 2013).

[58] See chapters on Gen George S. Patton Jr, LtGen Arthur G. Trudeau, and RAdm Walter C. Ansel in Daugherty, *Pioneers of Amphibious Warfare*, 298–400.

[59] *Landing Operations on Hostile Shores*, FM 31-5 (Washington, DC: War Department, 1941), II.

[60] Russell F. Weigley, *The American Way of War: A History of United States Military Strategy and Policy* (Bloomington: Indiana University Press, 1977, 1973), 264.

[61] For a recent analysis of the implications of A2/AD for amphibious warfare, see Sam J. Tangredi, *Anti-Access Warfare: Countering A2/D2 Strategies* (Annapolis: Naval Institute Press, 2013).

and preparations to breach obstacles, establish beachheads, and maintain logistical networks. All these missions can only be achieved under an umbrella of air superiority and a cordon of naval (surface and underwater) superiority that reach several hundred miles in all directions. These require truly "joint" operational capabilities.[62] From the defensive perspective, the key elements include disruption of enemy assault forces and logistical support efforts. Indeed, so effective have improvised explosive devices been on land, that they will doubtlessly be utilized to impede ship-to-shore transit and on-shore maneuver by amphibious assault forces. Just as was the case in the 1920s and 1930s, so too it is evident in 2015 that mastering the offensive side of amphibious warfare necessitates an equally clear understanding of the defensive side.[63]

The author of this article believes that the need for projecting military force and humanitarian assistance from the sea will not diminish, especially the fact that the majority of the world's population lives within a couple of hundred miles of major bodies of water. This statement is all the more relevant because of the pivot toward the Pacific Rim by the United States military now and in the future. The twenty-first century amphibious operational environment certainly requires the type of doctrine, force structure, and equipment that only the Marine Corps is in any position to develop. The Corps' amphibious mission—whether executing an assault under fire, landing to support humanitarian efforts, or defending a shoreline against enemy invasion—are different from the Second World War or even Gallipoli in degree, but not in kind. Indeed, it is arguable that the Marine Corps is in better shape now in the twenty-first century than in the previous century because the Corps possesses an integrated and tested force structure platform, albeit in ground combat deployments conducting counterinsurgency operations, in the Marine Air-Ground Task Force concept. Admittedly, it remains to be seen whether new, effective amphibious assault vehicles will keep pace with the requirements for speed, capacity, and agility in the contemporary operating environment. It is also uncertain how the amphibious mission should best be balanced against the Corps' other missions such as in counterinsurgency operations. Nevertheless, some 80 years hence, the foundational doctrines still ring true in *Tentative Manual for Landing Operations* (1934) and the *Tentative Manual for Defense of Advanced Bases* (1936) that helped the Corps prepare to fight the Pacific War.

Finally, the habits of mind of Thomas Holcomb, John Lejeune, Earl Ellis, Holland Smith, and others like them were needed to drive doctrinal development, equipment procurement, and force structure creation for meet challenges of amphibious operations during the Second World War. Their habits of mind included not only solving problems in the amphibious lane, but also ensuring institutional survival and maintaining strategic relevance during times of constricted resources in interwar periods.

[62] See, for example, *Amphibious Operations*, Joint Publication 3-02 (Washington, DC: Joint Chiefs of Staff, 2014).

[63] The December 2012 issue of the *Marine Corps Gazette* contains several articles on the current state and future prospects for the Corps' amphibious missions. More recently, see Trevor Howell, "Traditional Amphibious Warfare: Wrong for Decades, Wrong for the Future," *Marine Corps Gazette* 98 (September 2014): 18–22.

THE ROLE OF MARINE AND SHORE-BASED NAVAL AIR AT GUADALCANAL

Some Lessons for Today

by Major Philip F. Shutler

Marine Corps Gazette, May 1989

By mid-October 1942, the situation of the 1st Marine Division on Guadalcanal had gone from miserable to desperate.[64] The Marines had come ashore on 7 August to seize the partially completed Japanese airfield on the island. Within two days, Japanese air and surface attacks had forced the U.S. transports to withdraw with half of their supplies still aboard. Since then, Marines had lived on two meals per day of captured Japanese food supplies. They had managed to complete the airstrip using captured Japanese equipment and had named it Henderson Field after a Marine killed at Midway. As soon as it was ready, a makeshift contingent of Marine, Navy, and Army Air Corps squadrons flew in to help defend it. They had repulsed one major counteroffensive in September, but another one was clearly on the way. On the night of 12–13 October, two Japanese battleships stood offshore and pounded the field with almost a thousand 14-inch shells, destroying more than half of the aircraft, and almost all of the aviation fuel.

The next day a Marine colonel briefed some pilots from the Army's 67th Pursuit Squadron:

> *We don't know whether we'll be able to hold the field or not there's a Japanese task force of destroyers, cruisers, and troop transports headed this way. We have enough gasoline left for one mission against them. . . . After the gas is gone we'll have to let the ground troops take over. Then your officers and men will attach themselves to some infantry outfit good luck and goodbye.*[65]

Today, of course, we know that the Marines held, but we sometimes forget how close the struggle for Guadalcanal really was. We also sometimes forget the crucial role that Marine and Navy carrier aircraft based at Henderson Field played in the campaign. William S. Lind, for example, has argued that:

[64] The original article came from Philip F. Shutler, "The Role of Marine and Shore-based Naval Air at Guadalcanal: Some Lessons for Today," *Marine Corps Gazette* 73, no. 5 (May 1989). Minor revisions were made to the text based on current standards for style, grammar, punctuation, and spelling.

[65] *Pacific Counterblow*, Wings at War Series no. 3 (Maxwell Air Force Base, AL: Center for Air Force History, 1992).

> *[if] Marine aviation is truly to focus on supporting the Marines on the ground, it needs to be reorganized to emphasize [close air support] at the expense of air-to-air and deep interdiction capabilities.*

Other *Gazette* authors have argued that (1) the [Northrup Grumman] EA-6B Prowler, with its sophisticated electronic warfare system, is admirably suited for defending ships against missile attack or supporting "war-at-sea" strikes but ill-suited to the needs of the Marine Corps; and (2) that while the [McDonnell Douglas] AV-8B Harrier II is ideal for close air support it offers little advantage in a naval campaign. Essentially, all of these arguments assume a dichotomy between the roles of Marine Corps aviation and naval aviation and conclude that, in order to meet future fiscal constraints, we should reduce interoperability between the two communities.

A reexamination of the Guadalcanal campaign leads to a much more comprehensive understanding of the role of Marine Corps aviation. It suggests that, in order to fully support the Marine on the ground in an amphibious campaign, *Marine air must remain shore-based naval air, capable of both striking ships and defending them against air strikes.* To explain this apparent paradox, we should first review the broad outlines of the struggle for Guadalcanal chronologically and analyze them in a functional perspective. We can then draw some conclusions for future amphibious operations.

Chronologically, the struggle for Guadal-

Combat artist Dwight Shepler depicts the Cactus Air Force on Henderson Field, with a parked Army Air Forces Lockheed P-38 Lightning in the foreground and Marine Grumman FRF-3 Wildcats swarming overhead.
Art Collection, U.S. Navy

1stLt Hugh Laidman, *At the Edge of Henderson Field.* When it became known that the Japanese were constructing a new airfield on Guadalcanal, the Joint Chiefs of Staff issued a directive calling for the capture of one or more locations in the southern Solomons. Landings were made at Tulagi and Guadalcanal on 7 August. One of the first objectives was the partially completed airfield, which was quickly named Henderson Field in honor of Maj Lofton R. Henderson, a Marine dive-bomber pilot shot down at the Battle of Midway.
Art Collection, National Museum of the Marine Corps

canal developed in three phases that ended in September, October, and November 1942. (The island was not completely secured until 9 February 1943, but after November the campaign was primarily an Army land offensive.) Neither the American theater commanders nor their Japanese adversaries had planned for the battle to develop as it did. Instead, the Japanese belatedly realized the threat that Henderson Field posed to their stronghold at Rabaul [New Guinea] and gradually shifted the focus of their main effort from New Guinea to Guadalcanal. They mounted three major counteroffensives to recapture the airfield, each more powerful than the one before. The first two of these counteroffensives culminated in massed infantry attacks on the Marine perimeter: the Battle of Edson's Ridge in September and the Battles of the Matanikau and Bloody Ridge in October. But the overwhelming infantry attack that the Japanese planned for November never materialized, thanks in part to the Henderson Field flyers.

From a functional perspective, the opposing theater commanders fought the battle for Guadalcanal by mounting a series of tactical "shields;" Marine, and later Army, ground forces shielded the airfield against Japanese infantry attacks. In a continuous battle of attrition, American flyers attempted to shield ground and naval forces against Japanese air strikes from Rabaul.

By 1943, Marine pilots were flying aircraft that were equal or superior to the planes being flown by the Japanese. This growing technological advantage, however, would have been far less significant had it not been for the tireless efforts of Marine maintenance personnel who worked around the clock to keep the new aircraft operationally ready. In *Henderson Field, Night*, 1stLt Hugh Laidman depicts mechanics working through the night to service a Vought F4U Corsair.
Art Collection, National Museum of the Marine Corps

Because there were no all-weather attack aircraft in those days, it was up to U.S. Navy surface gunfire ships to shield the Marine ground forces against Japanese naval gunfire bombardments during the night. Initially, they were unsuccessful. Carrier task forces on both sides attempted to shield transports with inbound reinforcements. The epic "symmetric" battles of carriers against carriers, aircraft against aircraft, and infantry against infantry have received most of the historians' attention during the years. But it was the cross-functional or asymmetrical attacks by shore-based naval aircraft against Japanese transports that provided the margin of victory for American forces.[66]

It was clear even before the campaign began that U.S. carrier task forces would be unable to provide the air shield for Guadalcanal. Vice Admiral Robert L. Ghormley, commander of the South Pacific theater, wrote to the Joint Chiefs of Staff on 8 July, "The Carrier Task Groups will be themselves exposed to land based air [from Rabaul] while unprotected by our land based aviation, and it is extremely doubtful that they will be able to retain fighter escort to the transport area." Vice Admiral Frank Jack Fletcher, overall commander of the Guadalcanal operation, echoed this grim assessment when his forces assembled in late July. Japanese air attacks against the transports on 8 August confirmed this expectation, and Fletcher promptly withdrew the carriers that night with Ghormley's approval. The withdrawal of the carriers and the stunning Japanese victory against the naval gunfire shield that same night forced the withdrawal of the American transports, which we noted earlier.

Admiral Fletcher has been much criticized over the years for his decision to withdraw the carriers. Certainly the Marines on Guadalcanal must have had unkind words for him as they watched their supplies disappear over the horizon. But much of this criticism misses the point.

The United States had entered the war with six battle line carriers. By August 1942, Fletcher had already lost two of them at the battles of the Coral Sea and Midway. Before November, three

[66] For a discussion of shield and symmetrical and asymmetrical forces, see the 1987 Schulze Memorial Essay, *MCG*, November 1987.

of the remaining four—[USS] *Saratoga* [CV 3], *Wasp* [CV 7], and *Hornet* [CV 8]—would be sunk or knocked out of the war for several months. The point is that carriers were both valuable and vulnerable, just as they are today. Fletcher should be criticized not for husbanding scarce combat assets but for failing to aggressively exploit the operational advantage offered by Henderson Field in the ensuing campaign.

However, Admiral Fletcher was not alone in his failure to grasp the opportunity presented by Henderson Field. In fact, only a handful of the senior commanders realized that the airfield was not only the prize to be won but also the means to win it.

Foremost among those who pushed to increase the air contingent at the beleaguered field were Major General Alexander A. Vandegrift, commanding the Marines, and Rear Admiral John S. McCain, commander of shore-based naval air in the South Pacific and later one of the Navy's great carrier commanders. In a message to Admiral [Chester W.] Nimitz, Commander in Chief of the Pacific theater, on 1 September, McCain requested reinforcements and replacements for the aircraft at Henderson Field. He went on to say:

> *No help can or should be expected of carrier fighters unless based ashore. With substantially the reinforcement requested Cactus [Guadalcanal] can be a sinkhole for enemy air power and can be consolidated, expanded and exploited to enemy's mortal hurt. The reverse is true if we lose Cactus. If the reinforcement requested is not made available, Cactus cannot be supplied and hence cannot be held.*

Rear Admiral Richmond Kelly Turner, the commander of the amphibious task force, also saw

F4F Wildcat planes parked on the fighter strip at Henderson Field.
Official U.S. Navy photo, Naval Aviations News, *January–February 1943*

the operational significance of the airfield. In a letter to Ghormley in early September, he described it as "an unsinkable aircraft carrier."

More by default than by design, Henderson Field eventually received the reinforcements that Vandegrift, McCain, and Turner had argued for. After *Saratoga* was torpedoed and *Wasp* and *Hornet* were sunk, many of their surviving aircraft were transferred to Guadalcanal. As Japanese transports approached the island in November, bearing troops for the third major counteroffensive, the damaged *Enterprise* [CV 6] deployed its entire air wing to Henderson Field while the carrier retired.

The continuous transfer of aircraft from ship to shore during the campaign was easy because Navy and Marine squadrons flew identical aircraft. When Navy aircraft landed at Henderson Field, they were met by maintenance crews that had the skills, tools, parts, and ordnance to keep them flying. Thus, for example, when a flight of Navy dive bombers diverted ashore from *Enterprise* in August, they were able to stay and fight for more than a month, although the

Running a gauntlet of antiaircraft fire, four Rising Sun bombers come in low at Guadalcanal, Solomon Islands, to attack U.S. transports, extreme left. Black bursts show the intensity of the American antiaircraft assault, 8 August 1942.
Collection of Clifton B. Cates (COLL/3157), Archives Branch, Marine Corps History Division

crews brought nothing with them except the flight suits they were wearing.

Once they arrived, Navy and Marine aircraft proved to be a decisive factor during each of the three phases of the struggle. On 24 August, Japanese and American carriers dueled in the Battle of the Eastern Solomons, but it was dive bombers from Henderson Field that turned back the Japanese transport group the next day. Forced to operate only at night after that, the Japanese could send in only a trickle of reinforcements. Even after they were bombarded by the Japanese battleships in mid-October, Henderson's flyers were able to attack six enemy transports that next day, forcing the Japanese to beach three of them and withdraw the others. Consequently, the Japanese ground forces that hit the perimeter 10 days later were greatly reduced in numbers and effectiveness. In the titanic surface gunfire battle of 12–14 November, U.S. Navy battleships prevented the Japanese from shelling the airfield as they had in October. As a result, a full contingent of Henderson-based dive bombers, supported by the entire *Enterprise* air wing, was able to sink 7 of the 11 Japanese transports bound for Guadalcanal on 14 November. The remaining four were bombed and destroyed while unloading the next day. Only 2,000 Japanese reinforcements got through. Consequently, the Japanese were never able to mount a ma-

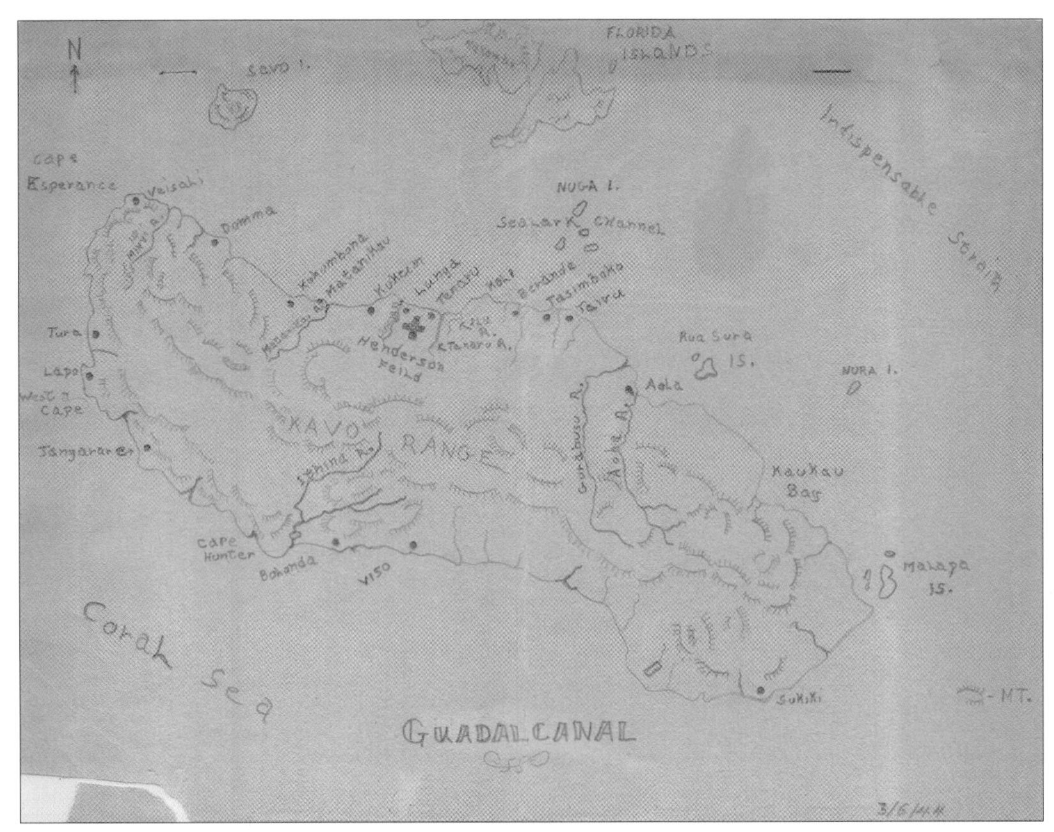

This hand-drawn map of Guadalcanal, dated 6 March 1944, features locations on the island including Henderson Field, Kaukau Bay, Tenaru, and the Coral Sea.
Guadalcanal Map (2010-2070), Archives Branch, Marine Corps History Division

jor ground attack to complete the third major counteroffensive in November. Henderson Field was finally secured.

Clearly, shore-based naval air power provided the margin of victory during the Guadalcanal campaign. Nevertheless, it is reasonable to ask whether innovations in technology have made the lessons of Guadalcanal irrelevant for future amphibious operations.

In fact, advances in technology have made those lessons even more relevant. With the introduction of tilt-rotor aircraft and air cushion landing craft, amphibious assaults can be conducted from much greater ranges than in the past. The landing force will depend upon fixed-wing air more than ever for the bulk of its fire support. The AV-8B and short airfields for tactical support (SATS) allow an amphibious task force commander to establish an air shield ashore in a much shorter time than it took to build Henderson Field.

At the same time, carriers are no less vulnerable to long-range bombers and submarines than they were in 1942. Reconnaissance satellites make it more difficult for carrier battle groups to evade detection, especially if they must remain within supporting range of a beachhead for long. Soviet naval aviation includes a substantial force of long-range bombers, including 140 [Tupolev TU-22M] "Backfires," whose primary mission is the destruction of U.S. carriers. According to unclassified sources, the Backfire has an operating radius of 2,160 nautical miles and can fire supersonic antiship missiles at targets 240 nautical miles away. Soviet *Oscar I*-class attack submarines and *Kirov*-class cruisers can fire antiship missiles from ranges up to 300 nautical miles. As early as 10 years ago, the Department of Defense Annual had concluded that the Soviets "can concentrate aircraft, coordinate attacks with air, surface, or submarine launched missiles, and use new technology to find our fleet units, jam our defenses, and screen their approach."

Of course, to counter the antiship missile threat, U.S. carrier battle groups will employ air defense systems like the AEGIS [ballistic missile defense] system on *Ticonderoga*-class cruisers, combat air patrols of fighters like the [Grumman] F-14 Tomcat, and multiple layers of electronic countermeasures. But how successful these measures will be against a cunning and determined attack cannot be predicted with certainty. It is safe to conclude that as long as we need aircraft carriers, we will also need the ability to establish "unsinkable carriers" ashore.

The lessons of Guadalcanal remain valid today. Theater commanders must have the capability to fight a campaign using carrier-based and shore-based naval aircraft in cooperation. The Marine Corps should not restrict its strike capability to close air support. Instead, it should continue to procure systems and develop tactics that will enable the landing force commander to conduct "war-at-sea" strikes. If we fail to maintain operational flexibility and interoperability between Navy and Marine aircraft, we may one day arrive at a situation like the one that almost developed during that desperate October on Guadalcanal when the most useful thing for the aviators to do will be to draw rifles and take up positions on the perimeter.

LETTER FROM SMITH TO CATES ON CHOSIN RESERVOIR

by Major General Oliver P. Smith

FROM THE COMMANDING GENERAL, 1st MARINE DIVISION TO THE COMMANDANT OF THE MARINE CORPS
17 December 1950[67]

At the present moment, I am in Masan. I sailed on the USS *Bayfield* [APA 33] from Hungnam on 15 December for Pusan. With the exception of certain shore party elements, elements of the AmphTrac [Landing Vehicle, Tracked] battalion, and NGF [naval gunfire] teams and TAC [tactical air control] parties, which are being retained by Corps at Hungnam for the time being, the entire division should close Masan today.[68] What our mission will be I do not know. When the remainder of the X Corps arrives in the Pusan area, the Corps will become a part of the 8th Army. [General] Lemuel C. Shepherd has made representations to Corps regarding the need for a period of time in which the division can integrate replacements, repair equipment, and be resupplied. The Corps is aware of this need, not only for us but also for the 7th Division, which lost practically en toto [*sic*] two infantry battalions and a field artillery battalion. However, Corps will not be calling the turns here.

You have probably read a lot of misinformation in the newspapers and it might be well to give you a factual account of what we have been doing for the past two weeks.

When I last wrote you, the 8th Army had not yet launched its attack. At that time, my mission was to establish a blocking position at Yudam-ni and with the remainder of the division to push north to the Manchurian border. As I explained to you, I did not press the 5th and 7th Marines, which had reached the Chosin Reservoir, to make any rapid advances. I wanted to proceed cautiously for two reasons. First, I had back of me 50 miles of MSR [main supply route],

[67] The original content came from Commanding General, 1st Marine Division letter to Commandant of the Marine Corps, subj Chosin, 17 December 1950 (MCHC, Quantico, VA). Minor revisions were made to the text based on current standards for style, grammar, punctuation, and spelling.
[68] Masan was the former capital city of South Gyeongsang Province, South Korea.

In front of the commanding general's quarters at Masan, 1950. From left: LtCol Raymond L. Murray, commanding officer, 5th Marines; Gen Oliver P. Smith; Col Lewis B. Puller, commanding officer, 1st Marines.
Oliver P. Smith Collection (COLL/213), Archives Branch, Marine Corps History Division

14 miles of which was a tortuous mountain road which could be blocked by bad weather, and I wanted to accumulate at Hagaru-ri at the southern end of the reservoir a few days supply of ammunition and rations before proceeding further [*sic*]. Second, I wanted to move [Colonel Lewis B. "Chesty"] Puller up behind me to protect the MSR and he had not yet been entirely released from other commitments.

By 23 November, both the 5th and 7th Marines were in contact with the CCF [Chinese Communist forces], the 5th to the east of the Chosin Reservoir and the 7th to the west thereof. The 7th was advancing to the blocking position assigned by Corps at Yudam-ni. In the 15-mile stretch of road between Hagaru-ri and Yudam-ni, the 7th had to traverse a 4,000-foot mountain pass and was impeded by the enemy, roadblocks, and snow drifts. Patrols of the 5th pushed to the north end of the reservoir.

On 24 November, the 8th Army's attack jumped off. With the attack came General MacArthur's communiqué, which explained the "massive compression envelopment" that was to take place. I learned for the first time that the 1st Marine Division was to be the northern "pincers" of this envelopment. At a briefing on 25 November, the details were explained. I was to make the main effort of the Corps in a zone of action oriented to the westward. I was

Commissioned by the Chosin Few Association to mark the dedication of the missile cruiser USS *Chosin* (CG 65), Col Charles H. Waterhouse's painting *Eternal Band of Brothers, Korea* depicts the U.S. military winding its way down Funchilin Pass.
Courtesy of the Waterhouse estate

to advance along the load from Yudam-ni toward Mupyong-ni, cut the road and railroad there, send one column on to the Manchurian border at Kuup-tong, and another column north to Kanggyeo. The 7th Infantry Division was to take over my former mission of advancing north up the east side of the reservoir and thence to the Manchurian border. The 3d Infantry Division was to take over the protection of the MSR up to Hagaru-ri. (This never transpired; and to the end of the operation, I had to retain one battalion of the 1st Marines at Chinhung-ni at the foot of the mountain and another battalion of the 1st Marines at Koto-ri at the top of the mountain. Otherwise, there would have been no protection for this vital part of the MSR). Under the plan, the Corps assumed responsibility for engineer maintenance of the MSR to Hagaru-ri. It also agreed to stock 10-days supplies at Hagaru-ri. I doubt if the Corps would have been able to do this. In any event, the enemy gave us no opportunity to prove whether or not it could be done.

D-day, H-hour for the attack to the westward was fixed by Corps as 27 November, 0800. By 26 November, [General Homer L.] Litzenberg, with all of the 7th, was at Yudam-ni. I decided to have him remain in the Yudam-ni area and pass the 5th through him for the attack to the westward. The 5th had not been in a serious engagement since the attack on Seoul.

The attack jumped off on schedule, but it was not long before both the 5th and 7th were hit in strength by the CCF. By 28 November,

Tank convoy of 1st Tank Battalion cross mountains on the way to Chosin Reservoir from Hamhung, Korea, 19 November 1950.
Defense Department photo (Marine Corps) A5343, Marine Corps History Division, courtesy of TSgt J. W. Helms Jr.

reports of casualties left no doubt as to the seriousness of the attack. At the same time, the 8th Army front was crumbling. No word was received from Corps regarding discontinuance of the attack or withdrawal. Under the circumstances, I felt it was rash to have [General Raymond L.] Murray attempt to push on and I directed him to consolidate on the positions he then held west of Yudam-ni. At the same time, I directed Litzenberg to open up the MSR between Yudam-ni and Hagaru-ri, which had been blocked by the Chinese, as had also the stretch of road between Hagaru-ri and Koto-ri. On this same day, 28 November, I moved my operational CP [command post] to Hagaru-ri. The movement was made by helicopter, the only feasible method in view of the cutting of the MSR. Fortunately, we had been able to get some vehicles and working personnel into Hagaru-ri before the road was cut.

Litzenberg's efforts to clear the MSR between Yudam-ni and Hagaru-ri were unsuccessful on the twenty-eighth. He reported he would make another effort with a battalion the following day, 29 November.

On 28 November, Puller organized Task Force Drysdale to open up the MSR between Koto-ri and Hagaru-ri. This force was under command of Lieutenant Colonel [Douglas B.] Drysdale of the RM [Royal Marine] Commandos. It consisted of the RM Commandos, 235 strong, G Company of 3/1 [3d Battalion, 1st Marines] coming north to join its parent unit at Hagaru-ri, and a rifle company of the 31st Infantry, which was moving north to join its parent unit east of the Chosin Reservoir. (The 7th Infantry Division had pushed north a battalion of the 31st, a battalion of the 32d, and a field artillery battalion to relieve the 5th Marines on the east side of the Chosin Reservoir.) In addition to the units enumerated, the Drysdale column included two companies of our M26 [Pershing] tanks, each less a platoon, and a truck convoy. The column was to move out on the twenty-ninth. I will cover its operations later.

During the night of 28–29 November, the enemy attacked Hagaru-ri in force. The attack started at 2130 and lasted all night. First the attack came in from the south, then shifted to the west, and then to the east. Our defense force consisted of 3/1, less G Company, and personnel of our Headquarters and Service units. Our casualties were 500, of whom about 300 were from the infantry and 200 from Headquarters and Service units. The Headquarters Battalion alone had 60 casualties.

Troops of Regimental Combat Team 7 held up just south of Hagaru-ri while Marine and naval close air support work over enemy positions with napalm, 6 December 1950.
Oliver P. Smith Collection (COLL/213), Archives Branch, Marine Corps History Division

We had at an early date realized the importance of Hagaru-ri as a base. On 16 November, [Lieutenant General] Field Harris and I had tentatively approved a site for a [Douglas] C-47 [Skytrain] strip at Hagaru-ri. Work was begun by our 1st Engineer Battalion on 19 November and the strip was first used by C-47s on 1 December, although at the time it was only 40 percent completed. This strip was essential for the evacuation of wounded and air supply in case our road went out either due to weather or enemy action. Hagaru-ri had to be held to protect this strip and the supplies that we were accumulating there. The movement of the Drysdale column from Koto-ri to Hagaru-ri would not only open the road, but would also furnish us needed reinforcements for the defense of Hagaru-ri.

The Drysdale column started north from Koto-ri on the morning of 29 November. About halfway to Hagaru-ri, it became engaged in a heavy firefight. Embarrassed as he was by a truck convoy, Drysdale was on the point of turning back to Koto-ri, but I sent him a message to push on through if at all possible. He started the truck convoy back toward Koto-ri under the protection of a company of tanks and some infantry, while the remainder of the column continued to fight its way toward Hagaru-ri. The truck convoy returning to Koto-ri was jumped by the Chinese, who had closed in on the MSR again. There was considerable mortar fire and tanks as well as trucks were pretty badly shot up before they got back to Koto-ri. There were also a considerable number of personnel casualties. Drysdale continued to fight on toward Hagaru-ri and toward evening arrived with about 150

John A. Groth's *Village near Hagaru* shows three Marines heavily clothed against the bitter Korean cold in November 1950.
Art Collection, National Museum of the Marine Corps

of his Commandos and G Company of the 1st Marines. The Army company never arrived, although some stragglers came in to Koto-ri. The conclusion was inescapable that a considerable force would be required to open up the MSR between Hagaru-ri and Koto-ri. We would not have any such force until the 5th and 7th Marines joined us at Hagaru-ri.

On 29 November, the 7th Marines started a battalion back along the MSR to open up it, but the battalion got nowhere. I then ordered Litzenberg to employ the entire 7th Marines on the following day, 30 November, to open up the MSR. At the same time, I ordered Murray to pull back his regiment to Yudam-nio. Late in the day of 29 November, I received a telephone call (radio link) from Corps stating that the whole scheme of maneuver was changed, that the Army battalions on the east side of the Chosin Reservoir, who were now cut off from us were attached to me and I was to extricate them, and that I was to withdraw the 5th and 7th Marines and consolidate around Hagaru-ri.

On 30 November, the Corps turned over to me command of all troops as far south as Sudong, which is four or five miles below the foot of the mountain. These comprised a battalion of the 31st Infantry, which was on its way up the mountain and miscellaneous engineer and service units.

During the day of 30 November, Puller was attacked rather heavily at Koto-ri but kept his perimeter intact.

On the afternoon of 30 November, General

Col Charles H. Waterhouse, *Road to Hagaru, North Korea*, acrylic on canvas.
Courtesy of the Waterhouse estate

[Edward M.] Almond flew up to see me. By this time, he had given up any idea of consolidating positions in the vicinity of Hagaru-ri. He wanted us to fall back in the direction of Hamhung and stressed the necessity for speed. He authorized me to burn or destroy equipment and supplies, stating that I would be supplied by airdrop as I withdrew. I told him that my movements would be governed by my ability to evacuate the wounded, that I would have to fight my way back and could not afford to discard equipment, and that, therefore, I intended to bring out the bulk of my equipment.

The problems of the 5th and 7th Marines could not be separated. By 30 November, between them, they had accumulated about 450 wounded who had to be protected. The only feasible thing to do was to pool their resources. The two regimental commanders drew up a joint plan (an ADC [assistant division commander] would have come in handy at this point) which was flown to me by helicopter and which I approved. Briefly, the 7th was to lead out from Yudam-ni and the 5th was to cover the rear. Artillery and trains were in the middle. The walking wounded were given weapons and marched in column on the road. Other wounded were loaded in trucks. The route these two regiments had to traverse was tortuous. From Yudam-ni, the road first led south up a narrow mountain valley and then turned eastward toward Hagaru-ri. At about the halfway point, the road crossed a 4,000-foot mountain pass and then descended toward Hagaru-ri. This last section of the road more or less followed the ridgelines and did not offer the same opportunities to the enemy to block the road as did the first part of the road out of Yudam-ni. As events transpired, the 7th and 5th did have a hard fight to get up to the pass, but the descent to Hagaru-ri, although opposed, was relatively easier.

During these operations, one company of the 7th Marines had a unique and remarkable experience. This was F Company. In his initial advance to Yudam-ni, Litzenberg had left E and F Companies in occupancy of high ground along the road to the rear. Litzenberg was able to extricate E Company, but could not reach F Company, which was in position at the top of the mountain. It was completely surrounded but held excellent positions. By pinpoint airdrops, we were able to keep the company supplied with ammunition and rations. It had 18 killed and 60 wounded but held out for more than three days when it was relieved by 1/7 [1st Battalion, 7th Marines] pushing back up the mountain from Yudam-ni.

During the night of 30 November–1 December, Hagaru-ri was again heavily attacked but the perimeter held. We were stronger this time as G Company of 3/1 and the Commandos had joined our defense force. The attacks were from the southwest and the east. The attack from the east fell on the sector manned by the Service Battalion. Lieutenant Colonel [Charles L.] Banks, an ex-[Edson] Raider, was in command of the Service Battalion. He did an excellent job in beating back the attack.

By 1 December, the situation with regard to care of casualties was becoming serious. Dr. [Navy Captain Eugene R.] Hering had at Hagaru-ri 600 casualties awaiting evacuation. These were being cared for by C and E Medical Companies. It was estimated 400 casualties would be brought in if the Army battalions east of the reservoir broke out. (Actually, we eventually evacuated more than 900 men from these battalions). We estimated the 5th and 7th would bring in 500 casualties. (Actually, they brought in 1,500.)

It was manifest that the only solution to our casualty problem was completion of the C-47 strip. (OYs [light observation planes] and helicopters could not make a dent in our casualty load.) Our engineers had worked night and day on the C-47 strip. On two nights, work had to be interrupted because of enemy attacks and the engineers manned their part of the perimeter near the field. The front lines were only 300 yards from the end of the runway. The strip was rather crude; 3,800 feet long, 50 feet wide, no taxiways, and a 2-percent grade to the north. The soil was black loam but it was frozen. Our equipment had considerable difficulty with the frozen ground. On 1 December, the strip, as I have described it, was considered to be 40 percent completed.

On the advice of the aviators, it was decided to bring in a C-47 for a trial run on the afternoon of 1 December. The plane landed successfully at about 1500 and took off 24 wounded. It takes about a half hour to load a plane with litter patients. Ambulatory patients go very much faster. At first, we could accommodate only two planes on the ground simultaneously. Eventually, as the field was improved, we were able to accommodate six planes on the ground without blocking the runway. Hours of daylight were from about 0700 to 1745 and use of the strip was limited to those hours. After the first plane landed, more planes came in. Five additional plane loads of wounded were taken out that afternoon. We would have gotten out more but an incoming plane, loaded with 105mm ammunition, collapsed its landing gear. The plane was too heavy with its load to push off the runway and we had to unload it, thus losing valuable time. (We attempted to have incoming planes loaded with ammunition and other needed supplies to supplement airdrops.)

I will complete the story of evacuation of casualties from Hagaru-ri out of chronology, as it is all one story and a very remarkable accomplishment when viewed as a whole. On the evening of 1 December, stragglers from the breakup of the Army battalions east of the lake began to drift in. During the day of 2 December, we evacuated 919 casualties by air, the majority of them from the Army battalions. During the morning of 3 December, the doctor cleaned out by air evacuation all his remaining casualties. This gave us an opportunity to fly out our accumulation of dead. The estimate of casualties of the 5th and 7th Marines had now risen to 900. At 1935, 3 December, the advance guard of the 7th Marines arrived at the perimeter. It was closely followed by the column of walking wounded. The column continued the movement during the night and each vehicle brought in more wounded, some on the hoods of jeeps. By morning, the doctor's hospital installations were full. On the day of 4 December, 1,000 casualties were evacuated by air. On the day of 5 December, 1,400 more casualties were evacuated by air. When we moved out from Hagaru-ri to Koto-ri on 6 December, we had no remaining casualties to evacuate.

I believe the story of this evacuation is without parallel. Credit must go to the troop commanders whose determination and self-sacrifice made it possible to get the wounded out, to the medical personnel whose devotion to duty and untiring efforts saved many lives, and to the Marine and Air Force [air crews] (including fatal accident[s] in spite of the hazards of the weather and a rudimentary landing strip.)

Breakthrough to Fox Hill, by Col Charles H. Waterhouse, portrays LtCol Raymond Gilbert Davis during the action for which he was awarded a Medal of Honor. Davis leads his battalion in the fourth attempt to rescue the beleaguered 1st Marine Division at Toktong Pass, who for six days and nights held off the sudden emerging forces of the Chinese armies. On the main supply route, adjacent to the Chosin Reservoir, the Toktong Pass was a lifeline. Davis led his battalion over three successive ridges in the dark and in blizzard snow at close to minus 75 windchill and in continuous attacks against the enemy.
Courtesy of the Waterhouse estate

To get back to the story of the operation in its proper chronological sequence. At 1335, 1 December, we got our first airdrop from Japan. These drops were known as "Baldwins." Each Baldwin contained a prearranged quantity of small arms ammunition, weapons, water, rations, and medical supplies. Artillery ammunition had to be requested separately. A Baldwin could be dropped by about six [Fairchild] C-119 [Flying Boxcar] planes. We were required to make request on Corps for the number of Baldwins desired, modified as desired. We usually requested Baldwins less weapons and water and plus given quantities of artillery ammunition.

Airdrop did not have the capability of supplying a Marine division in combat. When the drops were started, the total capability of the

Far East Air Force was 70 tons a day. This was stepped up to 100 tons a day. But to support an RCT [regimental combat team] in combat requires 105 tons a day. What gave us some cushion was the fact that, with our own transportation, before the roads were cut, we had built up at Hagaru-ri a level of six days rations and two units of fire. The airdrops continued until we left Hagaru-ri and were also made at Koto-ri, where Puller had to be supplied and where we had to accumulate supplies in anticipation of the arrival of the bulk of the division there. The drops were not always accurate, and we had personnel and materiel casualties as a result of inaccurate drops; however, we owe a considerable debt of gratitude to the Air Force for their efforts.

During the afternoon of 1 December, a deputy chief of staff of the Corps arrived and gave me the outline of the latest plan. Under this plan, the 3d Infantry Division was to move elements to Majong-dong (about 10 miles south of the foot of the mountain) and establish a covering force through which I would withdraw. Upon withdrawal, I was to occupy a defensive sector west and southwest of Hungnam and the 7th Division was to occupy a sector northeast and north of Hungnam.

Toward evening on 1 December, some 300 stragglers of the cutoff Army battalions up the reservoir drifted into camp, having made their way in over the frozen surface of the reservoir. They continued to drift in during the night and for three or four days thereafter. I have never found out exactly what happened. Apparently, the two battalions that had holed up at Sinhung-ni started south and had made some progress, with the support of a considerable amount of Marine aviation (10 planes on either side of the road). Then the acting regimental commander was killed and the column must have fallen apart and men made the best of their way out to the lake and thence down the lake to our perimeter. For some unknown reason, the Chinese did not do much firing at people on the surface of the lake. We evacuated some 900 men of the two infantry battalions and artillery battalion. There remained with us some 385 more or less able-bodied men whom I had the senior Army officer present form into a provisional battalion. We brought these out with us.

During the day of 2 December, Lieutenant Colonel [Olin] Beall and other volunteers conducted a remarkable rescue operation on the lake. Air cover was provided. They drove jeeps, often towing improvised sleds, as far as four miles over the surface of the reservoir, and picked up wounded and frostbitten men. Although the Chinese did not often fire on the wounded on the lake, they did fire at the jeeps. During the day, 250 men were rescued by these jeeps. Operations were continued the following day, but a lesser number were found. Beall was awarded the DSC [Distinguished Service Cross] by the Corps commander.

The 5th and 7th made some progress up the mountain during 2 December. Enemy opposition was still strong.

On 3 December, Litzenberg reached the top of the mountain between Yudam-ni and Hagaru-ri. However, there was still a buildup of enemy between him and us and he was running short of gasoline. In a slow-moving column, there is considerable idling of motors and in any event, in cold weather, motors have to be started up frequently. All this consumes a large quantity of gasoline. At Litzenberg's request, we made a pinpoint drop of gasoline to the head of the truck column. Unfortunately, he did not request diesel fuel, a lack of which later was responsible for the loss of several artillery pieces.

During the day of 3 December, Litzenberg continued to push over and down the mountain. At 1630, we sent out tanks with the Commandos to clean out the Chinese who were on the road near camp. At 1935, the advance guard of the 7th Marines arrived at the perimeter. Movement continued during the night, the 5th Marines following in after the 7th. In the darkness, it takes a long time to get units in from covering positions and on to the road. When they were only a few miles from Hagaru-ri, some of the tractors drawing the 155mm howitzers ran out of diesel fuel. This stopped the column. The Chinese closed in with mortar and automatic weapon fire. Some of the tractors were disabled. We later sent a column back with diesel fuel, but not all the guns could be gotten out because of disabled tractors. We lost 10 out of 18 155mm howitzers and 4 out of 30 105mm howitzers. The guns were spiked and later an air strike was put down on them. Despite the losses, it was still a remarkable feat to bring out three battalions of artillery minus these guns.

The last elements of the 5th and 7th Marines did not arrive at Hagaru-ri until about noon of 4 December. I was considerably relieved to have these two regiments rejoin. I considered that the critical part of the operation had been completed. Even with two depleted RCTs, I felt confident we could fight our way to Koto-ri where we would gain additional strength. The terrain was not as difficult, it lent itself well to air support, and we were able to lay down preparatory artillery fires all the way to Koto-ri. Artillery emplaced at Hagaru-ri could reach halfway to Koto-ri and Puller's artillery at Koto-ri could reach back to meet our fires.

After their grueling experience, the regiments were not in condition to continue the advance on 5 December. Also, we wanted to be sure that all our casualties were evacuated. Our order, therefore, provided for an advance on Koto-ri at first light on 6 December.

This blown bridge at Funchilin Pass blocks the only way out for U.S. and British forces withdrawing from the Chosin Reservoir in North Korea during the Korean War. Air Force C-119 Flying Boxcars dropped portable bridge sections to span the chasm in December 1950, allowing men and equipment to reach safety.
Official U.S. Air Force photo

The order for the advance on Koto-ri provided for an advance in two RCT [regimental combat team] columns. RCT 7 led out. The RCT was normal as to combat troops, with the provisional Army battalion attached. In addition, Litzenberg had within his column his own regimental train and Division Train No. 1. RCT 5 was to follow RCT 7. Its composition was normal except for the attachment of 3/1. Murray also had within his column his own regimental train and Division Train No. 2. He was to hold the perimeter until RCT 7 had gained sufficient distance to permit him to move out on the road.

The embarrassing part of this move was the trains. More than a thousand vehicles were involved. We carried two-days rations and two units of fire. We brought out all usable equipment and supplies, including tentage and stoves. Even the engineer pans were used as trucks to carry tentage.

Litzenberg had not advanced more than two miles before he ran into trouble. Using maximum air and artillery support, it required until 1400 to break through. Peculiarly enough, all the opposition came from the east side of the road.

At 1420, I moved my operational CP by OY plane and helicopter to Koto-ri. My radios, vans, and working personnel were mostly in Division Train No. l.

By 1800, 6 December, Litzenberg had reached the halfway point and was progressing satisfactorily. However, during the night, the Chinese cut into the train in two places. There was confused and close range fighting. We lost men and vehicles but remarkably few vehicles.

The column continued to move during the night and by 0590, 7 December, the leading elements of the 7th Marines began to arrive at Koto-ri.

The 5th Marines did not clear Hagaru-ri until 7 December. Murray had quite a rear guard action at that place, but came off with 200 prisoners. His last elements did not close Koto-ri until 2135, 7 December.

The advance from Hagaru-ri to Koto-ri cost us more than 500 casualties. Puller had an OY strip only. However, Field Harris agreed to land TBM [turboprop] planes, of which he had three, on this strip. During the day of 7 December, between OYs and TBMs, 200 casualties were evacuated. However, there were still 300 more casualties to evacuate. The aviators stated that, if 400 feet [were] added to the strip, it would be possible for C-47s to land. Therefore, during the night of 7–8 December, our engineers lengthened the strip by 400 feet. Unfortunately, the strip was periodically under enemy fire. On 8 December' C-47s began to land and we soon completed evacuation of our casualties.

Koto-ri is about two miles north of the lip of the mountain. From the lip of the mountain the road descends tortuously to Chinhung-ni about 10 road miles distant. At Chinhung-ni was Puller's 1st Battalion. On 7 December, the Corps had moved an Army battalion to Chinhung-ni in order to free 1/1 [1st Battalion, 1st Marines]. Theoretically, the road was open from Chinhung-ni to the south.

Our plan for getting down the mountain was simple. (However, it must be borne in mind that the enemy surrounded Koto-ri as they had closed in behind our columns.) The 5th and 7th Marines were to seize and hold the commanding ground to about the halfway point. 1/1 was to push up from Chinhung-ni and seize and hold commanding ground about halfway up the mountain. The 1st Marines, which had regained 3/1 from Hagaru-ri and additionally had a battalion of the 31st Infantry attached, was to hold the perimeter at Koto-ri until the trains cleared when it was to follow out (We now had 1,400 vehicles as a result of the addition of Puller's train and Army vehicles.) Once the commanding ground was seized, it was our intention to push the trains down the mountain. As the trains cleared, infantry would leave the high ground and move down the road. The last vehicles in the column were the tanks. We realized that if an M-26 ever stalled or threw a tread on a one-way mountain road, it would be very difficult to clear it out of the way.

In all this planning, there was one serious catch. The Chinese had blown out a 24-foot section of a bridge about one-third of the way down the mountain. They could not have picked a better spot to cause us serious trouble. At this point, four large pipes, carrying water to the turbines of the power plant in the valley below, crossed the road. A sort of concrete substation was built over the pipes on the uphill side of the road. A one-way concrete bridge went around the substation. The drop down the mountainside was sheer. It was a section of this bridge, which was blown. There was no possibility of a bypass.

[Lieutenant Colonel John H.] Partridge, our engineer, got together with the commanding officer of a Treadway Bridge unit, which was stranded at Koto-ri, and they devised a plan. This involved dropping by parachute at Koto-ri the necessary Treadway Bridge sections. These were dropped on 7 December. As a precaution, additional sections were spotted at Chinhung-ni at the foot of the mountain.

Marines on the road between Funchilin Pass and Chin-hung-ni, 1950. The weather was a constant enemy.
Oliver P. Smith Collection (COLL/213), Archives Branch, Marine Corps History Division

At 0800 on 8 December, the 7th Marines jumped off to seize Objectives A and B at the lip of the mountain; then it pushed on to Objective C further along. The 5th moved out and captured Objective D above the bridge site. 1/1 moved up the mountain and captured Objective E. All this was not accomplished as easily as it is described. There were delays and casualties. The bridging material did not get to the bridge site until 9 December. The bridge was completed at 1615 that date. In anticipation of completion of the bridge, the truck column had been moved forward and the leading truck was ready to cross as soon as the bridge was completed. Unfortunately, another block developed farther down the mountain where the road passed under the cableway. This block was caused partly by enemy fire and partly by additional demolition. This block was not opened until 0600, 10 December.

What we had feared regarding the tanks occurred. As I explained previously, we had placed them last in the column. As they were proceeding down the mountain, the brake on the seventh tank from the tail of the column locked. The tank jammed into the bank. Efforts to bypass the tank or push it out of the way were fruitless. To complicate matters, the Chinese closed in with mortar fire and thermite grenades and mingled with the crowds of refugees following the column. The tankers dismounted and fought on foot with the Reconnaissance Company, which was covering the tail of the column. There were casualties. Finally, the tankers did their best to disable the seven tanks and moved down the mountain. Next morning, an air strike was put in on the tanks as well as the bridge, which we had laboriously constructed.

During the day of 10 December, both Division Trains Nos. 1 and 2 cleared Chinhung-ni at the foot of the mountain and leading elements of the trains began arriving at Hamhung that afternoon. After the trains cleared the road, empty trucks were sent up for troops.

At 1300, 11 December, the last elements of the division cleared Chinhung-ni. The 3d Division was supposed to keep the road open south of Chinhung-ni, but Puller's regimental train was ambushed near Sudong. He lost a couple of trucks and had some casualties. However, Puller arrived at his assembly area with more vehicles than he had started down the mountain with. He had picked up and towed in some vehicles he had found at the scene of a previous ambush of Army trucks. Puller's last elements arrived in the assembly area at 2100, 11 December. This completed the move of the division from the Chosin Reservoir area.

Our rear echelon had set up 150 tents with stoves for each regiment. Hot food was available when the troops arrived.

While Puller was closing his assembly area on 11 December, the 7th Marines was embarking in the MSTS *Daniel I. Sultan* [T-AP 120]. The 5th Marines embarked 12 December and the 1st Marines on 13 December. Loading out of the division was completed about midnight 14 December, and the last ship of the convoy sailed at 1030, 15 December.

An approximation of the casualties from the date (27 November) we jumped off in the attack to the westward until we returned to Hungnam (11 December) is as follows:

KIA [killed in action]	400
WIA [wounded in action]	2,265
MIA [missing in action]	90
Total Battle	2,755
Non-Battle	1,395
	(mostly frostbite)
Grand Total	4,150

This is not the complete picture as there are many more frostbite cases, which are now being screened.

I am understandably proud of the performance of this division. The officers and men were magnificent. They came down the mountains bearded, footsore, and physically exhausted, but their spirits were high. They were still a fighting division.

EQUITATUS CAELI

by Colonel Keith B. McCutcheon

Marine Corps Gazette, February 1954

The first of 12 helicopters descended steeply to the sharp ridge near the top of the mountain, hovered momentarily, and landed.[69] Five fully equipped Marines jumped out. The time, 20 September 1951. The place, Korea. Thus began the first airborne assault by helicopter in the history of warfare. Operation Summit—it was called by the men of the 1st Marine Division who planned and executed it.

Helicopters were not exactly new to Korea; they had been used from the beginning [of the conflict], but not in this way and definitely not on this scale. Craft of the required size had not been available in sufficient numbers. It had taken time to plan, prepare, and get ready for this particular operation. In fact, it took more than five years.

The first atomic bomb tests at Eniwetok [Atoll, Marshall Islands,] had caused forward-looking Marine Corps planners to analyze critically the concept of amphibious operations. Not that the bomb made such operations obsolete. Not at all. In fact, the Corps and the Navy developed the doctrine for amphibious operations with the reverses at Gallipoli still fresh in their minds. Now, it was time to study them again in the light of lessons learned at Bikini [Atoll].

These planners reasoned that an amphibious task force of World War II proportions would constitute a profitable target for an enemy with an atomic bomb capability. A method was needed to reduce that profit. Dispersion, mobility, and speed needed to be injected into the task force and one solution seemed to lie in the use of helicopters.

With that in mind, the Marine Corps commissioned Marine Helicopter Squadron One (HMX-I) at the Marine Corps Air Facility Quantico, Virginia, in December 1947. The primary mission of the squadron was to develop tactics and techniques for the use of helicopters in an amphibious operation—a responsibility assigned to the Corps by the National Defense Act of 1947.

[69] The original article came from Col Keith B. McCutcheon, "Equitatus Caeli," *Marine Corps Gazette* 38, no. 2 (February 1954). Minor revisions were made to the text based on current standards for style, grammar, punctuation, and spelling.

Landing Zone, Korea, by Col H. Avery Chenoweth, depicts Sikorsky HRS-1 Chickasaw helicopters of HMR-161 ferrying Marines to the front. The artist served as an infantry platoon leader in Korea in 1951.
Art Collection, National Museum of the Marine Corps

For most of the next three years, the squadron busied itself with experiments. Utilizing the Sikorsky HO3S helicopter and the Piasecki HRP [Rescuer], the unit experimented with wire laying, cargo hauling, troop lifts, carrier operations, communications, maintenance, and all the other aspects of the overall problem. It was not a large squadron; helicopters were not plentiful and there were a lot of bugs to be worked out. But the program did succeed in training a small group of pilots and maintenance personnel and it did pioneer the use of rotary-wing aircraft in large-scale military operations.

Then came Korea. When the 1st Marine Brigade landed at Pusan on 2 August 1950, it had attached a small number of Sikorsky HO3S helicopters and a handful of pilots and mechanics from HMX-1. In the next few months, that unit made a name for itself. So successful were they in their operations that they completely sold all military men on their usefulness and necessity. Demands for helicopters far exceeded the supply. Expansion of production facilities followed and so did the interest of all potential military and civilian users. The Marine Corps not only accelerated its existing plans, but it also succeed-

A Sikorski HRS-1 helicopter of HMR-161 provides foxhole relief with a fresh load of replacements for the 1st Marine Division at an advance base in Korea, 22 November 1951. The troop drop was into the front lines, while Marine Corsair fighter-bombers furnished air support to keep enemy guns quiet on a nearby hill position.
Official U.S. Marine Corps photo NH 97100, All Hands Collection, Naval History and Heritage Command

ed in getting approval for an expanded program.

One of the types of squadrons to come out of the new funds was the Marine helicopter transport squadron [HMR]; and the first such unit was HMR-161, which was commissioned at the Marine Corps Air Station El Toro, California, in January 1951.

Under the leadership of Lieutenant Colonel George W. Herring, the squadron began the time-consuming process of organizing. Personnel were joined, equipment procured, aircraft accepted, pilots trained, and the whole unit was prepared for a prospective movement overseas.

On 7 April, the first helicopter arrived—a Sikorsky HRS-1. It was a three-bladed, single main rotor configuration with a single tail rotor to compensate for torque. Theoretically, it could carry 10 passengers in addition to the two pilots, but only for short distances.

Except for four pilots who had experience at HMX, the remainder had just received transition flight training to helicopters within the past five months. Most of the mechanics were new at the game too. Due to foresight, however, quite a number had been given on-the-job training at HMX pending arrival of the squadron's own helicopters. Pilots likewise had been trained either at Quantico or at the Navy's school at [Naval Air Station] Pensacola, Florida.

An intensive syllabus was conducted to introduce all the pilots to the types of operations that were expected to be conducted in Korea. Emphasis was placed on mountain flying up to 6,000 feet altitude. This in itself was a totally

Col Keith B. McCutcheon, commanding officer of HMR-161 in Korea, prepares for a reconnaissance flight in an HRS helicopter. McCutcheon was an innovator and theoretician as well as a doer, and like his hero MajGen Roy S. Geiger, he commanded both air and ground units in combat.
Official U.S. Marine Corps photo 127-N-A132705, Still Pictures Division, National Archives and Records Administration

new experience for all pilots as practically all of their previous experience had been at sea level.

Finally, in August 1951, the squadron loaded aboard ship and departed for Korea.

Upon arrival, it was attached operationally to the 1st Marine Division commanded by Major General Gerald C. Thomas. Camp was set up in a few days and the squadron was eager to go to work.

On 13 September, Colonel Herring attended a conference with the division chief of staff, Colonel Victor H. Krulak. At that morning conference were members of the division staff. The purpose of the meeting was to tee up a helicopter mission for the supply of a frontline battalion that afternoon. Operation Windmill I was born.[70]

It was a coordinated effort on the part of all arms-infantry, artillery, and air. It went off smoothly from the initial reconnaissance flight to the last supply run. Landing spots had to be developed in the rough terrain so the helicopters could land, communications had to be maintained between the various units and surveillance had to be maintained over the area to see what, if any, reaction the enemy would have.

Six days later, a similar mission was executed. The pilots gained invaluable experience in terrain appreciation, low-level navigation in unfamiliar terrain, and flying with external loads.

The flying crane, external hoist technique was used in these operations. It was a technique developed back in the early days of HMX and was now paying dividends under combat conditions. Cargo nets were loaded with the badly needed supplies and slung by means of hooks beneath the helicopters. This method permitted the aircraft to deliver the loads rapidly to small areas and cut down the loading and unloading time. It also reduced the time the helicopters would be vulnerable to enemy fire in the forward areas and provided a means for jettisoning the load quickly in the event of an emergency.

But it was Operation Summit that the squadron was looking forward to—the first trooplift. It was not long in coming, only one week after Windmill.

A reinforced reconnaissance company was to be airlifted to the front to relieve a unit of the ROK [Republic of Korea]. From the outset,

[70] The operation focused on getting one day's supplies to 2d Battalion, 1st Marines, more than seven miles away. In two-and-a-half hours, the helicopter crews delivered 18,848 pounds of cargo and evacuated 74 casualties.

it was obvious that landing sites were nonexistent in that razorback-like terrain that rose up to 3,000 feet above sea level. There were places, however, below the crest of the highest hill that could be developed into suitable landing sites within a reasonable period of time.

It was accomplished by hovering the helicopters over the selected points and letting specially equipped Marines climb down from them hand over hand by the use of knotted ropes, another technique dreamed up back on the banks of the Potomac [River]. In about an hour, the vegetation had been cut down and the razorlike ridge excavated, built-up, and flattened into an area 50 feet by 50 feet so that one HRS could land comfortably and the troops could disembark. Then, in a continuous column, the reinforced company was shuttled from the rear to the front. They were placed on high ground fresh and ready to fight. A new technique in the book of warfare had been demonstrated and successfully executed. To provide communications between the company and other units to the rear, two wire lines were laid by helicopter. In a matter of minutes, the wire was laid over terrain that would have required a patrol on foot hours to accomplish.

It was not all easy pickings. On one occasion, a pilot took off with a man still on the rope, and when the crew chief called up and said, "Sir, the man is still on the rope," the pilot recovered his momentary loss of balance, made another approach, and let the man down. The squadron later preserved the incident for history in its squadron song:

They were hovering on the slope
While the man came down the rope.
They still had lots of power to spare.
So before he reached the ground

A helicopter approaches the landing zone on Hill 812 with part of the Reconnaissance Company of the 1st Marine Division, as seen from another helicopter, 20 September 1951.
Naval History and Heritage Command, NH 97101

They took off and flew around
While the man was dangling freely in the air.

As he hung there in the breeze
From his thousand-foot trapeze,
He knew his chances must be pretty slim;
But they made another pass
And dropped him ___ ___ ___.

There were still skeptics of these new twirly birds though. They said the helicopters were too vulnerable, they could not fly at night or under conditions of low visibility.

Sure helicopters are vulnerable. But so is a tank, a ship, or an infantryman. All combat units expect to and do take losses. Techniques must be worked out to reduce the vulnerability. So far

1st Marine Division leathernecks move out after disembarking from an HRS-1 helicopter in a Korean War painting by combat artist Col H. Avery Chenoweth.
Art Collection, National Museum of the Marine Corps

the rotary-wing aircraft have proved that they can operate in and forward of the front lines if they are employed intelligently.

And they can operate at night.

Operation Blackbird proved that. The squadron was ordered to lift a reinforced company of the division reserve from its assembly area to a position near the division's left flank at night.

But there was a lot to do before the event came off. Liaison was established with the parent battalion; reconnaissance flights between the embarkation point and landing sites were made in an effort to determine compass headings, landmarks, time and distance checks, and altitudes; the engineers had to clear the prospective landing zones of mines; and night indoctrination and familiarization flights were made to ensure that all the pilots who were to participate were thoroughly checked out and oriented with the terrain.

In order to fly the designated route, the pilots took off initially from a dry streambed, climbed and cruised through two mountain passes, and let down into a valley to the landing spot. The return trip over a different route to ease the traffic problem required that three passes be negotiated with a final letdown of 1,000 feet to the streambed loading site. Just to keep the pilots on their toes, they were ordered to avoid several friendly artillery positions. To further complicate matters, there was no moon and the sky had a high, thin overcast that reduced visibility.

Blackbird proved that helicopters can operate at night even under adverse conditions, provided certain other conditions exist, such as daylight reconnaissance of the area, familiarity of the pilots with the locality, and the existence of some prominent terrain landmarks to guide the pilots. There is a great deal of development to be done yet before operations such as this become routine.

Perhaps one of the most publicized helicopter missions in Korea has been the evacuation of wounded from the front lines to rear areas where prompt, adequate medical attention was assured, including lifesaving surgery if required. HMR-161 came in for its share of such flights also, although this was not the primary mission of the squadron. There were other helicopters available from another unit for this purpose. Occasionally, however, the number of casualties required the use of a larger aircraft or the presence of a medical officer with the patient in the craft was deemed necessary, so HRSs were assigned to the mission.

In its first six months of operations overseas, the squadron evacuated a couple of hundred casualties, many of them from frontline company positions. Quite a few of them were flown to the coast and landed aboard the Navy's hospital ship USS *Consolation* [AH 15], the first hospital ship to have a helicopter landing platform installed. This ship provided a floating hospital.

During the period of her stay, the squadron made a number of landings aboard by day and night in weather that often prevented the operation of small boats between the ship and the beach. The ship believed that the HRS was very useful for this purpose because of the load it could carry and the fact that it could and did operate by day or night, seemingly without regard to weather.

All helicopter pilots received a great deal of satisfaction in transporting evacuees, as they realized the importance of this most humane mission. Literally thousands of American boys owe their very lives to these "flying angels of mercy."

HMR-161 has continued on with the pioneering efforts in the field of rotary-wing aircraft. Perhaps as a symbol of the future, they chose as their squadron slogan the Latin phrase *Equitatus Caeli*, Cavalry from the Sky. To date, they have performed many of the missions of the old horse cavalry and then some for full measure. They can very well become the eyes of the ground commander and provide him with visual protection of his front, flanks, and rear. In addition, they can move his battle elements to positions where they can do the most good. They can give the force, the speed, mobility and dispersion that are essential in this modern age of warfare. The techniques may be new, but the tactics are still those of the Confederate cavalry leader [Nathan Bedford] Forrest, who reputedly said, "Git there fustest with the mostest."

Equitatus Caeli!

THE "AFLOAT-READY BATTALION"
The Development of the U.S. Navy-Marine Corps Amphibious Ready Group/Marine Expeditionary Unit, 1898–1978

by Colonel Douglas E. Nash Sr.

Marine Corps History, Summer 2017

As any student of naval and maritime history knows, *sea power* is the ability of a nation to use and control the sea and to prevent an opponent from using it.[71] Merely having a fleet is not enough; any nation that wishes to control the sea must be able to project its power in real or concrete form. According to current U.S. Navy doctrine, *power projection* in and from the sea includes a broad spectrum of offensive operations to destroy enemy forces or to prevent enemy forces from approaching within range of friendly forces. History shows that there are generally three ways to accomplish this goal: amphibious assault, attack of targets ashore, or support of sea control operations.[72] The United States is, of course, the world's leading maritime power; a key component of its maritime power projection capability is the U.S. Navy and Marine Corps' Amphibious Ready Group/Marine Expeditionary Unit (ARG/MEU), a force that is increasingly relevant in today's complex operating environment. Understanding how the ARG/MEU concept evolved is an excellent example of how the Marine Corps has successfully adapted throughout its history to changing political and military circumstances.

Whenever a Marine Air-Ground Task Force (MAGTF) consisting of a battalion landing team, composite air squadron, and combat logistics battalion is embarked aboard a Navy Amphibious Squadron (PhibRon), an ARG/MEU is created. Up to three can operate continuously in the areas of responsibility assigned to the Geographic Combatant Commanders (GCC), including the Pacific, Central, African, and European commands. These versatile units provide the president of the United States, acting in his capacity as the commander in chief of the U.S. Armed Services, and the GCC commanders with credible deterrence and response capability across the range of military operations. ARG/MEUs serve as forward-deployed, flexible sea-based

[71] The original article came from Col Douglas E. Nash Sr., "The 'Afloat-Ready Battalion': The Development of the U.S. Navy-Marine Corps Amphibious Ready Group/Marine Expeditionary Unit, 1898–1978," *Marine Corps History* 3, no. 1 (Summer 2017): 62–88. Minor revisions were made to the text based on current standards for style, grammar, punctuation, and spelling.

[72] *Naval Operations Concept 2010: Implementing the Maritime Strategy* (Washington, DC: Department of the Navy, 2010), 51.

MAGTFs—an *afloat-ready force*—a force capable of conducting amphibious operations to respond to a crisis, conduct limited contingency operations, introduce follow-on forces, or support designated special operations forces at a moment's notice. ARG/MEUs are characterized by their sea-based forward presence, expeditionary nature, ability to plan for and respond to crises, combined arms integration, and interoperability with joint, combined, and special operations forces in support of theater requirements.[73]

However, the ARG/MEU concept did not simply spring into existence overnight. Its inception as an afloat-ready force dates back to the late 1800s and reflects a confluence of three factors: policy (i.e., the political-military need for afloat-ready forces by the U.S. government, and by extension, the U.S. Navy); the maturation of the Marine Corps' expeditionary doctrine that featured the ARG/MEU as its centerpiece; and the technological development of aircraft and amphibious assault shipping that enabled the MAGTF to operate in its maritime environment. This article will lay out the historical milestones of this concept, including its early origins, and show how policy, doctrine, and technology have contributed to the evolution during the past 118 years of the force deployed around the globe today.

HISTORICAL ORIGINS OF THE AFLOAT-READY FORCE

Since its inception in 1775, the U.S. Marine Corps has contributed a detachment of Marines, numbering anywhere from 6 to 60 Marines, to nearly every major warship's complement, from sloop to frigate, until the turn of the nineteenth century. Serving as "naval infantry" when needed, as marksmen in the "fighting tops" of sailing ships during sea battles, and as the ship's guard, they also were ready to enforce shipboard discipline when necessary. Should a landing party be ordered to go ashore to fight or land for less warlike purposes as part of a naval expedition, Marines would make up a portion of the party, but would usually be outnumbered by Navy bluejackets, who were part of the ship's normal complement.

As a rule, large numbers of Marines would not normally be embarked on a Navy ship, especially in cases where a fleet or flotilla might sail on missions lasting weeks or even months. There was simply no reason for them to do so, unless embarked on a troopship where they would be landed as part of a land campaign led by the U.S. Army. Exceptions were made should a large-scale amphibious landing be contemplated, such as at Veracruz, Mexico, in 1847, or Fort Fisher, North Carolina, in 1864, but Marines did not ordinarily embark to serve as a fleet's contingency landing force to be landed if and when a commodore saw fit. There was simply no room aboard contemporary warships for anything larger than a detachment of 10 to 50 men.

Despite this record, at least one naval officer during this period advanced the idea of having an embarked landing force sailing with the fleet at all times. The officer, Navy Commander Bowman H. McCalla, had recorded his suggestion in an after action report about the U.S. Navy and Marine Corps expedition of April 1885 to the Isthmus of Panama, then still part of Columbia. Noting how readily the brigade of Marines restored peace and prevented an insurrection once ashore, McCalla wrote that "in future naval operations an additional number of seamen and marines, organized in naval brigades, will

[73] *Amphibious Ready Group and Marine Expeditionary Unit: Overview* (Washington, DC: Headquarters Marine Corps, 2013), 1.

be carried in transports accompanying the battle ships." Though the seeds of an idea had been sown, the Navy Department did not concur and would continue to adhere to existing practice of forming ad hoc landing forces when needed.[74]

That policy changed in 1898, when the United States declared war on Spain. Confronted by a maritime enemy with naval and land forces stationed around the globe defending various overseas colonies, such as Cuba, Puerto Rico, and the Philippines, the U.S. Navy was challenged by the enormous distances involved in simply closing the distance to do battle. Another aspect of naval warfare that had changed since the Marine Corps' inception was the introduction of steam powered warships, which had completely replaced wooden sailing ships by the end of the nineteenth century. Instead of being driven by inexhaustible wind power, ships were now dependent upon coal to fire their steam plants, which enabled them to travel faster and at a steadier pace than with sail power. However, steel-hulled steam-powered warships could not carry enough coal, the fuel of choice, to travel 8,000 miles or more to reach some of Spain's far-flung possessions, where they presumably would do battle with the Spanish fleet once they arrived. Therefore, coaling stations and advanced bases located along the way were necessary and in fact became of strategic importance to the Navy.

While ships could and often did take on coal at sea, this was a slow and hazardous process that exposed a warship to danger while it had come to a complete stop and "hove to" alongside a fleet collier, unlike in today's Navy, where underway replenishment is a common procedure. A coaling station in a protected harbor or port was thought to be far more preferable. However, a protected harbor would most likely have to be taken from the enemy, who might be using it for the same purpose. While, in theory, sailors could (and occasionally did) fight as part of a landing party, the only infantry the Navy had of any strength was the fleet's few embarked Marines who actually had trained for ground combat as their stock-in-trade. To be effective, such an expeditionary landing force would have to be at least of battalion size (several hundred men), including artillery, which could embark and remain on board as an afloat-ready battalion and land whenever the naval commander deemed the situation required boots on the ground (in modern parlance) or when U.S. foreign policy dictated that they land. And therein lies the true genesis of the fleet's "ready reserve" force, the forerunner of today's Amphibious Ready Group/Marine Expeditionary Unit.

HUNTINGTON'S BATTALION

The Marine Corps' first ready reserve force or afloat battalion was "Huntington's Battalion," which was activated for expeditionary service during the Spanish-American War on 16 April 1898. Composed of Marines recruited from nearly every shipyard and naval installation detachment on the East Coast of the United States, it was created by the Colonel Commandant of the Marine Corps, Colonel Charles Heywood, in anticipation that the Navy would ask for such a force, but without knowing exactly how, when, or where it would be employed. This ad hoc organization, known officially as the 1st Marine Battalion (Reinforced), consisted of 654 Marines and one Navy surgeon.

It was organized into five infantry com-

[74] Bowman H. McCalla, *Report of Commander McCalla upon the Naval Expedition to the Isthmus of Panama, April 1885* (Washington, DC: Navy Department, 1885), 43–81.

Col Robert W. Huntington as a major in the 1870s.
Naval History and Heritage Command, NH48984

Marching off to war in the late afternoon on Friday, 22 April 1898, the battalion, preceded by the New York Navy Yard band playing the popular "The Girl I Left behind Me," is led down Navy Street in Brooklyn, NY, under the command of LtCol Huntington astride Old Tom (Capt George F. Elliot's charger).
Archives Branch, Marine Corps History Division

Marine officers who landed with 1st Marine Battalion (Reinforced) at Guantánamo, Cuba, on 10 June 1898. From left: 1stLt Herbert L. Draper, adjutant; Col Robert W. Huntington, battalion commander; and Capt Charles L. McCawley, assistant quartermaster.
Defense Department photo (Marine Corps) 514827

panies and one artillery battery equipped with four 3-inch rapid-fire guns and a battery of four Colt-Browning M1895 machine guns.[75] There was, of course, no aircraft to support this modest force, since the Wright brothers' pioneering flight was still five years out. Having received no definite mission from the Navy's Atlantic Fleet, the battalion commander, Lieutenant Colonel Robert W. Huntington, was ordered to have his men board the converted transport USS *Panther* (1889) in New York City on 22 April 1898. While underway, they learned that they were bound for the naval blockade of Cuba.

The *Panther* was hardly suited as an attack transport. It was old and crowded, having been purchased with the intent of carrying only half the number of Marines that were actually embarked. A former South American banana

[75] John J. Reber, "Huntington's Battalion Was the Forerunner of Today's FMF," *Marine Corps Gazette* 63, no. 11 (November 1979).

USS *Panther*, ca. 1902–3.
Naval History and Heritage Command, NH68336

freighter, its hasty conversion to a troopship failed to address many of the amenities taken for granted today, such as adequate ventilation and heads (toilets) and galley (kitchen) spaces. Given the time constraints, it was the best the Navy could do. After nearly two months in limbo, half of the time being spent ashore at Key West, Florida, and the other half afloat, Huntington and his battalion finally landed at Guantánamo Bay, Cuba, on 10 June 1898 at the site the Atlantic Fleet had selected for a protected coaling station.[76]

For Huntington and his Marines, the landing could not have come soon enough. Besides having to cope with crowded and uncomfortably hot living conditions aboard the *Panther*, a variety of command-related issues had arisen between Huntington and the ship's captain, Commander George C. Ritter, since embarking in April. One well-known example involved Ritter's order forbidding his crew to assist the Marines in landing their supplies and equipment, forcing the Marines to do it by themselves, thus prolonging the landing operation. Additionally, the Marines were not allowed to land all of their rifle ammunition, since Commander Ritter claimed he needed it kept aboard to serve as ship's ballast.

Moreover, Ritter, following Navy custom, insisted on establishing his authority over the Marines on every matter, large or small. While this certainly was his prerogative in regard to a normal Marine Corps ship's detachment, Huntington believed that this authority was overstated in regard to an embarked Marine battalion,

[76] Incidentally, the same bay is still in use by the U.S. Navy 118 years later.

The USS *Marblehead* steams ahead on its way to Guantánamo, Cuba.
Official U.S. Navy photo

Cdr Bowman H. McCalla, captain of the USS *Marblehead* at Guantánamo Bay, Cuba.
Naval History and Heritage Command, NH72745

which was under the command of its duly appointed commander. Timely intervention at one point by the overall flotilla commander, Commander McCalla of the warship USS *Marblehead* (CL 12), ensured the cooperation of both the ship's captain and commander of the landing force for the duration of the operation.

Nevertheless, Huntington's Battalion was successfully landed on 10 June with all of his men, guns, tents, and equipage and they immediately went about securing the heights surrounding the bay. The Spanish defending force was resoundingly defeated at the Battle of Cuzco Wells on 14 June, leaving the battlefield to the Marines. Not only did the Marines fight ashore as an independent, all-arms force for the first time, new techniques in ship-to-shore communication, fire support, and inter-Service cooperation also were established, if not perfected. With the heights secure and the Spanish bottled

The first bloody engagement of U.S. troops on Cuban soil. U.S. Marines going ashore at Guantánamo with their Krag–Jørgensen rifles in June 1898.
Official U.S. Marine Corps photo

Group of Marine officers at Portsmouth, NH, immediately after the Spanish-American War and their return from Cuba. Col Huntington (front row, fifth from right) with his line and staff officers, August 1898.
Defense Department photo (Marine Corps) 515613

up safely in the town of Caimanera, McCalla's flotilla sailed into the excellent harbor and used it continuously for the next several months, which was finally established as a permanent U.S. naval base by treaty when the war was over. Following the war's conclusion, Huntington and his Marines sailed back to the United States, arriving at Portsmouth, New Hampshire, on 26 August 1898.[77]

[77] Surprisingly, 98 percent of the men had been unaffected by any tropical disease, compared to the Army contingent in the Cuban campaign, which suffered inordinately from diseases such as yellow fever. Their good fortune was attributed to the fact that, for most of the campaign, the Marines had been embarked aboard a ship away from the swampy lowlands, and while they were ashore had practiced rigorous field sanitation procedures.

Col John H. Russell Jr., commander of Russell's Battalion, shown here as a major in 1902, future Major General Commandant of the Marine Corps. *Historical Reference Branch, Marine Corps History Division*

Back on American soil on 19 September 1898, Colonel Commandant Heywood ordered the battalion paraded and then had it disbanded, with its Marines being sent back to the various East Coast barracks and naval installations from whence they had come.[78] Although Huntington's Battalion had successfully accomplished its mission, Heywood did not contemplate this expeditionary adventure becoming a standing requirement. Instead, the Colonel Commandant saw it as a distraction from the Marine Corps' traditional role, which he felt was continuing to serve as ship's detachments and guarding the various naval installations throughout the United States. Whether he or the Marine Corps cared for the concept or not, the afloat-ready battalion had proven itself in practice, and the U.S. Navy took notice.

THE AFLOAT-READY BATTALION CONCEPT REVIVED BY THE NAVY

The next incarnation of the afloat-ready battalion came four years later in the form of Russell's, Haines's, Pope's, and Lejeune's Battalions. At the beginning of September 1902, the USS *Panther* once again embarked a Marine battalion (16 officers and 325 enlisted men) at the request of Secretary of the Navy William H. Moody, who had stated his desire the previous July to have such a battalion ready for training with the fleet, as well as to be on hand to serve in an expeditionary capacity and ready to land anywhere the fleet deemed it desirable to do so.[79]

Commanded by Lieutenant Colonel Benjamin R. Russell, this first afloat-ready battalion was composed of men from the Marine Barracks Brooklyn Navy Yard and Philadelphia Navy Yard. Hastily formed for service in what they were told would be Western Caribbean waters, the battalion sailed on 14 September 1902. Upon arrival off the coast of Columbia, the *Panther* would serve as a station ship, able to launch an expeditionary battalion-size landing force anywhere in the region at a moment's notice. It and its three successor battalions would protect American interests during ongoing unrest in Honduras and Panama for the next 16 months, serving as an important tool of U.S. national policy in the region.

The Marines did not have to wait long. On

[78] The origination of military parades harkens back to military formations during close-order maneuvers. More recently, the actions became strictly ceremonial in nature, particularly during the nineteenth and twentieth centuries when military units were returning from deployments or as a means to demonstrate the military might of a nation.

[79] Allan R. Millett and Jack Shulimson, *Commandants of the Marine Corps* (Annapolis: U.S. Naval Institute Press, 2004), 140.

USS *Prairie* in a harbor, while she was fitted with sailing rig for training ship service, ca. 1901–5.
Naval History and Heritage Command, NH105835

Col Percival C. Pope (shown here as a major in 1890), commander of Pope's Battalion, 1902.
Naval History and Heritage Command, NH85788

23 September, on orders from Rear Admiral Silas Casey III, commander of naval forces in the Caribbean, Russell and his Marines landed at Colón, in what is now modern-day Panama, to protect U.S. interests during a period of civil unrest between Colombia and the United States, which exercised governmental authority over the region where the Panama Canal was being built. The landing of a disciplined battalion of well-armed and -equipped Marines, and its visible presence throughout the city, was enough to convince the warring parties—loyalists and separatists—to stand down and cease their violent acts against the local government in Colón and American businesses.

After two uneventful months of patrolling and supporting the local police, the battalion once again embarked on board the *Panther* on 18 November and sailed for the advanced naval base at Culebra, an island off the coast of Puerto Rico, where the Marines disembarked and conducted training ashore.[80] By 30 November 1902, most of Russell's men had become sick from various tropical diseases incurred after two and a half months of service in the Caribbean, forcing the weakened battalion to return to the United States, where it was immediately disbanded. Despite the lingering effects of the various tropical illnesses, the presence of an armed and well-trained battalion of Marines embarked aboard a station ship had proved its worth.

Meanwhile, once again at the behest of the Navy, another Marine battalion was formed on 5 November that same year in Norfolk, also for service in the Caribbean. This battalion, commanded by Colonel Percival C. Pope, was sent directly to Culebra aboard the transport USS *Prairie* (AD 5) to train with Russell's battalion, since the immediate need for troops in Panama had seemingly passed. Discovering that Russell's battalion had been forced to return to the United States for health reasons, the flotilla commander decided to keep Pope's battalion on station aboard the *Prairie* instead.

Pope was no stranger to service afloat. He had served on the staff of Huntington's Battalion

[80] Millett and Shulimson, *Commandants of the Marine Corps*.

at Guantánamo Bay and was a good choice to lead the new battalion, which was nearly twice as large as Russell's. It was a balanced force, consisting of 600 men organized into six companies, along with artillery, machine guns, and rudimentary signal equipment. However, in fleshing out this battalion, the East Coast was effectively denuded of nearly every able-bodied Marine who had not deployed with Russell five months earlier. It also forced the Colonel Commandant to delay his plans to create an advanced base defense force, which had become the Marine Corps' primary focus since the Spanish-American War.

EMPHASIS SHIFTS TO ADVANCED BASE FORCE

With its traditional role of serving as ship's detachments threatened by the increasing modernization of the Navy, which felt that it no longer needed such a seemingly anachronistic body of troops on board its ships, Marine Corps leaders belatedly realized that the advanced base force was where its future lay.[81] After its victory in 1898, the United States had acquired a far-flung overseas empire with coaling stations located all around the globe that needed to be defended or seized if the president deemed it necessary. The ensuing deployments of not only Pope's battalion in an expeditionary capacity, but of two subsequent ones, forced Heywood to delay his plans for creating such a force for at least two more years, since it was patently obvious that nothing could be done until the Navy overcame its desire to keep large numbers of Marines embarked on station ships or until it became impractical to continue doing so.

Nevertheless, Admiral George Dewey, commander of the Atlantic Fleet and hero of the Battle of Manila Bay, remained enthusiastic about the utility of an embarked ready battalion. Shortly after the Spanish-American War had concluded, he commented that "If there had been 5,000 Marines under my command at Manila Bay, the city would have surrendered to me on May 1, 1898, and could have been properly garrisoned."[82]

The further utility of the Marines for service in the Caribbean was evinced by Dewey's deputy, Rear Admiral H. C. Taylor, in a letter to the secretary of the Navy, in which he asserted that the Marines served two purposes: one of "being ready for service anywhere," and the other "that of improving the base and harbor" of Culebra as "a most valuable adjunct."[83] It was not an entirely negative development for Pope's Marines, who gained valuable experience in constructing and defending an advanced base during a lengthy exercise carried out by the Navy that ended on 3 January 1903.

That same month, Colonel Pope handed over command of his battalion in Culebra to Major Henry C. Haines, who then was ordered to transfer his Marines back aboard the creaking *Panther* later that month. They would serve aboard this station ship as part of the Atlantic Fleet's newly activated "Caribbean Squadron" until late July 1903. Finally landed in Maine to take part in Army-Navy joint maneuvers at the end of that month, Haines and his battalion sailed to the Philadelphia Navy Yard in August 1903, where he was relieved of command by Major John A. Lejeune in October.

[81] At the time, an *advanced base force* was understood to be a coastal and/or naval base defense force designed to establish mobile and fixed bases in the event major landing operations would be necessary beyond U.S. shores.

[82] James D. Hittle, "Sea Power and the Balanced Fleet," *Marine Corps Gazette* 32, no. 2 (February 1948): 57.

[83] Millett and Shulimson, *Commandants of the Marine Corps*, 141.

THE FLOATING BATTALION OF THE ATLANTIC FLEET

Lejeune's Battalion, now known by the Navy as the "Floating Battalion of the Atlantic Fleet," then embarked aboard the transport USS *Dixie* (1893) and sailed once more to the Caribbean to take part in the upcoming 1903–4 winter maneuvers.[84] That exercise never came to pass because Lejeune and his men were diverted from Culebra to Panama instead, where they went ashore at Colón on 5 November to discourage Colombian forces from invading. Joined two months later by Brigadier General George F. Elliott's provisional Marine brigade, Lejeune and his men participated in the Panama Canal crisis of 1903–4, but did not see combat. After Colombia backed down from its threats to invade Panama, mainly due to the presence of Elliott's brigade, peace was restored and Panamanian independence was formally recognized. No longer needed, Lejeune's Battalion returned to the United States in February 1904, where it was finally disbanded.[85]

BGen George F. Elliot (shown here as a major general while serving as the 10th Commandant of the Marine Corps, 1903–10), commander of Elliott's Brigade.
Marine Corps History Division

Though Pope's, Russel's, Haines's, and Lejeune's battalions had satisfactorily served as precursors for the Navy's forces afloat concept, the Marine Corps recorded its objections to the overall concept, feeling that it was a diversion from what it saw as its evolving primary mission of serving as the fleet's nascent base defense force. In addition to this objection, Colonel Commandant Heywood complained in 1903 that this unfunded program came out of the Marine Corps' thinly stretched budget and was not compensated for by the Navy, that it used "borrowed" manpower needed elsewhere, and that the proper onboard equipment and small boats needed to receive, store, and land supplies were lacking on the ships used to carry the Marines. There was also the issue of the ships themselves—the USS *Panther*, *Prairie*, and *Dixie*—which were never intended to serve as troopships and had undergone inadequate conversion to prepare them for that role. They were cramped, poorly ventilated, and lacked adequate space for the embarked Marines to exercise or perform any sort of drill.

Another issue that continually raised its head was the never-ending conflict of authority between the successive Marine battalion commanders and each ship's captain. In many cases, not only did the ship's captain insist on enforcing

[84] Spencer C. Tucker, *Almanac of American Military History: 1000–1830*, vol. 1 (Santa Barbara: ABC-CLIO, 2012), 1210.
[85] Tucker, *Almanac of American Military History*, 1213.

his writ upon every Marine on board, circumventing the Marine Corps chain of command, but some ship's captains also attempted to give precise instructions on the employment of the Marines once they had gone ashore. However it might vex Colonel Commandant Heywood and his successor Brigadier General Commandant Elliott, there was little either of them could do about it, since they had no authority over their Marines from Washington, DC, once they were sailing as part of the fleet, unless, like Elliott, he sailed with his Marines to command them in person as commander of the provisional brigade sent to Panama.

While Heywood or Elliott could complain to the Navy about this practice, both had to confront the admiral's belief, deeply rooted in tradition, that anyone embarked on a U.S. Navy warship was subject to the captain's authority. Heywood, when he first confronted this assertion, countered that this was nonsense, given that the *Panther*, *Prairie*, and *Dixie* were mere troopships, which by naval custom gave the commander of the landing force authority over his own men. Unfortunately for Heywood, the Navy's counterargument that the presence of a few small-caliber cannon on board these converted freighters buttressed its contention that these were indeed warships, which practically ended all discussion of the matter during the rest of his and Elliott's tenure.

Despite the ineffective resistance of the Marine Corps, which was in any case subordinate to the Navy, the latter Service still wanted to continue the practice. After the success in Panama and elsewhere, the Navy believed that an embarked battalion of Marines enhanced the Navy's expeditionary capability. However, events conspired to end the practice altogether for nearly 43 years. During the first decade of the twentieth century, the United States quickly discovered that its new overseas empire needed to be policed and that the numbers of troops on hand, both Army and Marine Corps, were insufficient for the purpose. The Marine Corps especially found itself pulled in every direction, having to send detachments to protect new naval bases in the Philippines, Guam, Puerto Rico, and Cuba, as well as the American legation in China. While the authorized size of the Marine Corps had increased, it still had not attained the minimum number of Marines that Colonel Commandant Heywood felt adequate—a total of 10,000 men—to meet all of the Corps' commitments, most especially when it was focused on the evolving advanced base defense force concept.

The dichotomy between the desires of the Navy, which wanted an expeditionary afloat-ready battalion, and the Marines Corps, which wanted an advanced base defense force, would continue unresolved until 1947. During the interval, both Services were consumed by a variety of challenges, including modernizing the fleet, fighting World War I, participating in a series of protracted counterguerrilla and nation-building operations in the Caribbean during the 1920s and '30s (the Banana Wars), experimenting with air-ground cooperation, and—most important from the Marine Corps' perspective—developing and maturing the advanced base force concept that included the concept of amphibious assault against a defended beachhead. These and other events, including successfully waging World War II, required the complete dedication and cooperation of both sea Services to achieve their goals.

Ambassador George F. Kennan in 1947.
Library of Congress Prints and Photographs Division

RAdm Bernhard H. Bieri, U.S. Navy.
Naval History and Heritage Command, NH80-G-701987

AFLOAT-READY BATTALION CONCEPT REDISCOVERED BY STATE DEPARTMENT

This is where things stood until December 1947, when the concept was resurrected at the beginning of the Cold War. On this occasion, it was not the Navy that called for an afloat-ready force, but the U.S. Department of State, which felt that the United States needed a variety of policy options to employ as a counter to what had become an increasingly belligerent and assertive Soviet Union. In April and again in December of that year, Ambassador George F. Kennan called for a scalable, highly mobile amphibious reaction force that could be based at sea and prepared to conduct a landing operation anywhere in the Mediterranean Sea to assist U.S. allies threatened by Communist expansion.[86] Instead of seeking a military confrontation with the Soviet Union, which was engaged in destabilizing several Western European nations and consolidating its control over Eastern Europe, Kennan believed that U.S. goals would best be achieved by containing the Soviet threat over a long period by using political, military, informational, and economic levers of power.

Kennan was serving at the time as Secretary of State George C. Marshall Jr.'s influential director of policy planning and was highly

[86] Kenneth W. Condit, *History of the Joint Chiefs of Staff: The Joint Chiefs of Staff and National Policy, 1947–1949*, vol. 2 (Washington, DC: Office of Joint History, Office of the Chairman of the Joint Chiefs of Staff, 1996), 6.

USS *Bexar* (APA 237) underway off San Diego, CA, ca. 1954.
Naval History and Heritage Command, NH66834

respected throughout the U.S. government for his depth of understanding of the growing Soviet menace. When the Soviet Union began to exert diplomatic and military pressure upon Greece and Italy throughout the summer and fall of 1947, the State Department was able to convince President Harry S. Truman that to assist these democratic governments, both threatened by Communist agitation, the Navy could help further the nascent "containment policy" and Truman Doctrine against the Soviet Union by conducting a variety of fleet exercises and amphibious demonstrations that would send a signal to Josef Stalin of the inadvisability of continuing his destabilizing actions.[87]

Consequently, in addition to sending additional ships of the Sixth Fleet to the eastern Mediterranean as ordered by the president, the commander of U.S. naval forces in the Mediterranean, Vice Admiral Bernhard H. Bieri, also requested that a battalion-size Marine Corps amphibious task force be deployed to bolster the fleet's striking power, which up to that point did not include any battalion landing teams. The request was duly approved and a chief of naval operations order dated 20 December 1947 directed the temporary assignment of a reinforced Marine battalion to augment existing Marine detachments on Sixth Fleet warships and to provide a ready landing force.[88] This order brought about the actual resurrection and implementation of the afloat-ready battalion concept, the first time since Pope's, Haines's, Russel's, and Lejeune's Battalions of 1902–4 that an amphibious expeditionary force would embark aboard Navy ships and remain on station, awaiting a

[87] Allan K. Henrikson, "The Creation of the North Atlantic Alliance: 1948–1952," *Naval War College Review* 32, no. 3 (May/June 1980): 12. Editor's note: the May/June 1980 issue of *Navy War College Review* was published bearing the incorrect volume number (32); the correct volume number for all 1980 issues was 33.

[88] John G. Norris, "Navy Places Its Top Strategist in Command of Area," *Washington Post*, 6 January 1948.

possible contingency order that could result in their landing on a foreign shore at a moment's notice.

Within days of receiving this order, a battalion landing team of 1,000 Marines from the 2d Marine Regiment (Reinforced), along with vehicles, tanks, artillery, and supplies, formed up and began loading on board the World War II-vintage U.S. Navy attack transports USS *Bexar* (APA 237) and USS *Montague* (AKA 98) in Morehead City, North Carolina.[89] Sailing from the East Coast on 5 January 1948, this force remained afloat with the Sixth Fleet for three months in the eastern Mediterranean, returning on 12 March 1948 after being replaced by a similar battalion from Camp Lejeune, North Carolina, a move that initiated a series of cruises that would normally last six months.

Unlike its lukewarm acceptance of the concept in 1903, the Marine Corps embraced this new mission enthusiastically. Embroiled as it was in the 1947–48 military roles and missions debate, which involved nothing less than the continuing survival of the Marine Corps as a Service, this type of mission was tailor-made for what it specialized in—expeditionary operations and amphibious assault—as part of the Navy's "balanced fleet."[90] Having proven its ability to carry out these kinds of assignments in the Pacific during World War II, the Marine Corps felt that it was uniquely suited for the afloat-ready battalion mission in the Mediterranean, as compared to the U.S. Army, which was almost fully committed to occupation duties in Germany, Japan, China, and Italy. The greatest obstacle to filling the Navy's requirement was that the number of existing battalion landing teams had been reduced to six, of which only half were considered to be available for service with the Sixth Fleet in the Mediterranean, the Caribbean, and elsewhere.[91]

Though it never fired a shot in anger, the first afloat-ready battalion to deploy to the Mediterranean participated in several amphibious exercises within close proximity of Greece and Italy, a move that the Soviets could not fail to notice. Combined with other political and military signals being sent by the U.S. government at the time, the presence of the amphibious force and the national resolve that it signified were enough to influence the Soviet Union to decrease its support to the Communist rebel movements in Greece, Italy, and Turkey, granting the governments of these countries the breathing space they needed to renew efforts to bolster their defenses against their respective insurgencies.[92]

EVOLUTION OF THE MEDITERRANEAN AFLOAT BATTALION 1948–60

This first afloat-ready battalion, though still a powerful unit by today's standards, was not a true combined Marine air-ground organization in the modern sense. The battalion landing team was not paired with an aviation component and lacked a commander and staff to exercise command and control of any Marine Corps air and ground units that might operate together. It was completely dependent on its troop transports for logistical support, having none of its own, rendering it unable to operate independently ashore for more than a few days. To compound command and control issues, neither the *Bexar* nor the *Montague* was equipped with the communications gear that would have allowed the

[89] Norris, "Navy Places Its Top Strategist in Command of Area."

[90] Hittle, "Sea Power and the Balanced Fleet," 59.

[91] Condit, *History of the Joint Chiefs of Staff*, 150.

[92] George F. Kennan, *Report by the Policy Planning Staff: Review of Current Trends—U.S. Foreign Policy*, Policy Planning Staff Paper no. 23 (Washington, DC: U.S. Department of State, 1948).

U.S. Marines and Lebanese Army personnel debark from LCTs with one of the Marine Corps' new M50 Ontos, light armored antitank vehicle. This unit was from the NELM Battalion assigned to the Navy's Sixth Fleet.
Defense Department photo (Marine Corps) A17399

Gen Lemuel C. Shepherd, 20th Commandant of the Marine Corps (1952–55).
Historical Reference Branch, Marine Corps History Division

battalion commander to exercise control over his forces while afloat. It was a stopgap, expedient solution but it was enough to send the right message of political will.

Though Marine fixed-wing aircraft were operating aboard aircraft carriers of the Sixth Fleet at the time, they fell under the Navy's control and were not considered to be part of the afloat-ready battalion's "force package" or authorized temporary organizational structure. The battalion and its equipment were not configured for an amphibious assault either, since neither of the two attack transports were accompanied by the necessary landing ships, tank (LSTs) nor did they carry any landing vehicles, tracked (LVTs) like those recently used during the war with Japan with such great effect. It also had no helicopters of its own, a newly introduced aerial system that had not yet gone far beyond the experimental stage but one that showed great future promise as a means of landing troops in support of an amphibious assault.

Nevertheless, this move initiated the Marine Corps' practice of maintaining an air and landing force with the Sixth Fleet in the Mediterranean on a recurring basis, a practice which, except for short-term breaks in continuity due to overwhelming requirements for troops elsewhere (e.g., the Korean and Vietnam Wars), has continued from 1948 to the present day. That same year, this afloat-ready battalion also was given its first name—the Naval Forces, Eastern Atlantic and Mediterranean Battalion, or NELM Battalion.[93] In 1960, the Sixth Fleet redesignated

[93] "Marines Are on Their Way," *Sunday Star-News* (Wilmington, NC), 13 January 1957, 8-A.

it as the Landing Force, Mediterranean, or LanForMed, but little else changed.

THE DOCTRINAL REVOLUTION OF THE 1950S

Except for the existing battalion landing team doctrine dating back to the late 1940s, the Marine Corps had yet to devise a tactical system or a way of thinking about how to incorporate all of the disparate elements needed to make such an air-ground force capable of operating in a nuclear environment complete. Even had there been doctrine, or helicopters advanced enough to carry troops and cargo, in 1948 there was as yet no ship suitable enough to serve as a floating base, though aircraft carriers did hold promise. Unfortunately, the Navy was reluctant to allocate its large fleet carriers or the funding for such a project, not convinced yet that the helicopter would prove itself as the panacea that the Marine Corps thought it was. Fixed-wing aviation continued to operate from aircraft carriers assigned to the various fleets.

Between January 1948 and early 1960, a succession of NELM Battalions continued sailing with the Sixth Fleet throughout the Mediterranean. However, a portent of the future gradually began to take shape upon the publication of a bulletin on 9 November 1954 written by General Lemuel C. Shepherd Jr., Commandant of the Marine Corps.[94] Weighing the increasing capability of the Marine Corps' rotary- and fixed-wing aviation elements, and foreseeing how they might work in concert with ground combat elements, Shepherd decreed that a new organizational structure—what he termed a Marine Air-Ground Task Force, or MAGTF—would be needed in the future to enable the Marine Corps to continue its amphibious warfare mission while at the same time leveraging new technology to make it a more lethal and agile force.

[94] Lemuel C. Shepherd Jr., *The Marine Air-Ground Task Force Concept* (Washington, DC: Headquarters Marine Corps, 1954).

Shepherd stated that the "future employment of Fleet Marine Force elements will normally involve organization as air-ground task forces in which air and ground units will habitually operate as a single operational command" and that this Marine Air-Ground Task Force should consist of a balanced all-arms team.[95] Shepherd was not prescriptive in the bulletin as to the actual makeup of the force, but he was clearly influenced by the all-helicopter amphibious assault concept, commonly referred to as *vertical envelopment*, which foresaw an even greater employment of the helicopter than was possible at the time. More importantly, Shepherd stressed the importance of all arms—air, ground, and logistics—being placed under the command of a single Marine commander not tasked with the additional duty of commanding one of the MAGTF's components.[96]

While Shepherd's bulletin was important, it was not yet settled as doctrine and commanders of Marine Corps units in service with the various fleets were not bound to follow it. A year later, however, Shepherd's thoughts were reinforced by *Concept of Future Amphibious Operations*, Landing Force Bulletin 17 (LFB-17), which did have the force of doctrine behind it.[97] This bulletin stressed that the MAGTF concept was uniquely suited toward the conduct of vertical

[95] While the development of MAGTF doctrine did have some influence on the continuing evolution of the afloat-ready battalion concept, particularly in regard to the integration of the aviation element, it is a separate concept that evolved along parallel lines and will be covered in greater detail in a future volume of this publication.

[96] Shepherd, *The Marine Air-Ground Task Force Concept*, 2.

[97] *Concept of Future Amphibious Operations*, LFB-17 (Washington, DC: Headquarters Marine Corps, 1955).

Marines from Golf Company, 2d Battalion, 2d Marines, landing in Lebanon, 15 July 1958. These were the first and second waves of Marines to hit the beach at the airport in Beirut.
Defense Department photo (Marine Corps) A17497, courtesy of TSgt Ed Scullin

envelopment as part of amphibious operations in a nuclear environment and that MAGTFs must leverage all of its elements to achieve success. However, the bulletin did not actually provide much guidance concerning how the doctrine was to be put into practice, leaving it up to the commanders to decide what a MAGTF actually was and what it would look like. Additionally, the Marine Corps was still without a suitable seagoing platform to carry such a force, even had there been a consensus with the Navy as to what it was to be. Though smaller escort carriers had been temporarily made available to the Marine Corps for training exercises and experimental purposes since 1948, no dedicated or purpose-built Navy ships yet existed that could transport and support the kind of MAGTF that Shepherd envisioned.

IMPACT OF NEW DOCTRINE ON THE FLEET MARINE FORCE

Armed with the knowledge that incorporating helicopters into the afloat-ready battalion concept was now expected to become standard practice, beginning in 1956, all three Marine Expeditionary Forces, or MEFs (I MEF, II MEF, and III MEF), began to experiment using the forces assigned to them; however, each MEF headquarters, faced with different challenges posed by its area of operations and the availability of amphibious shipping, approached the matter differently. One of the first prototype MAGTFs that deployed consisted of 6th Marine Regiment Headquarters, with one battalion landing team, joined by two Marine Medium Helicopter Squadrons (HMM)—HMM-261 and

USS *Thetis Bay* (LPH 6), the first Marine Corps landing platform, helicopter.
Naval History and Heritage Command, 19-N-69574

HMM-262. This force left Morehead City for NELM Battalion duty with the Sixth Fleet in the Mediterranean on 20 August 1957, making it the first standing MAGTF to serve in that capacity on a rotating basis.[98] The regimental headquarters served as the overall command and control element of the MAGTF, but the limitations of existing amphibious shipping meant that its helicopters had to embark aboard fleet carriers, leaving the MAGTF commander with little authority over their employment until they could be landed and joined with the rest of the MAGTF ashore, the same procedure that governed employment of fixed-wing aircraft.

Throughout the late 1950s and early 1960s, there were many examples of the NELM Battalion/LanForMed being used to conduct noncombatant evacuation operations and to support humanitarian assistance operations, such as the Suez Crisis of 1956, the Lebanon Crisis of 1958, and the Cyprus Crisis of 1965, among others. On 14 July 1958, President Dwight D. Eisenhower ordered the NELM Battalion, along with two additional battalion landing teams, to land in Lebanon to evacuate American citizens and to forestall a coup of the democratically elected government. Joined by elements from the 2d Marine Division that were already afloat in the Mediterranean with the Sixth Fleet for an exercise, the NELM Battalion was quickly landed and began conducting operations nearly a week before the U.S. Army's airborne task force, entirely dependent on airlift, arrived from Germany.[99] The ability of the afloat-ready battalion with its associated aviation element to land and begin conducting operations within 48 hours of notification was a powerful testament to the

[98] Ralph W. Donnelly, Gabrielle M. Neufeld, and Carolyn A. Tyson, *A Chronology of the United States Marine Corps, 1947–1964*, vol. III, Marine Corps Historical Reference Pamphlet (Washington, DC: Historical Division, Headquarters Marine Corps, 1971), 34.

[99] Donnelly, Neufeld, and Tyson, *A Chronology of the United States Marine Corps*, 36.

U.S. Marine Corps HUS-1 Seahorse helicopters lift off the USS *Boxer's* (LPH 4) flight deck during operations off Vieques Island, Puerto Rico, with the 10th Provisional Marine Brigade on 8 March 1959.
Naval History and Heritage Command, NH97288, courtesy of Grantham

utility of the concept, once again proving the usefulness it first demonstrated in 1898.

NEW AND MODIFIED SHIPS MAKE THEIR APPEARANCE

Meanwhile, a new class of Navy ships pointed toward the possibilities of the modern MAGTF. On 20 July 1956, the USS *Thetis Bay* (CVE 90), a converted World War II escort carrier, was recommissioned by the Navy as a landing platform, helicopter (LPH) ship, which was the Marine Corps' first amphibious assault ship able to embark both the troops from the battalion landing team and 12 aircraft from a composite helicopter squadron, combining the functions of both an aircraft carrier and attack transport, changing the way afloat-ready battalions would operate forever.[100] Though the *Thetis Bay* primarily served as a training platform, it saw many operational deployments as the Marine Corps worked out the technical details of the vertical envelopment concept. On 10 November 1958, the first permanent Marine aviation detachment afloat was activated for service and would ulti-

[100] "'Copter Carrier Commissioned," *Naval Aviation News*, September 1956; and "Fuji Feels Marine Assault," *Naval Aviation News*, December 1957, 36.

mately serve on board the USS *Boxer* (LPH 4), a converted World War II *Essex*-class fleet carrier, then undergoing conversion at Norfolk, Virginia.[101] The unit was activated to provide supply, maintenance, and flight deck control to Marine helicopter squadrons and troops assigned to the *Boxer* once the ship was placed back into service.

This trend accelerated in 1959, when the USS *Boxer* was finally recommissioned on 30 January. Twice as large as the *Thetis Bay*, the *Boxer* carried up to 30 helicopters (21 on deck and 9 in the hangar deck) as well as nearly 2,000 Marines of an embarked battalion landing team.[102] Two other converted *Essex*-class carriers, the USS *Princeton* (LPH 5) and USS *Valley Forge* (LPH 8), soon followed, joining the *Boxer* and the USS *Thetis Bay*.

Finally, the Marine Corps had the major pieces of what would constitute the future amphibious force, but these converted carriers had their limitations. For example, while they could embark troops via helicopters, they had no surface landing craft of their own, forcing the Navy and Marine Corps to continue to rely on LSTs and landing ship, docks (LSDs) to carry the ship-to-shore craft that would transport the bulk of the battalion landing team and the logistics elements ashore. These vessels were much slower than the *Essex*-class ships, often requiring the LPHs to sail separately.

With four LPHs on hand between 1959 and 1964, the Marine Corps focused on merging the MAGTF concept with the afloat-ready battalion concept. The half-formed doctrine still lagged behind the evolution of landing craft and helicopters. Realizing this, the Navy and the Marine Corps began working closely together in both the Atlantic and Pacific Fleets as their experimentation progressed, though again each fleet approached the challenge differently. For example, in 1960, the Sixth Fleet in the Mediterranean Sea announced the initiation of the Fast Amphibious Force (FAF) concept.

According to an article in the *Marine Corps Gazette*, which represented the unofficial voice of the Marine Corps' leadership, the FAF consisted of an afloat-ready battalion, a composite helicopter squadron, and a small logistics element embarked aboard the ships of a Navy amphibious squadron, consisting of fast amphibious ships (including an LPH) capable of steaming at 20 knots that would allow them to avoid slower Soviet submarines. The concept stressed that both the Marine and Navy elements of the FAF must train and operate in concert with one another to boost proficiency and overall effectiveness. Although the FAF concept was intended to be implemented in both Pacific (with the Seventh) and Atlantic (with the Sixth) Fleets, it only seems to have been put into effect under that title in the Atlantic.[103]

THE FLEETS EXPERIMENT WITH NEW CONCEPTS

As the outlines of future amphibious doctrine began to take hold, the concepts for Marine Expeditionary Unit (MEU), Marine Expeditionary Brigade (MEB), Marine Expeditionary Force (MEF), and Marine Expeditionary Corps (MEC) had become common usage within the Marine Corps by 1960, though they had not yet been

[101] Like many of the ships from this period, the *Boxer* saw a great deal of change during its service. Originally classified as an aircraft carrier (CV 21) in 1945, it was repeatedly reclassified, first as an attack carrier (CVA 21) then as an antisubmarine carrier (CVS 21).

[102] LtCol Eugene W. Rawlins, *Marines and Helicopters, 1946–1962*, ed. Maj William J. Sambito (Washington, DC: History and Museums Division, Headquarters Marine Corps, 1976), 87–88.

[103] R. A. Stephens, "Fast Amphibious Force," *Marine Corps Gazette* 45, no. 1 (January 1961): 46–47.

U.S. Navy Seventh Fleet Amphibious Ready Group underway in March 1965 (from left): USS *Bexar* (APA 237), USS *Princeton* (LPH 5), USS *Thomaston* (LSD 28), and USS *O'Bannon* (DD 450). Sikorsky UH-34D Seahorse helicopters of HMM-365 fly above the ships while *Princeton*'s crew spells out the task group designations, "TG 76.5/79.5," on the flight deck.
Official U.S. Navy photo 1142349

encapsulated in doctrine.[104] This led to misunderstandings throughout the Marine Corps regarding their exact usage and composition, whether it applied to the NELM Battalion (redesignated in 1960 as the Landing Force, Mediterranean, or LanForMed) or the FAF.

For example, the 24th MEU (the first recorded use of that designation) was activated at Cherry Point, North Carolina, on 15 November 1960. It consisted of a brigade headquarters, a battalion landing team from the 2d Marine Division, and a provisional Marine Aviation Group consisting of a light helicopter transport squadron and an ordnance-laden attack or jet fighter squadron embarked separately on an aircraft carrier with the Sixth Fleet.[105] Technically speaking, it was a MAGTF, but not quite like the MEU as they are known today, and it did not deploy to the Mediterranean for LanForMed Battalion duty, but deployed only for a series of training exercises. One possible reason is that the Sixth Fleet did not perceive the FAF as a permanent organization; its primary purpose appears to have been to serve as a means to train and familiarize Marines with the emerging doctrinal concepts.

The Sixth Fleet's FAF went through a number of permutations over the next several years, as different units rotated in and out within its structure, but the FAF itself, which at one point included the provisional 16th MEB, never existed for more than three months at a time.[106] By the time the Vietnam War began in 1965, it appears to have disappeared altogether, as the

[104] "Fleet Marine Force," *Marine Corps Gazette* 44, no. 7 (July 1960): A-1.
[105] Donnelly, Neufeld, and Tyson, *A Chronology of the United States Marine Corps*, 43.
[106] "Marine Expeditionary Brigade Returns from Mediterranean Area Maneuvers," *Camp Lejeune Globe* (Jacksonville, NC), 8 June 1961, 6.

demands for ships and battalion landing teams outweighed all other considerations, though the LanForMed Battalion deployments appear to have continued unabated throughout the 1960s.

The introduction of the FAF concept evolved along similar lines with the Seventh Fleet during the early 1960s, but with some typical differences in the operational style between the Atlantic and Pacific Fleets. In 1961, the Seventh Fleet, under the guidance of its commander, Admiral Harry D. Felt, designated its FAF equivalent as the Amphibious Ready Group or ARG. Felt, with Marine Corps support, first proposed the organization of such task forces to Chief of Naval Operations Admiral Arleigh A. Burke, who concurred. This force, the Navy and Marine Corps' first ARG, was designated Task Force 76.5 and based out of the U.S. Naval Base Subic Bay in the Philippines.[107]

Gen David M. Shoup, 22d Commandant of the Marine Corps (1960–63).
Historical Reference Branch, Marine Corps History Division

The prototype ARG consisted of an amphibious squadron with three to four "fast" ships (one LPH, one LSD, and/or an attack cargo ship or AKA) and a Special Landing Force (SLF) instead of a MEU, consisting of a battalion landing team from the 3d Marine Division based in Okinawa and a composite helicopter squadron that included both utility and heavy-lift aircraft. Thus combined, the ARG/SLF would rotate its embarked Marine units every six months, remaining at sea "on station" in support of various Southeast Asia contingencies involving Laos, Cambodia, Thailand, and Vietnam between 1960 and 1964, but was not deployed ashore. That would change during the summer of 1965, when the United States stepped up its involvement in South Vietnam.[108]

The first MAGTF to sail the Atlantic with its own aviation element was built around battalion landing teams from 3d Battalion, 8th Marine Regiment, and 1st Battalion, 6th Marine Regiment, in February 1961 and included aviation elements from the 2d Marine Aircraft Wing. With an 83-man headquarters element provided by 2d Marine Division, this force was designated as 4th MEB. It was joined shortly thereafter by the 24th MEU, which was already at sea sailing as part of the aforementioned FAF. Combined, the 4th MEB included the USS *Boxer* and the fast

[107] Edward J. Marolda and Oscar P. Fitzgerald, *The United States Navy and the Vietnam Conflict: From Military Assistance to Combat, 1959–1965*, vol. II (Washington, DC: Naval Historical Center, Department of the Navy, 1986), 42.

[108] Marolda and Fitzgerald, *The United States Navy and the Vietnam Conflict*, 474, 529.

Marines from the Pacific Fleet's ARG/SLF wade ashore near Da Nang, Republic of Vietnam, ca. 1965.
Defense Department photo (U.S. Navy) 1110983

attack transports from the Navy's Amphibious Squadrons 2 and 8.[109]

This was the first time that a balanced, self-contained, brigade-size MAGTF had participated in a routine afloat mission in the Atlantic and Caribbean. It had not been activated for any specifically designated contingency, such as those brigade-size task forces that had been quickly created or stood up for the Lebanon or Cyprus crises. Instead, like the FAF before it, the MEB served as an enormous sea-going laboratory for amphibious warfare. Though much larger than a MEU, the 4th MEB allowed the Marine Corps to experiment with both its new ships and doctrinal concepts during Exercise LantPhibEx 1-61 before the brigade was deactivated after three months at sea.

A cursory examination reveals that both the Pacific and Atlantic Fleets composited or assembled their afloat-ready forces differently. Those activated for service in the Atlantic, the Caribbean, or in the Mediterranean as LanForMed, tended to be somewhat larger than the special landing force in the Pacific, usually approaching a Marine Expeditionary Brigade (as shown above in the case of the 4th MEB) in size versus that of a battalion landing team-size Marine Expeditionary Unit. Thus, even as late as 1962, it appears that the Marine Corps had still not completely settled the argument about what exactly a MAGTF or a MEU was and what were its constituent elements.

[109] "Marines at Work: 4th MEB," *Marine Corps Gazette* 45, no. 4 (April 1961): 4–5.

USS *Okinawa* (LPH 3) underway in the South China Sea in January 1969 with several CH-34 Seahorse helicopters parked on her flight deck.
Naval History and Heritage Command, NH107670, courtesy of PHC A. L. Smith

MARINE CORPS ORDER SETTLES THE DOCTRINE DEBATE

While the evidence indicates that nearly every senior-level Marine (e.g., lieutenant colonel and above) were in general agreement about the overall MAGTF concept, opinions differed widely as to their size, mission, composition, and other important topics. On 27 December 1962, the MAGTF debate was settled once and for all when *Marine Corps Order (MCO) 3120.3* was issued by Headquarters Marine Corps after extensive consultation with the Navy, which, after all, would be providing the amphibious warfare ships to carry them. This order, signed by Commandant of the Marine Corps General David M. Shoup, formally codified a MAGTF's composition in doctrine and specifically enumerated the four types of MAGTFs based on the size of the command.[110]

The order stated that a MAGTF, regardless of size, would henceforth consist of a ground combat element (GCE), a command element (CE), an aviation combat element (ACE), and a combat service support element or CSSE (now called logistics combat element, or LCE). Additionally, the order specified that a MEU would be based on a battalion landing team, just as the first battalion landing team had been in 1898, and would be augmented by a composite heli-

[110] *MCO 3120.3, The Organization of Marine Air-Ground Task Forces* (Washington, DC: Headquarters Marine Corps, 1962).

copter squadron and a dedicated logistics battalion. Though the doctrine was quickly accepted and the terminology agreed upon, the Seventh Fleet, demonstrating its independent streak, continued to use the term *special landing force* for the MEU sailing with the ARG in the Pacific.[111]

In line with the declaration laying out the composition of a MEU, the order further stated that a MEB was to be based on a reinforced regimental combat team, a composite air group, and a logistics regiment. The MEF would be based on a Marine division, air wing, and appropriately sized logistics elements. The MEC, though the term was never used in practice due to the U.S. Army's objections, was to be based on two or more Marine divisions, with an appropriately sized air wing and logistics element. In practice, however, the MEF has effectively functioned as a corps-size headquarters, demonstrated by the performance of I MEF during Operation Desert Storm in 1991 and during Operation Iraqi Freedom in 2003.

THE MARINE CORPS INTERNALIZES THE CONCEPT

Simply stating "Let there be MEUs, MEBs, MEFs, and MECs" was not the same as creating these organizations. A great deal of learning had to be accomplished and many different subordinate doctrinal publications, tables of organization, tables of equipment, and reams of Service regulations had to be written. The best way to try out new doctrine was through actual practice using real Marines and real ships, where lessons could be learned and modifications made. New organizations had to be created out of thin air, so to speak, since they may not have existed before or even been contemplated in the original 1961 order.

For instance, MEU headquarters generally did not exist in the early 1960s. The usual practice was to take a Marine division or regimental headquarters and either increase its capability with more staff and equipment to make a MEU headquarters or strip down a division headquarters to the bare essentials to create a MEB headquarters. For a more permanent solution, MEU headquarters had to be designed and built into future budgets so the manpower spaces could be allocated, equipment purchased, and funding for training programmed.

An excellent example of how each MEF worked through the task of incorporating the new doctrine can be seen in how the Pacific Fleet's SLF evolved. The first SLF created in 1960 lacked a separate command element. Instead, the commander of the battalion landing team served in a dual capacity as both battalion commander and SLF commander. With the addition of an aviation element consisting of a mixed rotary-wing squadron, this quickly proved to be an unworkable arrangement, since the battalion landing team commander lacked the expertise and communications means to command and control it. Consequently, III MEF in Okinawa authorized the activation of a small permanent SLF command element in 1965.[112] A year later, this staff had evolved into a true MAGTF headquarters approximately the size and capability of an infantry regiment's staff.[113]

As the demand for more troops to support

[111] *MCO 3120.3*, enclosures 1–2.

[112] Jack Shulimson and Maj Charles M. Johnson, *U.S. Marines in Vietnam: The Landing and the Buildup, 1965* (Washington, DC: History and Museum Division, Headquarters Marine Corps, 1978), 196, 200–1.

[113] Jack Shulimson, *U.S. Marines in Vietnam: An Expanding War, 1966* (Washington, DC: History and Museum Division, Headquarters Marine Corps, 1982), 297.

USS *Raleigh* (LPD 1) underway at Guantánamo Bay, Cuba, on 1 February 1963. The ship carries two CH-34 Seahorse helicopters parked on her afterdeck.
Naval History and Heritage Command, NH107694, courtesy of PH3 Houchins

the Marines in Vietnam increased, the ARG/SLF was increasingly employed ashore, where it took part in numerous ground combat operations, so much so that the Navy decided it needed another ARG/SLF activated just to ensure that the Seventh Fleet still had a reserve landing capability should other emergencies arise in the Pacific Rim not involving Vietnam. Accordingly, a second ARG/SLF, or ARG-B, was created to complement the original ARG, now designated ARG-A. From 1965 to 1969, both ARGs rotated between service with the fleet and ashore in support of III MAF in South Vietnam. During this period, both ARGs carried out 62 amphibious landing operations in Vietnam while taking part in dozens of cruises with the Seventh Fleet.[114]

[114] Benis M. Frank and Ralph F. Moody, "SLF Operations in Vietnam" (unpublished paper, History and Museums Division, Headquarters Marine Corps, 1972), Section VII, 4.

The only change that occurred during this period worth noting involved the renaming of Marine Expeditionary Units, which, due to perceived South Vietnamese sensitivity to the term *expeditionary* (with the attendant negative connotations of French Colonial rule), were redesignated as Marine Amphibious Units or MAUs in 1965. Between 1965 and 1990, a variety of former MEUs carried over this term until all existing MAUs reverted to their former naming convention of MEU. Except for name change, everything else remained the same. Finally, even the Seventh Fleet's special landing force was redesignated as a MAU in 1969, then once again as a MEU by 1988.[115]

[115] Jonathan D. Geithner, *Historical ARG/MEU Employment* (Arlington, VA: CNA, 2015), 3–4; and *All Marines Message 023/88, Change of Marine Corps Task Unit Designations* (Washington, DC: Headquarters Marine Corps, 5 February 1988).

USS *Tarawa* (LHA 1) underway in the Pacific Ocean in March 1979. Note the light-colored Douglas A-4 Skyhawk parked between two AV-8 Harriers on the ship's starboard after flight deck.
Naval History and Heritage Command, NH107654

U.S. NAVY SUPPORT TO THE ARG/MEU CONCEPT

The final piece of the afloat-ready battalion concept involved related activities of the U.S. Navy, which had continuously coordinated its ship design initiative with the Marine Corps. While the ARG/MEU concept had become embedded in Navy practice, if not in doctrine, purpose-built ships to replace the World War II-vintage ships did not arrive in the fleet until 26 August 1961, when the first LPH was built as such from the keel up, and the USS *Iwo Jima* (LPH 2) was commissioned. Six others of its class soon followed. These ships were capable of carrying 26 helicopters and nearly 2,000 Marines of a reinforced battalion landing team, equating to 193 officers and 1,806 men.[116]

However, as impressive as these vessels were in terms of their vertical envelopment capability, they lacked a well deck, thus forcing the ARG/MEU to rely on older LSTs and LSDs to carry the ship-to-shore "connectors" (LVTs, LCVPs, LCUs, etc.) for over-the-beach amphibious capability. Fortunately, by 1962, a new type of amphibious warship, the landing platform, dock or LPD (known today as amphibious transport docks), entered service, easing the reliance on the World War II-era vessels. It had both a well deck and purpose-built helicopter landing platforms, giving it a versatility that the older ships lacked.

By 1970, composition of ARGs, at least as far as the Navy was concerned, had become a settled issue due to the retirement of the rest of the few remaining World War II-era amphibious ships and the construction of enough new ones to replace them. The "standardized" ARG/MEU was now composed of an amphibious squadron

[116] "*Iwo Jima* (LPH-2), 1961–1993," Naval History and Heritage Command, 10 November 2015.

An AV-8B Harrier, assigned to Marine Medium Tiltrotor Squadron 163 (VMM-163) (Rein), hovers above the flight deck during a vertical takeoff from the amphibious assault ship USS *Makin Island* (LHD 8) on 23 August 2016. As the flagship of the Makin Island ARG, the ship is deployed with the embarked 11th MEU to support maritime security operations and theater security cooperation efforts in the U.S. Navy's Fifth Fleet area of operations.
Official U.S. Navy photo, courtesy of PO3 Devin M. Langer

(PhibRon) consisting of one LPH (later replaced by the amphibious assault ship, general purpose, or LHA), one LPD, and one LSD combined with a MEU of 2,200 Marines. However, fixed-wing aircraft were still embarked aboard Navy fleet carriers, which were not part of the ARG/MEU combination. War plans directed that, once a beachhead had been taken, airstrips would then be seized or constructed, allowing fixed-wing aircraft to land, where they would then revert to ARG/MEU control. This unsatisfactory situation would not change until 1979.

The slower *Newport*-class LSTs and attack cargo ships, or AKAs, were relegated to other amphibious squadrons before they were phased out entirely by 2000, being superseded by the new classes of ships being commissioned. Attack transports (APAs), the last vestige of the World War II-type attack transport, also were completely phased out by 1980. Though the ARG/MEU composition had been settled for nearly 40 years, it was not until 2010 that it was finally codified in the U.S. Navy's Operational Instruction *OPNAV 3501.316B* on 21 October.[117]

[117] *OPNAV 3501.316B, Policy for Baseline Composition and Basic Mission Capabilities of Major Afloat Navy and Naval Groups* (Washington, DC: Department of the Navy, 2010).

EVOLUTION INTO TODAY'S ARG/MEU

The last significant development occurred on 29 May 1976, when the first *Tarawa*-class LHA was commissioned. The USS *Tarawa* (LHA 1) was the first LHA with a well deck for carrying and launching landing craft, utility (LCU), amphibious assault vehicles-personnel 7 (AAV-P7), and landing craft, air cushion (LCAC). There were five of these enormous ships built, each capable of carrying as many as 41 helicopters or a balanced mix of Boeing Vertol CH-46 Sea Knights, Sikorsky CH-53 Sea Stallions, Bell AH-1W Super Cobras, and Bell UH-1 Iroquois, as well as 1,903 troops.[118] The USS *Tarawa* deployed on its first Western Pacific cruise in March 1979 and for the first time operated with a McDonnell Douglas AV-8B Harrier vertical short takeoff and landing (VSTOL) jet squadron, in addition to an embarked helicopter squadron in a successful experiment to determine the feasibility of VSTOL aircraft operating from an LHA.

With the addition of the AV-8B Harrier, the ARG/MEU combination finally had its own organic fixed-wing aircraft squadron, capable of providing combat air patrol coverage as well as close air support to the MEU. Today, ARG/MEUs often deploy as part of an Expeditionary Strike Group (ESG) consisting of an aircraft carrier or other surface warfare combatants. Since 2015, the more-capable USS *Wasp*-class LHDs and the new USS *America*-class LHAs have completely replaced the *Tarawa*-class LHAs. Both classes of ships now operate with the new Bell Boeing MV-22 Osprey and will soon host the newly introduced Lockheed Martin F-35 Lightning II VSTOL aircraft, which is replacing the AV-8B. One thing has not changed, however. Conventional Marine fixed-wing aviation assets, such as the McDonnell Douglas F/A-18 Hornet, still operate from the decks of Navy fleet carriers. Once suitable airfields are constructed ashore, they will deploy as part of the ARG/MEU, though the Naval Task Group commander still retains the option of having them operate under his control.

INTRODUCTION OF THE MEU (SPECIAL OPERATIONS CAPABLE) CONCEPT

Another change to the ARG/MEU concept occurred in December 1985, when the 26th MAU (redesignated as a MEU in 1988) received the special operations capable, or SOC, designation, becoming the 26th MAU(SOC). Though the actual organization of the MAU itself did not change, its mission profile did, based upon an increasing awareness within the Department of Defense that the growth of terrorism around the world required an effective military response that went beyond traditional capabilities, bordering on those ordinarily possessed by special operations forces (SOF). The addition of a SOC designation to its title signified that a MAU had been issued certain equipment "enhancements" and had trained to a rigid standard prior to deploying. Once it had arrived on station, a MAU (SOC) might be called upon to accomplish special operations-like missions, such as *in extremis* hostage rescue or noncombatant rescue operations and antiterrorist operations.[119] By 1987, all deploying MAUs were required to train to MAU (SOC) status.

Marine Amphibious Units from that point

[118] *Amphibious Ships and Landing Craft Data Book*, Marine Corps Reference Publication 3-31B (Washington, DC: Headquarters Marine Corps, 2001), 3–5.

[119] *Report of Examination of Marine Corps Special Operations Enhancements* (Norfolk, VA: Fleet Marine Force, Atlantic, 1985).

onward would only receive the MAU(SOC) designation prior to deployment after they had met special operations certification requirements; otherwise, when not deployed, they would retain the normal MAU title (in 1988, they were once again redesignated MEUs and became MEU [SOCs]). This concept remained in effect from 1985 until 2005, when the newly activated special operations companies of Marine Corps Special Operations Command (MARSOC) began to assume the mission and MEUs finally dropped the SOC appellation.[120] Currently, MEUs can only use the SOC designation if a Marine Corps special operations component is attached to carry out specific special operations-related missions, though in practice this rarely occurs due to the high demand for their services within the U.S. Special Operations Command.

CONCLUSION

While current operational concepts such as a disaggregated or split-based ARG/MEU have recently been put into practice, the core concept of the ARG/MEU remains unchanged and will probably stay that way for the foreseeable future. As this article has shown, during the past 118 years, a progression of changes in national security policy, Service doctrine, and technology have combined to provide today's afloat-ready force the capability that Lieutenant Colonel Huntington could only dream about. Though the modern expeditionary amphibious force, with its warships, aircraft, and landing craft, is far removed from the afloat-ready battalion that saw its debut during the Spanish-American War, the concept itself—that of having an embarked self-sustaining battalion-size force ready to be landed anytime, anywhere at the order of the U.S. government—has hardly changed at all. Though debate may swirl around the notion that amphibious warfare has become obsolete, one thing is certain—as long as there is a U.S. Marine Corps, there will be an ARG/MEU at sea somewhere, ready for the call to carry out the nation's bidding.

[120] Frank L. Kalesnik, "MARSOC: U.S. Marine Corps Forces, Special Operations Command, The First Decade, 2006–2016" (PhD dissertation, Marine Corps Special Operations Command, 2016), 6–7. According to *MCO 3120.9B*, the MEU(SOC) was required to demonstrate interoperability with the embarked Naval Special Warfare Task Unit (SEALs) prior to deploying. Other sources suggest that the SEALs stopped deploying with MEUs shortly before 2001.

MERRITT AUSTIN EDSON BIOGRAPHY

"There it is. It is useless to ask ourselves why it is we who are here. We are here. There is only us between the airfield and the Japs. If we don't hold, we will lose Guadalcanal."

~Lieutenant Colonel Edson Merritt A. Edson,
Lunga Ridge, September 1942

Major General Merritt Austin Edson was born in 1897, he joined the Marine Corps Reserve on 26 June 1916, and was commissioned on 9 October 1917. He served with the 11th Marines in France and the occupation army of Germany. After the war he became a Marine Corps pilot, serving in the Pacific until physical reasons forced him to give up his flying status. He saw extensive action in Nicaragua in 1928–29, and was awarded his first Navy Cross. He returned to the states for several training assignments; in 1935 and 1936, he led the Marine Corps national rifle and pistol teams as they won the national trophies both years. He served in Shanghai, China, from 1937 to 1939, observing the Japanese military at firsthand.

He took command of 1st Battalion, 5th Marines, in June 1941 and in January 1942, began the process of transforming into the 1st Raider Battalion. Edson led his battalion through the landings on Tulagi, Solomon Islands, and the subsequent fighting on Guadalcanal. He was awarded his second Navy Cross for the Tulagi assault. On the night of 13–14 September 1942, he led his battalion and the 1st Parachute Battalion in the defense of Lunga "Bloody" Ridge. He was awarded the Medal of Honor for that action.

Following the conclusion of the Solomons campaign, he served as chief of staff for the 2d Marine Division during the Battle of Tarawa, for which he was awarded the Legion of Merit. He was promoted to assistant division commander for the landings on Saipan and Tinian, for which he was awarded the Silver Star. He went on to serve as chief of staff and then commanding general, Fleet Marine Force, Pacific.

The President of the United States takes pleasure in presenting
the MEDAL OF HONOR to
COLONEL MERRITT A. EDSON
UNITED STATES MARINE CORPS
for service as set forth in the following

CITATION:

For extraordinary heroism and conspicuous intrepidity above and beyond the call

William H. V. Guinness, *BGen Merritt Austin Edson*, oil on panel.
Art Collection, National Museum of the Marine Corps

of duty as Commanding Officer of the 1st Marine Raider Battalion, with Parachute Battalion attached, during action against enemy Japanese forces in the Solomon Islands on the night of 13–14 September 1942. After the airfield on Guadalcanal had been seized from the enemy on August 8, Col. Edson, with a force of 800 men, was assigned to the occupation and defense of a ridge dominating the jungle on either side of the airport. Facing a formidable Japanese attack which, augmented by infiltration, had crashed through our front lines, he, by skillful handling of his troops, successfully withdrew his forward units to a reserve line with minimum casualties. When the enemy, in a subsequent series of violent assaults, engaged our force in desperate hand-to-hand combat with bayonets, rifles, pistols, grenades, and knives, Col. Edson, although continuously exposed to hostile fire throughout the night, personally directed defense of the reserve position against a fanatical foe of greatly superior numbers. By his astute leadership and gallant devotion to duty, he enabled his men, despite severe losses, to cling tenaciously to their position on the vital ridge, thereby retaining command not only of the Guadalcanal airfield, but also of the 1st Division's entire offensive installations in the surrounding area.

Franklin D. Roosevelt

From left: LtGen Thomas A. Holcomb, Col Merritt A. Edson, and MajGen Alexander A. Vandegrift caught in a candid shot during the lieutenant general's inspection on Guadalcanal, December 1942. *Official U.S. Marine Corps photo 50891, courtesy Koepplinger*

A superb combat commander and staff officer, Major General Edson epitomized the Old Breed Marines who fought through the Banana Wars, brought the Corps' amphibious doctrine to life, and led the young Marines of the Second World War across the coral and sand to victory in the Pacific.

Sgt Richard L. Yaco, *Beyond the Rice Paddies,* acrylic on board.
Art Collection, National Museum of the Marine Corps

CHAPTER THREE

The Manpower Renaissance

by Paul Westermeyer

World War II and the Korean War established the Marine Corps as one of the world's elite fighting forces. Dedicated to its amphibious warfare mission as well as being the nation's "First to Fight," it had unquestionably become the preferred "ready force" available for deployment by the president at a moment's notice. Additionally, the National Security Act of 1947 provided the Marine Corps with a long-needed statutory protection; its existence had finally been enshrined in law. However, this could not protect the Corps from budget cuts or the possibility of operational irrelevance in the nuclear age.

The Cold War, which began in earnest after 1948, introduced new geopolitical realities that required the Marine Corps to rethink its doctrines and concepts. On the one hand, the Corps needed to be prepared to support the global needs of the Navy with ready amphibious forces, while on the other it need to find a place for itself within the framework of the North Atlantic Treaty Organization's war plans. Doctrinally, this demanded the continual refinement of air-ground task force organization developed over the decades and the hurried adoption of evolving technologies that would allow the Corps' to maintain its expeditionary edge and naval mind-set.

This steady evolution continued during one of the most dynamic periods of the Marine Corps' history, even as it supported the Army in the decade-long Vietnam War, which spanned the spectrum from traditional counterinsurgency warfare to large-scale conventional operations. To successfully fight this increasingly unpopular kind of war, the Corps drew on its experiences from World War II and the Banana Wars, testimony to its flexibility and willingness to adapt to arising challenges.

As its techniques and tactics evolved to ensure that the Corps would remain relevant and effective in the changing military-technological environment of the twentieth century, it was also perfecting its ability to make Marines. Following "a form of unfailing alchemy," as Lieutenant General Victor H. Krulak describes it in his polemic, *First to Fight: An Inside View of the U.S. Marine Corps*, drill instructors transform the young men and women of character who arrive at Parris Island, San Diego, or Quantico into Marines, "whose hands the nation's affairs may safely be entrusted."[1]

Despite this fabled alchemy, making Marines has never been easy for the Corps. Incidents such as the notorious Ribbon Creek, South Carolina, tragedy in 1956 illustrate that the Service can never take the process of making Marines for granted. Doing so requires constant refinement and attention.

With the end of Selective Service and rising racial tensions in the wake of the Vietnam War, the decade of the 1970s introduced new challenges into the process of making Marines. Nevertheless, the veterans of Korea and Vietnam faced these challenges with the same fortitude and flexibility they had demonstrated during these complex conflicts and were able to preserve the character of the Corps, something their predecessors had done unfailingly since 1775.

[1] Krulak, *First to Fight*, xv.

THE IMPACT OF PROJECT 100,000 ON THE MARINE CORPS

by Captain David A. Dawson

INTRODUCTION[2]

On 23 August 1966, Secretary of Defense Robert S. McNamara announced that, starting 1 October 1966, the military would begin accepting men previously rejected for military service.[3] McNamara wanted to "salvage tens of thousands of these men each year, first to productive military careers and later for productive roles in society." He held out the hope that these men would "return to civilian life with skills and aptitudes which for them and their families will reverse the downward spiral of human decay."[4] The armed forces would take in 40,000 of these disadvantaged youths the first year, and 100,000 every year thereafter, hence the name "Project 100,000." McNamara dubbed the previously ineligible men accepted under Project 100,000 "New Standards" men. He also required the military Services to accept a minimum portion of their new recruits from men with low, but previously acceptable, test scores. Project 100,000 lasted until December 1971, bringing roughly 38,000 New Standards men into the Marine Corps.

A Confluence of Interests

McNamara's proposal to use the military for social purposes resulted from the confluence of two separate approaches to military manpower. Army officers were interested in developing effective ways to train "marginal" men, so they could be used effectively if a major war required the full mobilization of the nation's manpower. Many political leaders, noting that the armed forces trained and cared for millions of men, saw the military as an excellent tool for correcting social problems. Both of these views grew out of the military's experience during World War II.

Impressed by the military's ability to train and care for millions of men during the war,

[2] The original content came from Capt David A. Dawson, *The Impact of Project 100,000 on the Marine Corps* (Washington, DC: History and Museums Division, Headquarters Marine Corps, 1995), 1–8, 181–95. Minor revisions were made to the text based on current standards for style, grammar, punctuation, and spelling.
[3] Although McNamara's plan included both low-score men and men with minor physical defects, the medical remedial program made up a very minor part of Project 100,000 and medical remedials are therefore ignored in this paper. Medical remedials, all volunteers, accounted for less than 9 percent of all Project 100,000 men. Of the medical remedials, 65 percent consisted of overweight men and underweight men made up another 20 percent. Generally, once these under and overweight men achieved a normal weight, they were indistinguishable from other recruits.
[4] Homer Bigart, "McNamara Plans to 'Salvage' 40,000 Rejected in Draft," *New York Times*, 24 August 1966, 18.

2dLt Gayle W. Hanley reloads a magazine with ammunition during a lull in action while participating in The Basic School Exercise, 20 April 1977.
Defense Department photo (Marine Corps) A454019, courtesy of SSgt Jan E. Fauteck

Recruits engage in swimming exercises at the Weapons Training Battalion area at Marine Corps Recruit Depot Parris Island, SC. All recruits were afforded an opportunity to qualify as a swimmer.
Official U.S. Marine Corps photo, courtesy of Hans Knopf

many political leaders began to view the armed forces as a potential tool for correcting social problems. The belief that military service fostered a variety of virtues, usually including strength, courage, and a sense of loyalty and responsibility to the appropriate political body, dated back to classical times. After World War II, President [Harry S.] Truman argued that universal military training, in addition to achieving the aforementioned goals, could correct the educational, intellectual, or physical deficiencies of disadvantaged Americans. President Truman was unable to implement universal training, but his vision of using the military to train the most disadvantaged members of society persisted.

Military officers opposed efforts to use the military for social purposes. Many officers were, however, interested in training men with poor academic skills. During World War II, manpower shortages forced the Armed Services to accept large numbers of men with low test scores. All of the Services provided remedial academic instruction to bring these men up to a minimum standard. After the war, the Army, Navy, and Air Force, realizing that future mobilizations would again force them to accept low score men, conducted experiments to develop better remedial training programs. The Army, which expected to experience the largest increase in the event of mobilization, showed the greatest interest in finding methods for training low score men.

Marine Corps Opposition to Project 100,000

The Marine Corps did not share the Army's interest in the problem of mobilizing men with low test scores and opposed Project 100,000 from the start. Initially, the Marine Corps objected to Project 100,000 on the grounds that

Sgt Eulas Talley Jr. observes recruits firing pistols in the mid-1970s.
Official U.S. Marine Corps photo

this program forced recruiters to turn away better qualified volunteers. When massive racial and disciplinary problems swept through the Marine Corps at the end of the Vietnam War, senior officers, including former Commandant General Leonard F. Chapman Jr., blamed them on Project 100,000.[5]

General Chapman's opinion is still widely shared throughout the Marine Corps. When the subject of Project 100,000 comes up, serving Marines familiar with "McNamara's Morons" almost invariably condemn the Marines enlisted under Project 100,000 as nothing but untrainable troublemakers.[6]

Marines condemning Project 100,000 thought that men with low test scores created the most disciplinary problems. Since Project 100,000 was forced on an unwilling Marine Corps by unpopular civilian Defense Department officials, this program provided a convenient way for Marines to blame the Corps' troubles on an outside influence beyond their control. Because the disciplinary problems experienced by the Marine Corps appeared shortly after the start of Project 100,000, many Marines simply assumed a cause-and-effect relationship. Their assumption was wrong.

The lowering of standards also raises ques-

[5] Gen Chapman was the 24th Commandant of the Marine Corps from 1 July 1968 to 31 December 1971. Gen Leonard H. Chapman intvw with Marine Corps Historical Center (MCHC) historians, 28 March 1979, 87; see also MajGen Lowell E. English intvw with Benis M. Frank, 13 June 1974, 74; MajGen Rathvon M. Tompkins intvw with Benis M. Frank, 13 April 1973, 93–94; and LtGen John E. McLaughlin intvw with Benis M. Frank, 19 October 1978, 149; all in MCHC Oral History collection.

[6] *McNamara's Morons* was a term that many officers and sergeants used to refer to low-IQ men who were taken into the military under the new program.

tions about the combat performance of New Standards men. Although many professional soldiers fail to grasp this point, tactics are partly determined by the capabilities of the troops in a unit. If soldiers cannot master certain skills, leaders have to simplify their tactics. On the surface, then, it would seem that New Standards men might have hampered Marine units fighting in Vietnam by forcing leaders to modify their tactics. They did not.

The Impact of Project 100,000

Critics of Project 100,000 ignore the Marine Corps' previous experience with low score men. During World War II and Korea, the Marine Corps accepted far more low score men than it did during Project 100,000. Many of these men had lower scores than the New Standards Marines. Yet there are no reports of rampant disciplinary problems in 1945 or 1953. Nor did the presence of low score men keep Marine units from earning a reputation as one of the world's finest fighting forces in both wars.

The Marine Corps' experience with New Standards Marines matched its experience with low score men in earlier wars. New Standards Marines were somewhat more likely to be punished for minor infractions, but only slightly so. There were not enough of them to account for more than a fraction of the discipline problems experienced by the Marine Corps at the end of the Vietnam War. In fact, the low point for Marine Corps discipline seems to have occurred sometime around 1974 or 1975, well after the last Project 100,000 Marine had been discharged or reenlisted.

In combat, their record is less clear. Some performed poorly, some performed well. In a few instances, the failures of New Standards men probably cost their lives and the lives of other Marines. But New Standards Marines did not force leaders to alter their tactics, nor do they seem to have lowered the overall fighting power of Marine units.

New Standards Marines did place an additional burden on the Marine Corps' training system. By 1965, the need to send a constant stream of replacements to Vietnam forced the Marine Corps to drastically reduce the length of training given to recruits. New Standards Marines were much more likely to need additional training to complete, or to fail, their basic training. Additional training required additional time, effort, and money. Failure wasted the Marine Corps' investment to that point. Both placed another strain on a system already stretched to the limit.

The Marine Corps, however, had no viable alternative. Marines condemning McNamara's Morons assume that the Marine Corps passed up better-qualified men to take New Standards Men. But even with Project 100,000, by the beginning of 1967, the Marine Corps had great difficulty finding qualified volunteers. By late 1967, the Marines were consistently exceeding its quotas for low score men. By late 1968, the Marine Corps needed draftees to fill its ranks. Barring a major change in draft deferment policy, the Marine Corps almost certainly would have lowered standards anyway and probably would have accepted about the same proportion of men in Mental Group IV that it took under Project 100,000.[7] If anything, Project 100,000 may have helped the Marine Corps by preventing the Air Force and Navy from taking only the highest scoring volunteers.

In the end, Project 100,000 had almost no

[7] Recruits in Mental Group IV scored between the 10th and 30th percentile on the Armed Forces Qualification Test and were required to complete additional aptitude tests.

Students of the 15th Officer Candidate Course, Marine Corps Schools, Quantico, VA, engage in mass calisthenics with rifles, 21 March 1956.
Defense Department photo (Marine Corps) A40273

impact on the Marine Corps. Project 100,000 did not significantly contribute to the Marine Corps' disciplinary problems or hamper combat operations in Vietnam. New Standards men placed a burden on the training system, but this was a burden that the Marine Corps would have had to bear anyway. Given the Marine Corps' inability to attract better qualified recruits, not long after McNamara's announcement, the Marine Corps would probably have followed the precedent of World War II and Korea, lowering standards to fill its ranks.

During Project 100,000 the armed forces, including the Marine Corps, followed the practice of previous wars. As in World War II and Korea, the increased need for recruits led to a lowering of standards. Vietnam differed from earlier wars in that the shortage was artificially created by a generous draft deferment policy. But if standards had been quietly lowered to meet end strength without Secretary McNamara's "Great Society" rhetoric, in all likelihood no one would have noticed it at the time or remember it today.[8]

THE IMPACT OF PROJECT 100,000

In a very real sense, Project 100,000 had little impact on the Marine Corps. The New Standards men created problems, but compared to the other problems faced by the Marine Corps, the burden created by New Standards men was slight. And this was a burden the Marine Corps would almost certainly have borne without Project 100,000. During this program, the percentage of low-score Marines increased dramatically, but it would have increased to about the same degree anyway due to the demands of the Vietnam War.

Project 100,000 had an equally limited impact on the New Standards Marines. Of course, military service undoubtedly had a profound effect on all of the New Standards Marines. Even two years of peacetime service has a profound impact on a person, immersing that person into a world that is more controlled, disciplined, and organized than anything found in civilian life. Many of the New Standards Marines served in combat, one of the most powerful human experiences. But the vast majority of the New Standards Marines were volunteers; if we accept that the Marine Corps would have lowered its standards without Project 100,000, it is reasonable to assume that most of these men would have volunteered to serve without Project 100,000.

[8] The concept for the "Great Society" came from Lyndon B. Johnson, "The Great Society" (speech, Ann Arbor, MI, 22 May 1964).

And Project 100,000 had almost no impact on the conditions of their service.

The Impact of Military Service on the New Standards Marines

Few New Standards Marines received any kind of training that might be considered vocational. McNamara admitted this after the fact, but still maintained that the military taught "discipline, self-reliance, and promptness . . . exactly the skills employers need."[9]

A study published in 1987, conducted by persons involved with Project 100,000, backed McNamara's assertion, finding that military service benefited the New Standards men. The authors of this study found that by 1974 two-thirds of the former New Standards men had used their GI Bill educational benefits, and that they were more likely to try to complete their high school education than similar nonveterans. Compared to their peers who did not serve, New Standards men had a lower rate of unemployment, better jobs, and earned more.[10]

The most recent study, published in 1989, reached the opposite conclusion. The authors of this study found that New Standards men were more likely to be unemployed, generally earned less if employed, had less education, were less likely to have received vocational training, and were more likely to be divorced than similar men who did not serve.[11]

Both of these studies had great difficulty tracking down New Standards veterans, and even more difficulty finding a suitable group of nonveterans for comparison, making their conclusions doubtful. Secretary McNamara was probably correct when he observed that, since the careful follow-up of these men he envisaged was never carried out, we will probably never know the real truth.[12]

It is still possible, however, to reflect on the probable benefits of military service. To do this, it might be instructive to consider the progress of a New Standards Marine. Of course, the experience of each one of the 38,000 New Standards Marines was unique, but these Marines also shared many experiences. A useful device for examining the experience of these Marines, therefore, might be to follow the career of a hypothetical "typical" Marine, which included the most common elements.

More than 90 percent of all of the New Standards Marines were volunteers, not draftees, and so let us make this hypothetical Marine a volunteer. Nearly half of all New Standards men were Southerners. Almost 60 percent of New Standards Marines were white, and 40 percent were black. The average age upon enlisting of black New Standards Marines was 19.3 years, of white New Standards Marines 18.1. This made him the same age as other Marine recruits and, like other Marine recruits, more than a year younger than recruits in other Services.

Only one-third of the New Standards Marines had finished high school, the lowest percentage of any Service. Sixty percent of other Marine recruits had completed high school, also the lowest percentage of any Service. Only a quarter of the white New Standards Marines had finished high school, compared to half the black New Standards Marines. The white New Standards Marines had completed an average

[9] Robert S. McNamara intvw with Capt David A. Dawson, 4 June 1991, hereafter McNamara intvw.

[10] Thomas G. Sticht et al., *Cast-off Youth: Policy Training Methods from the Military Experience* (New York: Praeger, 1987), 62–64.

[11] Janice H. Laurence, Peter F. Ramsberger, and Monica A. Gribben, *Effects of Military Experience on the Post Service Lives of Low-Aptitude Recruits: Project 100,000 and the ASVAB Misnorming,* Final Report 89-29 (Alexandria, VA: Human Resources Research Organization, 1989), 161–63.

[12] McNamara intvw.

Table 6-4

Unadjusted Reenlistment Rates for Marine Regulars by Fiscal Year, 1961-1976

	Marine Corps wide 1st Term Regular reenlistment rate	Inf, Gun Crews & Allied Specialists 1st Term Regular reenlistment rate	Marine Corps wide Career reenlistment rate	Inf, Gun Crews & Allied Specialists Career reenlistment rate
FY 61	18.3	*	78.7	*
FY 62	20.0	*	83.1	*
FY 63	15.5	*	84.6	*
FY 64	14.4	*	85.7	*
FY 65	16.3	15.7	84.5	88.3
FY 66	16.3	15.6	88.6	90.2
FY 67	10.6	9.2	77.9	76.1
FY 68	11.9	10.3	76.0	62.0
FY 69	7.4	6.2	74.5	59.8
FY 70	4.7	3.1	78.0	72.5
FY 71	7.9	4.5	81.8	77.6
FY 72	12.3	11.5	82.6	75.8
FY 73	13.0	5.1	81.7	59.5
FY 74	15.3	9.8	79.6	68.7
FY 75	20.4	15.1	73.3	60.1
FY 76	26.4	22.8	75.7	71.6

* DoD changed its occupational categories in 1965, and breakdowns for similar occupations are not readily available.

Source. DoD, Selected Manpower Statistics, 1968-77

Table 8-1

Marine Corps Annual Rate of Selected Disciplinary Indicators Per 1,000 Enlisted Strength, 1965-1977

Fiscal Year	Unauthorized Absence	Desertion	Courts Martial	Non-Judicial Punishment
1965	NA	18.8	45.3	157.0
1966	NA	16.1	39.0	142.8
1967	77.8	26.8	43.5	142.7
1968	87.6	30.7	44.0	186.2
1969	120.3	40.4	55.0	181.0
1970	174.3	59.6	57.5	228.6
1971	166.6	55.8	64.7	239.9
1972	170.0	46.4	64.9	251.6
1973	241.0	64.8	68.4	304.0
1974	291.4	88.3	68.0	370.6
1975	298.3	99.8	67.8	335.2
1976	213.9	69.1	52.5	309.0
1977	103.8	47.1	NA	NA

Source: Manpower Dept, HQMC (in Deserters, Vietnam Manpower Statistics Files, Reference Section, MCHC).

of 10 years of school, the blacks 11.1 years of school. Both blacks and whites, however, read below the sixth-grade level, and could compute at just above the sixth-grade level. Other Marine recruits had completed 11.4 years of school and could perform at roughly the 10th grade level.[13]

Our imaginary "typical" Marine then was a white, Southern, 18 year old, who completed the 10th grade but could only perform at a 6th-grade level. He was the same age as his peers, but had less schooling and considerably poorer academic skills.

He was a volunteer, and probably would have volunteered without any draft pressure.

[13] Data from this section is taken from, "Project One Hundred Thousand: Characteristics and Performance of 'New Standards' Men: Final Report" (unpublished paper, Assistant Secretary of Defense, Manpower and Reserve Affairs, 1971), tables B-1–B-10.

Table 8-2

Percent of Enlisted Marines with High School Educations, 1950-1978

Year	USMC Recruits % HSG (FY)	Male Civ. 18-19 % HSG (March)	USMC total Enlistd % HSG (CY)	Army total Enlistd % HSG (CY)	DoD total Enlistd % HSG (CY)	Male Civ. 20-24 % HSG (March)
1950		40.3				48.7
1952		46.8				52.2
1957		46.5				61.4
1959		46.9				63.2
1960*	57		62.8	63.5	66.1*	
1962		56.5				68.3
1964		52.0				72.3
1965		53.2	70.5	77.1	81.6	72.7
1966	66.4	56.1				75.0
1967	65.3	63.3	71.7	79.1	82.7	76.4
1968	57.4	54.1				77.2
1969	56.2	54.6	65.9	79.8	82.0	77.6
1970	55.1	57.0	67.0	83.5	85.2	80.1
1971	49.9	56.7	67.4	85.0	85.6	81.3
1972	50.8	58.9	64.2	76.3	81.3	83.4
1973	49.6	59.8	65.1	83.4	86.2	83.8
1974	54.2	60.1	66.4	84.7	86.7	83.8
1975	59.1	58.7	71.5	85.3	87.4	84.8
1976#	69.3	63.7	73.6	85.6	87.8	84.1
1977#	75.5	62.1	78.9	87.8	86.2	83.8
1978#	76.8	56.3	82.2	87.9	89.4	83.5

*This is the only year for which the Marine Corps did not have the lowest percentage of High School Graduates. The service with the lowest percentage of High School Graduates was the Navy, with 51.4 percent

#Part of the increase in these years reflect an increase in the percentage of female recruits, all of whom were required to have a high school diploma. However, since the percentage of females in the Marine Corps as a whole increased from slightly less than one percent to just over three percent, their impact on the overall rate is negligible.

Note: Marine Recruit HSG includes GED holders

Source: Vietnam Manpower Statistics file, Reference Section, MCHC; Census Bureau, Current Population Reports: Educational Attainment; DoD, Selected Manpower Statistics.

Table 9-1

Quality of Marine Corps Enlistees, FY 60-FY 75.

	Mean AFQT/ASVAB for enlistees (CY)	% High School Grads (FY)
1960	52.0	54.3
1961	57.2	50.9
1962	61.1	55.0
1963	58.4	60.1
1964	58.1	60.4
1965	57.8	59.6
1966	56.6	66.4
1967	52.2	65.3
1968	51.7	56.8
1969	49.2	54.1
1970	49.4	52.8
1971	49.1	49.9
1972	48.3	50.8
1973	51.0	49.6
1974	56.7	50.2
1975	59.3	52.7

Source: CMC Reference Notebook 1975 subj: Recruit Attrition, encl (2) 24Jul75; Plag et al., Predicting Effectiveness, p. 6.

He almost certainly had never heard of Project 100,000, and probably did not expect to learn any skills that would carry over into civilian life.

The Marine Corps considered New Standards Marines poor candidates for rank and responsibility. After 1 October 1967, all New Standards Marines were limited to two-year enlistments, so our imaginary Marine also entered on a two-year enlistment.

Like all Marines, he went to boot camp. There was no special literacy training, or any other special training, for New Standards men in the Marine Corps. Like all recruits, he was told exactly what to do every minute of the day. Only one-third of the New Standards Marines needed remedial training in boot camp, so our imaginary Marine probably graduated with his platoon. His fellow recruits, however, probably helped him keep up. Before 1970, his drill instructors were under considerable pressure to graduate 90 percent of their recruits. His drill instructors were very good at making sure his entire platoon suffered if individuals lagged.

After boot camp, our new Marine went to a brief school to learn a military specialty. He probably became an infantryman; almost half of all Marine New Standards men did. Even if his test scores had not precluded his assignment to the more advanced technical courses, his two-year enlistment barred him from all but the shortest courses. Even those New Standards Marines assigned to technical sounding fields, such as combat engineers or supply, almost invariably were assigned to jobs that required far more brawn than brains. In fact, the job most frequently held by New Standards men in the supply field was and is referred to as box kicker by other Marines; today's Marine Corps does not bother with any formal schooling for this assignment. But three-quarters of New Standards Marines went into combat arms assignments, and half went into the infantry, so let us make this Marine an infantryman.

A platoon of recruits stands at attention during Women Marines training at Marine Corps Recruit Depot Parris Island, SC, 24 April 1974.
Official U.S. Marine Corps photo 0160843674

After he completed a few weeks of infantry training, this young Marine was granted around 10 days leave. His recruiter, following a common practice, dropped in on him, in part, to ensure that he understood his orders. When it came time to leave for his next assignment, his recruiter took him to the station, just as he had when sending our Marine to boot camp. The recruiter was not worried that the Marine might try to desert; rather, he was worried that this new Marine might have some difficulty dealing with ticket windows, or getting on the right bus, or any task that involved reading or writing.[14]

Almost all Marines who joined the Marine Corps before the end of 1969 went to Vietnam for their first assignment. Almost none of those who enlisted after 1969 went to Vietnam. Since three-quarters of the New Standards Marines joined before 1970, our "typical" Marine should be a combat veteran. After a three-week train-

[14] Conversations with former recruiters indicate that this was a common practice.

ing period at the aptly named Staging Battalion at Camp Pendleton, California, and probably no more than five months from the day he first stood in the yellow footprints at boot camp, our Marine boarded an airplane and went to war.

In Vietnam, he did the things the Marine Corps expected of him. Although formal discipline was far more relaxed than any thing he had experienced so far, there was still a clear chain of command telling him what to do and when to do it.

After 13 months in Vietnam, he returned to the states. At this point, he had less than six months left to serve. He might have tried to reenlist, but his low scores on the entry tests probably made him ineligible. Like most Marines on two-year enlistments, the Marine Corps probably offered him an early release, and he probably took it. If he had stayed in the extra six months, he would have been eligible for a program called "Project Transition," [which] arranged training in civilian occupations for servicemen nearing their discharge date. But like almost every New Standards Marine, he did not avail himself of this opportunity.

So he became a civilian again. He had been a Marine for less than 20 months. His military experience consisted of boot camp, a few weeks additional training in a purely military skill, and then a year in Vietnam. Throughout the entire time, someone was responsible for him, and someone constantly checked on him to make sure he did everything he was supposed to, including bathing, eating, and getting up in the morning.

Before passing judgment, consider this story from the perspective of Headquarters Marine Corps. There was a war on. The Marine Corps recruited our young man, trained him in a skill the Marine Corps needed, and sent him to play his part in a job the Marine Corps was assigned to do. When he returned, the Marine Corps needed to make room for another man to do the same job, so it offered him a chance to leave early. He eagerly accepted this offer. Throughout his time on active duty, the Marine Corps made sure that he was housed, clothed, fed, paid, and generally cared for. This was the story not only of the New Standards Marine, but of most Marines who served during the Vietnam War.

The Impact of New Standards Men on the Marine Corps

New Standards Marines did not hamper combat operations in Vietnam. Nor did they significantly add to the massive disciplinary problems experienced by the Marine Corps at the end of the war. New Standards Marines did strain the Marine Corps training system. By late 1965, the demand of the Vietnam War had already stretched the training establishment to its limits. Men with low test scores, by needing additional instruction, recycling, or failing, stretched the training establishment further.

This strain, however, cannot be blamed on Project 100,000. Despite the Commandant's repeated complaints about the Marine Corps' Mental Group IV quotas, the Marine Corps needed these men. By late 1967, the Marine Corps could not attract enough high score volunteers to fill its ranks. It would almost certainly have lowered standards even if McNamara had not instituted Project 100,000.

New Standards Men in Combat

New Standards Marines did not hurt the war effort in Vietnam. A small minority of Marines recalled serious problems with slow learners in combat. A larger number recalled that some of their best Marines had low test scores. Others noted that Marines who could not master more complex tasks could perform mundane but necessary tasks, such as "ammo humper."[15]

There is no way to determine if New Standards men caused additional casualties. The foolish mistakes of some Marines unnecessarily cost lives, but foolish behavior has never been confined to the poorly educated or those with low test scores. Nor was it usually possible, given the chaotic, confusing nature of combat, to differentiate between mistakes and bad luck. And often the distinction between foolish behavior and heroism was equally blurry. In any case, New Standards Marines could not have been exceptionally foolish, since they were no more likely to be killed than other Marines.

Attempts to decide if a New Standards Marine caused an "unnecessary" death are inherently futile. Every death is a tragedy, but death is a part of war. The best that can be said is that the presence of New Standards men did not significantly increase the overall casualty count.

Thomas O'Hara, *Bivouac, Elliots Beach, Parris Island*, acrylic on board.
Art Collection, U.S. Navy

The Impact of Project 100,000 on Discipline

Persons regarding New Standards men as inherently unfit for service ignored the fact that hundreds of thousands of men with test scores below those of most New Standards men served during World War II and tens of thousands with scores as low served during Korea. During these wars, men with low test scores or poor educations were somewhat more likely to be formally disciplined or rated as poor performers, but their presence did not create a disciplinary crisis, nor did they receive large numbers of unfavorable discharges.

The service of the New Standards Marines followed the same pattern. They were more likely to be formally punished, receive poor performance ratings, and receive less than honorable discharges, but only to a small degree. New Standards men accounted for only a tiny part of a huge disciplinary problem.

The Added Cost of Low-Score Men

Despite McNamara's claim that New Standards men could be trained at no additional cost, these Marines did cost more. In June 1969, Irving M. Greenberg, the director of Project 100,000, estimated that New Standards men cost the military about $200 more than other men. The added costs came from remedial training, higher attrition, hospitalization for physical marginals, and requirements for data collection.[16] The additional $200 might not seem a large sum, but in June 1969, the typical first-term Marine, a lance corporal (paygrade E-3) with less than two years of service, was paid $137.70 a month.[17]

[15] *Ammo humper* is military slang for a Marine who carries ammunition.

[16] I. M. Greenberg, "Project 100,000: The Training of Former Rejectees," *Phi Delta Kappan* 50, no. 10 (June 1969): 574.

[17] *A History of Armed Services Pay Scales* (Camp Lejeune, NC: Disbursing Instructional Section, Marine Corps Support School, 1981), 8.

The 300-yard line on "B" Range at Camp Matthews with Platoons 13 and 156, 14 February 1952.
Defense Department photo (Marine Corps) A219081

Desperately short of Marines due to budget-driven limits on end strength and forced to drastically shorten its basic training program, during the Vietnam War the Marine Corps could not afford even a slight drain of its resources.

During the first few months of [the project], Project 100,000 created a drain, as recruiters turned away men with higher scores to meet their Mental Group IV quotas.

By late 1967, however, Project 100,000 could no longer be counted as a burden. New Standards men still cost more to train, but the Marine Corps could no longer truthfully blame their presence on the Defense Department quotas. Unable to attract recruits with higher scores, the Marine Corps needed these men to fill the ranks. Recruiters, desperate for volunteers, consistently exceeded their Project 100,000 quotas. By late 1968, the Marine Corps could not find enough volunteers, even by exceeding their Project 100,000 quotas, forcing the Marine Corps to resort to the draft for recruits. When Congress finally abolished Mental Group IV quotas, the Marine Corps continued to sacrifice recruit quality to meet end strength. The New Standards Marines' added cost would have been incurred if there had been no Project 100,000.

Criticism: Based on a False Assumption

At the heart of the criticism of Project 100,000 was the belief that an absolute standard for military service existed, and that no one falling below that standard should have been allowed to serve. Many Marines agreed with the assessment of Louise B. Ransom, a counselor for imprisoned veterans: "these guys should never have been in the military."[18]

This belief was false. Standards for military entry were not absolute. For all military specialty courses, a certain portion of the persons achieving a given score would fail. Higher entry standards resulted in a smaller portion of those accepted for training failing to complete any given course of instruction; lower standards resulted in a larger portion failing. To minimize the number of failures, the military tried to set minimum scores at the highest level that would still allow enough people to pass to fill the ranks.

If the number of persons needed increased or the pool of applicants decreased, the only way to get more graduates was to lower the cutoff score. This would produce more graduates, since many of the individuals previously reject-

[18] Myra MacPherson, *Long Time Passing: Vietnam and the Haunted Generation* (Garden City, NY: Doubleday, 1984), 643.

ed were always capable of passing the course. At the same time, the proportion of persons beginning training who failed to complete the course would increase. Thus, military recruiting obeyed the economic law of supply and demand, with test scores substituting for cost. As demand (willing applicants) decreased in proportion to recruits needed, recruiters were forced to lower the price (test scores). Marine Corps manpower experts understood this principle well before Project 100,000 began.[19]

The Marine Corps' experience in Vietnam followed this economic law. Unable to attract enough volunteers, the Marine Corps was forced to lower standards. The low score men brought in did not perform as well. A higher proportion of New Standards men required additional training or failed basic training. In general, the New Standards Marines who passed basic training did not perform as well as other Marines.

Most New Standards Marines, however, graduated from boot camp and rendered useful service. As a group, they may not have been as good, but the Marine Corps needed them to perform its mission.

The Legacy of Project 100,000

Critics of Project 100,000 forget that the Marine Corps of the Vietnam era contained the best-educated Marines, with the highest average test scores, that ever fought a major war. Project 100,000 did not hurt the Marine Corps. In the absence of McNamara's program, the Marine Corps would almost certainly have lowered standards to roughly the same level to fill its ranks. Nor did the presence of low score men create or significantly exacerbate disciplinary problems.

Project 100,000 had its greatest impact on the Marine Corps after it ended. It taught the Marine Corps a false lesson. By coinciding with one of the Marine Corps' darkest hours, Project 100,000 convinced a generation of career Marines that men with low test scores should not be enlisted under any circumstances.

[19] Deputy Assistant Chief of Staff, G-1, LtGen Samuel Jaskilka to Deputy Manpower Coordinator for Research and Information Systems, 9 May 1967, file 1510 HQMC Central Files 1967. See also the G-1 comments at the General Officer's Symposium, 1964, discussing the Marine Corps' ability to lower the proportion of Mental Group IV recruits due to the favorable recruiting climate.

PATHBREAKERS
Dealing with Race–The 1970s

by Fred H. Allison, PhD, and Colonel Kurtis P. Wheeler, USMCR

The Vietnam War had a range of effects on race relations in the Marine Corps.[20] On one hand, African American Marines served with great distinction, valor, and heroism in combat, and more black officers served in leadership positions than before. On the other, racial tensions escalated to unprecedented levels. Black troops defied authority figures they believed were racist and struggled to project their racial identity through afro hairstyles and special greetings like the "dap."[21] Racial tension that originated during the 1960s continued and, in some ways, intensified during the early 1970s.

The Marine Corps in the 1970s undertook measures that significantly increased the number of African American officers in its ranks. Because there was a larger group of them than there had been of their predecessors, these officers, on a broader level, cleared the way to make the presence of black officers a norm in the Corps. The wave of black officers that entered the Marines as a result of the big push during the decade served across the Marine Corps landscape in different specialties, rising through the ranks into positions of increasing importance.

As young officers in the 1970s, they faced different challenges. There were many black enlisted men, and as noted, racial tension at the time was prevalent. As African Americans, they naturally could identify and sympathize, but the challenge was determining how far they could carry the empathy while also trying to remain unbiased and maintain good order and discipline. They also faced issues off base as well.

Frank E. Petersen Jr.

We weren't dealing with an organized prejudicial system within the Corps. We were dealing with individuals in command and authoritarian positions who were prejudiced. Only 30 years before, blacks were just entering the Corps. . . . We were looking at officers trained some 20 years before, and others with significant rank trained no less than 10 years before. There were no officers who could exude even a modicum of understanding when managing black troopers. Beyond that, even when some black officers

[20] The original content came from Fred H. Allison and Col Kurtis P. Wheeler, *Pathbreakers: U.S. Marine African American Officers in Their Own Words* (Washington, DC: History Division, 2013), 115–43. Minor revisions were made to the text based on current standards for style, grammar, punctuation, and spelling.

[21] The term *dap* refers to a physical form of greeting, which can include a whole series of hand motions besides just the fist bump.

came on board, some of [them] refused to become involved. It was a put-down to speak out, they thought. Some acquiesced, caught between pride and the demands of some commanders that problems with blacks be solved in the old, traditional ways.

These officers, both white and black, represented the focal points, I felt, at which to begin corrective action. Not many jumped on my bandwagon, some preferring to pinpoint the "problem" as a reflection of what was going on in civilian arenas, where black social consciousness and civil rights activity were on the rise. . . . I prefer[red] saying, unlike the civilian community, that we have a strong but small, highly disciplined microcosm of the civilian community, and because of that, we can solve the problems if they are present. It's a little tricky when you get into this stuff to say that we have the same problems or that our problems are a "reflection" of the civilian community, because they are really two different communities.

We'd blown it badly when it came to our decrees about the approved length of a trooper's haircut—failing to account for what Caucasian hair versus black hair looked like at two or three inches in length. . . . We went a little ballistic in the way we regarded the "dap"—the special way that black troopers greeted one another during Vietnam—probably one of the biggest mistakes we ever made. We couldn't understand, or did not know, when we demanded the closely shaved head that some black men suffered from a condition in which, if they shaved too closely, the hairs curled on themselves, grew back into the skin, and caused painful pimples, pustules, cysts, scars, keloids, and infection if not addressed properly. . . . In lieu of medical attention and researched methodology for a solution, we simply discharged black Marines with this condition, saying they were unfit to serve—not the greatest of morale boosters.

2dLt Frank E. Petersen Jr. flew with Marine Fighter Squadron 212 in Korea in 1953, one of only two African American Marine officers to serve in combat during that conflict. This photograph was taken in November 1963 just after his promotion to major.
Defense Department photo (Marine Corps) A42086

To some, these were small things indeed. However, add to them all those other things involving prejudicial attitudes, misinformation, or just plain ignorance on the part of many in authority, and little human engineering time bombs were set to go off all around the Corps.

The point was to disarm them by conducting as many open discussions about cultural differences as we could. To try to make whites understand that there were blacks who hated whites; it wasn't just whites who hated blacks. It was to make them understand that the officers who were leading at the unit levels were not equipped to understand the culture and mores of blacks and the Hispanics and their cultural differences. To help them understand the fact that country and western music at the enlisted club may not be the choice of some of the minority troops, that the regulation that decreed length of hair needed to be amended to allow for cultural input, that the inability to swim like a fish hinged on the absence of facility, not because minorities simply couldn't swim.[22]

[Petersen later turned to discuss the impact of racial issues on recruitment, training, and retention.]

A major cause of the Marine Corps' problems was the need to satisfy manpower requirements during the Vietnam conflict, which brought into the Corps floods of black youth who had been exposed to the rising hue and cry of militancy and nationalism in the ghettos. . . . Our self-congratulation on our handling of racial problems within the Corps had begun to wane. A major riot had erupted at Camp Lejeune in July 1969. Marine enlisted men fought one another in San Diego, Hawaii, and Camp Pendleton, not to mention tension between the forces still remaining in Vietnam.

By now, the Marine Corps had firmly implanted a human relations program, which went a long way toward convincing officers and NCOs [noncommissioned officers] that they would have to take positive action to stop interracial tension and allay the fears of black Marines that they would be victimized by "The Man." White Marines had to be convinced that not all black Marines were potential thieves and muggers and that violence among Marines would not be tolerated. . . .

There needed to be, in my view, selection of blacks for more accelerated promotions. A policy was needed that required the Corps to look at the career patterns of selected black officers to ensure that they were on track and getting the right assignments. Something needed to be done regarding interpretation of fitness reports, not only for minorities, but for all officers. At that time, it was still a problem.

Another look at The Basic School needed to be taken. Why, for example, were so many black officers coming out of Basic School with supply, transport, and service military occupational specialty (MOS) designations? I still remember an old survey we ran in which we wanted to discover how many blacks were in command billets. The answer was that, of the 300 or so black officers, only 7 were in command billets. That was a pretty grim statistic. The black Marines knew the score, and they were becoming more and more verbal about it. . . .

I didn't want to become a general officer because I thought I was going to solve the problems all alone. I knew I couldn't do that. But perhaps I could make a positive impact on the problem. So the decision was to stick around and make a try for general officer rank if only because there was a grave need for a show of faith on the part of the Marine Corps.[23]

Brigadier General George H. Walls Jr.

Hutson: Can you discuss your experiences as Marine detachment commander in 1970–71 on the USS *Franklin D. Roosevelt* (CVA 42)?[24]

Walls: That was quite an interesting experience for me because it was the first time I had served with the Navy. I remember reporting to the ship and walking up to the quarterdeck. The officer of the day is there, and there is another officer kind of in my peripheral view who looks like he slept in his uniform and had maybe five or six days' worth of beard. It was the captain of Marines I was relieving. . . .

The Marines aboard that ship had the worst

[22] Frank E. Petersen Jr. with J. Alfred Phelps, *Into the Tiger's Jaw: America's First Black Marine Aviator* (Novato, CA: Presidio Press, 1998), 197–99.

[23] Petersen and Phelps, *Into the Tiger's Jaw*, 227–28.

[24] This oral history interview was completed by CWO-3 William E. Hutson on 17 February 2012.

BGen George H. Walls Jr. served for more than 28 years, retiring in 1993.
Courtesy of BGen George H. Walls Jr.

BGen Walls in Cuba with Haitian refugees while he commanded the joint task force for Operation Guantánamo in 1991–92.
Courtesy of BGen George H. Walls Jr.

reputation for discipline and personal appearance, and it was a bad situation I went to, but it was a good situation for me because there was nowhere to go but up. I met the captain of the ship. . . . We sat and talked, and he was very candid about what had happened with the Marine detachment and what he expected me to do with the Marine detachment. That was my welcome to the *Roosevelt*. . . .

Hutson: Were any of the problems with the detachment racially oriented, or was it just bad Marines?

Walls: It was a combination. There were some young Marines there who would go on liberty and get drunk out of their minds. . . . The other situation that we had on that ship, and I'm sure it was true on other Navy ships, went on after the "Z-grams" [or directives] came out. Admiral [Elmo R.] Zumwalt [Jr.] liberalized dress codes. Sailors could grow beards, they could do all sorts of things that they weren't allowed to do before. And there were a significant number of black sailors on the ship. I probably had half a dozen black Marines or more in the detachment. And again, the civil rights thing was still going on. There was a group of black sailors led by a petty officer first class whose name I don't remember. They were constantly demanding to change things, to do things, to the point that a couple of times it got where the captain was concerned about unrest on the ship. I'm talking about physical kinds of things happening where he called out the Marine detachment to be on standby in case these kinds of things happened. Fortunately, it never got to the point where it boiled over to where the Marines had to engage with the sailors. But it came close on a couple of times. And really, the only way that it got quelled was that the chaplain on the ship was black. His name was Carroll [R.] Chambliss. He retired as a captain. His son, Chris Chambliss, played baseball for the New York Yankees and some other teams. But between Carroll Chambliss and I, we were able to, in most cases, calm these young sailors down to the point where it didn't become a confrontation other than a lot of talk.

So there were those kinds of situations, and then there were just young people going ashore doing stupid stuff that got them in trouble. But after we kind of weeded out the problem children in the detachment, at least from my standpoint, things got [better].

Major General Charles F. Bolden Jr.

Allison: In 1970, you received your wings as a naval aviator and were assigned to Marine Attack Squadron 121 [VMA-121]. What were your experiences as a new aviator joining a gun squadron? There could not have been many other black aviators in the squadron.[25]

Bolden: When I got to VMA-121, I was fortunate because I was not the first black. [Richard] Dick Harris was a former enlisted BN [bombardier-navigator]. Dick had flown [Douglas] C-117s [Skytrains] as an air crewman. He and another aviator, Brewster, were running buddies. Brewster was about as close to being a redneck you could be, but they loved each other. They were like brothers. I think Brewster was from Tennessee or somewhere. He and Dick Harris had enlisted in the Marine Corps at the same time, they'd come through the ranks together, had both gone through ECP [Enlisted Commissioning Program] or something and had gone to flight school. They had been in [Douglas] EF-10s [Skyknights], then gone into flying [McDonnell-Douglas] RF-4s [Phantoms], and then they had transitioned to [Grumman] A-6s [Intruders]. So Harris was in 121 when I checked in.

Allison: What did that mean to you to have a fellow black officer as an aviator?

Following his final shuttle flight in 1994, Charles F. Bolden left NASA and returned to active duty as the Deputy Commandant of Midshipmen at the U.S. Naval Academy in Annapolis. In July 1998, he was promoted to his final rank of major general. *Official U.S. Marine Corps photo, Marine Corps History Division*

Bolden: It was phenomenal because we actually lived near each other at Cherry Point [North Carolina]. We all lived in what was called MOQ [married officer's quarters], and we did a lot of stuff together. I want to say there were three or four black aviators on the base—total—a couple of [Lockheed] C-130 [Hercules] guys, Dick, me, and a [Douglas] A-4 [Skyhawk] guy named [Clarence L.] Clancy Davis. Before he retired, he was the second black commander of a squadron in the Marine Corps. Clancy took VMA-214; incidentally, the call sign of that squadron was the Black Sheep. Clancy was interesting. For his change of command, when he came out, they played the theme song from the movie *Shaft* as an introduction before he came out. That's the

[25] This oral history interview was completed by Dr. Fred Allison on 7 May 2008 and 27 February 2012.

MajGen Bolden's 34-year career with the Marine Corps included 14 years as a member of NASA's Astronaut Office. He traveled in orbit four times around the space shuttle between 1986 and 1994, commanding two of the missions and piloting two others. His flights included deployment of the Hubble Space Telescope and the first joint U.S.-Russian shuttle mission, which featured a cosmonaut as a member of his crew.
Official NASA photo, Marine Corps History Division

kind of guy Clancy was. Clancy was very controversial.

But Dick Harris and all those guys, they were tough in that day. Dick had been through a lot, so he didn't take a lot of crap. He wore his feelings on his shoulder, and you better have your act together if you said something to him. Segregation wasn't very long ended down there in Cherry Point, and it didn't make any difference to him. He'd just as soon fight as anything. So he was sort of a revolutionary black officer; but that's who I came in the squadron behind, so I didn't have to do anything.

Because Dick was a senior BN, we flew together quite a bit because I'm the new guy. He got the new guy. I got a chance just to see how he handled himself and stuff like that. We did a lot of cross countries together, which brought a lot of stares when we rolled into an air base, particularly because most places we flew into were below the Mason-Dixon Line. I can remember how it worked: you'd land at an air base, and a guy would come over from the transient line and look at me and then go around to the other side to see who was in charge of the airplane. He looked at Dick, who was darker than I was, and the guy would be just baffled: "This airplane must be stolen because there is no way in the world that that there are going to be two black guys in this Marine Corps airplane!" We enjoyed it whenever we went on cross countries. So I didn't have any trouble when I got into the squadron.

Allison: Then you were assigned to VMA-533

[Marine Attack Squadron 533] and went to the western Pacific. This was in 1972; these were dark times for the Marine Corps. What was the situation in the 1st Marine Aircraft Wing at that time?

Bolden: That was a real bottoming out time, whether it was drugs, race riots, and stuff. It was probably the worst time in the history of the Marine Corps. But General Petersen, I think he was a lieutenant colonel at that time, he was the first African American to command a fighter squadron. He came to Iwakuni [Japan] and down to where we were flying from, Nam Phong, Thailand, with several of what they call human relations teams. So I had a chance to meet him, and talk to him, and get to know him a little bit. From then on, I stayed in touch with him, and he sort of became a mentor and a role model.

Major General Clifford L. Stanley

Andrasi: Tell us about your experiences once you got out of Officer Candidates School in 1969. Then it was to The Basic School [TBS]. What did you experience there?[26]

Stanley: When I got to TBS, we actually started at Camp Upshur, which is pretty spartan. We were all tight. I mean, my classmates, when we would go home from OCS [Officer Candidates School], even Camp Upshur [Quantico] initially, I had white classmates going home with me. I still get letters from some of them. So these weren't issues for me, personally, right then. But then folks started making it an issue, particularly in The Basic School. I remember the first time it came up. The photographer came to take a picture of, you guessed it, how many black students are there. I didn't ask the photographer to come. I'm sitting in class like everybody else, trying to stay awake, and the guy zooms in on me, and he zooms in on somebody else. There were folks upset about that; they were mad. But they weren't mad at the photographer, they were mad at me. Not that I was new to race issues, but it was just that I'm experiencing, now, classmates mad at me for having my picture taken and an article was done on folks. I said, "This is messed up," because I didn't ask for this. I'm sitting here, minding my business, doing the same thing they're doing, and that kind of stuff happens.

Before I joined the Marine Corps, though . . . [I was] the only one in my class [at South Carolina State] that joined the Corps. Then af-

MajGen Clifford L. Stanley served in numerous command and staff positions during his 33-year career. In 1993, he assumed command of the 1st Marine Regiment, Camp Pendleton, CA, making him the first African American to command a U.S. Marine Corps infantry regiment. He is pictured here at the rank of brigadier general.
Official U.S. Marine Corps photo, Marine Corps History Division

[26] This oral history interview was completed by LtCol Mark D. Andrasi on 28 February 2012.

A breakthrough assignment was that of then-Maj Stanley to the prestigious Marine Barracks Washington, DC, known for its evening parades and ceremonies. Stanley and his parade staff are pictured in front of the Marine Corps War Memorial in Arlington, VA.
Official U.S. Marine Corps photo

ter I joined the Corps, at Officer Candidates School, as I'm graduating, I ran across another guy who later became a general, Arnie [Arnold] Fields. He and I went to the same school, and he finished like a semester before me. I had no idea . . .

But again, my experience in OCS and The Basic School, it didn't seem overly traumatic.

I guess the other thing was MOS selection. I didn't know that much about the Corps, just being a Marine, that's it. I chose it, but I think it might have been my first choice. I was a supply officer initially, but I wanted to be a Marine. . . . I saw something I wasn't too comfortable with. There were a couple other black officers over there [when assigned to Okinawa], and they were in supply and things like that. I made a decision that if I was going to stay in, if I even thought about staying in, I was going to be infantry. And so, fast forward as I finished with the tour at Okinawa, I came back to the states, was at Quantico, applied for augmentation, and I also applied for an MOS change. Unheard of. I had no idea. I just said, "I want to be a Marine officer, I want to stay in. I'd like to be an infantry battalion commander one day. I don't see how I can get there with a 3002 MOS." And believe it or not, I was augmented, and a small group of people were augmented, and they changed my MOS to 0302, and I was transferred to Camp Lejeune [North Carolina]. I was still young enough in this to be a platoon commander. Actually, in that case, because I'd already been a platoon commander up at Quantico . . . I went to Lejeune, had a company—Mike Company 3/8 [Company M, 3d Battalion, 8th Marines].

The issue of race, though, was never that far away. But for me, it was less of an issue than a lot of folks because I was very comfortable with my skin, who I was and who I am.

Andrasi: As that infantry platoon commander, what was the relationship between you and your Marines? Was there any type of racial tension there from any of your subordinates?

Stanley: I wasn't naïve, but not that I know of. I mean, the tensions came from maybe officers, and not necessarily senior officers. . . . I found there were times in the Corps where there were people with hang-ups, and I saw it, but I was actually focused on being a Marine. . . .

When I went to the Naval Academy, I was a captain. This was after my MOS was changed. I'm an 03, I'm at Camp Lejeune in Mike Company 3/8. They moved me to the four [S-4] shop, and then I had orders to the Naval Academy. Then shortly after that, my wife was shot, and it was a racially motivated shooting. She's paralyzed. And today she's living it, and that's what happened.

Andrasi: Where did that happen?

Stanley: That happened in Wheaton, Maryland, and I was a captain stationed at the academy. I'd been at the academy for a few months. You've got to keep in mind [that] we're still, even though you're serving in the Marine Corps, you're still living in a world that's still a little mixed up. This guy was shooting black targets of opportunity. . . . He killed two people and wounded five, all black in a white area. He was walking around, and we just happened to be driving down the street. We weren't even walking. He killed my uncle, who was in the car in front of us, maybe a block separated, because we were separated by light. . . . My wife was shot, and she was the most seriously wounded of those who survived. There were, like I said, four other people shot and wounded. Two killed and five wounded.

I was stationed at the Naval Academy then, and I was with a lot of folks who were real pros. Even the Commandant at that time just took good care of us and reached out. The Marine Corps, that's when my relationship with the Marine Corps took a different bend because I saw how the Marine Corps took care of its own.

Major General Arnold Fields

Fields: I reported into 1st Battalion, 6th Marines [1/6 at Camp Lejeune in 1970] . . . . I was the only black lieutenant in the battalion.[27] I may have been the only black lieutenant in the regiment back then. But I was welcomed as a fellow officer and Marine by my fellow platoon commanders and my company commander. In fact, one such gentleman and I are the closest of friends and have maintained that friendship over all of the years subsequent to the experience at Camp Lejeune [and] 1/6. So I had a good relationship with my fellow platoon commanders.

I had a good relationship with the staff NCOs. My first staff NCO was a black staff NCO, Staff Sergeant Harris. And all of these folks, all of the staff NCOs and almost all the NCOs, had already had at least a tour, if not two tours, in Vietnam under their belts. So I thought they responded to me well, given the fact that I had not been to Vietnam, I was a brand-new second lieutenant, and I was black, which probably was an experience that they did not have in any of their previous contributions to the Marine Corps.

But again, I don't feel that there were any issues race-wise that were significantly outstanding above and beyond what one might expect from a group of Marines who are now

[27] This oral history interview was completed by LtCol Mark E. Wood on 16 February 2012.

under the leadership of this brand-new, inexperienced second lieutenant. I was treated well by the battalion commander. I don't think I was not given any opportunities that I was not due as a lieutenant. So it was a good experience.

Wood: Following that assignment, where did you go?

Fields: After I left 1/6, I reported directly into 2d Battalion, 4th Marines, where I was assigned duties as the 81mm platoon commander. I was very proud of that, because back then the mortar platoon consisted of 96 Marines, the largest platoon in the infantry regiment. Because we had the big tubes, I felt a little macho about that. I was really proud to be commander of that platoon, and I had great Marines working for me. No racial issues per se in that platoon to the best I can recall.

But there was a race issue all around. This, again, is the early '70s. We're now talking 1971, where strife, racially, was almost omnipresent within the Marine Corps and certainly in the organizations of which I was a part. Okinawa was particularly an area in which race was very much a polarizing aspect of the Marine Corps society. It was a considerable challenge. I wound up in the midst of an expectation on behalf of my black colleagues, expecting that I would be doing and acting like some of the black community would act, not all badly, but there were certain things that were done, such as we called it "passing the key." . . . It has to do with a greeting. There was a thing that was done with the fist, which was a means of communicating friendship and brotherhood with a fellow black person. Those kinds of things were frowned upon if not against the expectation of the Marine Corps. I did see some of my black colleagues doing it, officer and enlisted, but it was not something that I did because it was not an expectation, I believe, being an officer, and being a part of the whole of the Marine Corps.

An early product of the Marine Corps' push to recruit African American officers, Arnold Fields was commissioned in 1969 and was a series commander as a first lieutenant at the Drill Instructor School, Parris Island, SC, at the time of this photograph, ca. 1972.
Courtesy of MajGen Arnold Fields

But race was very much an issue, and it was not uncommon for there to be fights and so forth breaking out with a race connotation to them. The mess hall was one such environment or venue within which there was a very high probability that something was going to happen in that regard.

Wood: How did you . . . how did the battalion deal with these issues? How did the Marine Corps deal with it in a bigger sense?

Fields: Well, the Marine Corps once again was

MajGen Fields (left) visits peacekeeping forces in Sierra Leone during his time as deputy commander of Marine Forces Europe (2001–3).
Courtesy of MajGen Arnold Fields

still at that time trying to align its present and future with the mood of the nation and the mood of the Department of Defense when it came to race relations. Some formal race relation programs had already gotten underway, and some of those programs were having a general impact on the Marine Corps. I felt really that the Marine Corps was trying to be serious about being as much of a leader in race relations as it was on the battlefield. . . .

I felt the Marine Corps dealt pretty well with it, but probably not at as fast a pace that the environment, I felt, demanded at the time. To put what I just said into perspective, when I reported into 2d Marine Division, I recall the density, if you will, of black lieutenants or officers in that division to be seven black officers in the division that consisted of about 20,000 Marines in general. So I think the [African American] officer population back then was something like one point something percent in the Marine Corps, if that high. . . . No black officer immediately comes to mind with whom I associated in 2d Marine Division, and similarly, when I arrived at 2d Battalion, 4th Marines [2/4]. However, one of the company commanders in 2/4 was a black captain, and a very good one. . . . So I was pleased by that. But I do feel that the Marine Corps was trying to make a concerted effort to turn things around. . . .

I only spent one year and maybe a couple months or two as deputy director of the DI [drill instructor] school because [I was picked] to go to Quantico because the Marine Corps had put together what it referred to back then as a leadership training branch. . . . The essence of this leadership branch was to be the Marine Corps' way of focusing on race relations. The Marine Corps was reluctant to refer to cleaning up our act as anything but cleaning up our leadership. The Marine Corps felt that race relations had more to do with the quality of leadership than it did with any other characterization one might wish to apply to race relations. So we didn't really call it race relations training; we called it leadership training. I was picked to go to Quantico to be on that leadership team. I did that for a full three years as an instructor. I instructed in all the schools—Command and Staff College, Amphibious Warfare School, The Basic School, the MP [military police] School—all of them, on leadership, with a reasonably heavy emphasis on a human relations component. It was, in fact, the principal leadership package that the Marine Corps was offering to the schools, but especially The Basic School.

Wood: Was that program stood up around the time that you joined the unit? Or was this already in motion when you came to the leadership branch?

Fields: Yes, It was brand new. I was on the ground floor, and we kind of put the thing together and developed the curriculum and the whole package. It was under the leadership of a Marine colonel. [We had] a couple of them during my tenure of the three years, one of whom had previously been a regimental commander, and the other had been the CO [commanding officer] of The Basic School. So the Marine Corps was putting a high value on the quality of leadership training that this branch would otherwise provide. . . .

Wood: What kind of impact do you think the leadership branch had?

Fields: Well, the appearance at the time was not a very positive appearance. Why? I feel that there was still a considerable rejection, for one reason or another, of the approach that the Marine Corps and the defense establishment had in dealing with race relations. So it was a challenge, actually, to instruct in that branch.

Major General Leo V. Williams III

Andrasi: Can you relate your experiences early in your Marine Corps career? You were commissioned in 1970, I believe.[28]

L. Williams: Being in the Marine Corps was very much like being at the Naval Academy. The numbers of black officers in the Marine Corps, we thought we could count them on one hand. There were very, very seldom more than two or three or four of us in any major command. In my Basic School company, there were four of us—three from the Naval Academy and one other guy. So the Marine Corps, especially the officer corps, has always been challenged in terms of the numbers of black officers they were able to both bring into the Marine Corps, and retain in the Marine Corps. It's always been a substantial challenge.

I could see that right from the beginning of my Marine Corps career, just looking at my Basic School class. But I did have a number of mentors right from the beginning. My company commander somehow took a special interest in me. He was an artillery officer, and we maintained a really close relationship for the next 20 years. I don't remember any racial incidents when I was in The Basic School. It was just the numbers were small, and we all thought that we were headed to Vietnam, and it was a time to pull together, not pull apart. My Basic School experience was really a pretty pleasant memory.

After graduating from the Naval Academy in 1970, Leo V. Williams III was commissioned in the Marine Corps. Following two tours in artillery units with the FMF, a tour on staff at The Basic School, and a tour at Manpower, Headquarters Marine Corps, he transferred to the Reserve and began a career at Ford Motor Company. He was promoted to brigadier general in 1997 and to major general in 2000.
Official U.S. Marine Corps photo, Marine Corps History Division

[28] This oral history interview was completed by LtCol Andrasi on 5 March 2012.

Pride, progress, and prospects in the form of four Marine general officers with a common heritage, and diverse Marine Corps backgrounds. From left: BGen Clifford L. Stanley, MajGen Charles F. Bolden Jr., BGen Leo V. Williams III, and BGen Arnold Fields.
Official U.S. Marine Corps photo

I was in an artillery firing battery initially at [Camp] Pendleton [California] when there was a very, very tense racial time because we had lots of guys who were draftees, black and white, who did not want to be in the Marine Corps. A lot of these guys were Vietnam vets. A lot of them were disgruntled. A lot of them were really racially sensitive on both sides. And the tension level was sky high. I mean it was a powder keg most of the time. What I found, though, was that because I was one of the few that the black enlisted Marines could relate to, I was able to calm a lot of situations that otherwise might have been explosive. What it speaks to is the importance of diversity in maintaining good order and discipline. . . . In a lot of cases in the Marine Corps, troops did not have access to someone who looked like them, who shared a common experience, and who could talk them down from some explosive situation. . . . I was the only black officer in the regiment at that time, in the entire 11th Marines.

Andrasi: Were you called upon by that regimental CO to act in the capacity that you just described? Or [is] that was something that just kind of happened?

L. Williams: No, it just happened. I was never called on by . . . [I was] occasionally by the company commander. He would say, "Hey, we've got a situation building here. Let's go in." Often we would go in together, and once again, you can't account for stupidity. You get a young lieutenant who thinks that because he's got a sidearm that he's got all the power. And there were a couple of times when one of these young second lieutenants in particular nearly got himself killed because he got into a situation that quickly got over his head. He was confrontational when confrontation was not the right position to take. He should have been in an advisory or a negotiating position, and he decided that he's the baddest mother in the valley. So a couple of times, the captain and I had to go in and defuse a situation . . . that never really needed to get out of hand in the first place.

A drill instructor corrects the posture of a Marine during personnel inspection.
Richard Spencer Papers (COLL/5233), Archives Branch, Marine Corps History Division

But it really does speak to how important it is to have sufficient diversity that people, no matter who they are in the unit, feel both culturally comfortable and environmentally comfortable, if that makes sense. When you are in a situation, when you look around and you're the only guy who looks like you do, it can sometimes be intimidating. As I think back, that's been my experience through most of my professional career, both in the Marine Corps and in the 25 years that I spent with Ford Motor Company. Most of the time, for whatever the department was that I was in, I was the only [black] guy there. A couple of times there were one or two others, but I think I was always the senior.

I had the great opportunity to be a headquarters battery commander at Camp Pendleton. I really appreciated the battalion commander for having confidence in me, as a first lieutenant, to take that job. . . . Then I got selected to go back to Headquarters Marine Corps as the first black in officer assignments in the history of the Marine Corps. That was in June 1974. Interestingly enough, the mentor who was able to put me in that assignment was now-Lieutenant Colonel [Edward L.] Ed Green, the same Ed Green who was the major at the Naval Academy who brought in Colonel Petersen and Colonel [Kenneth H.] Berthoud [Jr.] to influence half of my black midshipmen Naval Academy class to go into the Marine Corps. So Ed Green has been looking over my shoulder for all of my adult life. Still does. But I came back to officer assignments, and it took me a little while to understand and appreciate, first of all, how significant an assignment that is, but also it took me a little bit of time to appreciate the caliber of the officers who were there.

Lieutenant General Ronald S. Coleman

Wheeler: Can you describe how you came to join the Marine Corps?[29]

Coleman: I was in Vietnam [in the U.S. Navy

[29] This oral history interview was completed by LtCol Wheeler on 12 August 2011.

from] '69 to '70. When I came out of Vietnam in '70, if you were coming out of Vietnam and you had nine months or less to do, you were released from active duty. So I got home in July and immediately got out of the [U.S. Navy]. I came out, by this time I'm married . . . [and I] wasn't sure what I was going to do. I had a mentor that said—I was going to go to school at night—and he said, "You'll never finish, just go to school during the day." So I used the GI Bill and went to then Cheyney State College, a historically black college in Cheyney, Pennsylvania, to be a teacher. . . . I went all summer every summer, took 15 hours every summer, so in three years, I graduated in '73. I taught school for a year and then it just, I wanted more. . . .

I went down to the Philadelphia Navy Yard and met my OSO [officer selection officer], Pierce [R.] King, who I still know today. His assistant was just a picture-poster Marine—that was him. I mean he had on his "mod blues" [modified dress uniform] and was just as sharp as he could be. They knew I was a college grad, and I told them I was interested. . . . I lived in Darby, and about two towns over was Drexel Hill, and that's where Pierce King lived. He was a white officer. He pointed toward my name and said, "Are you one of the Colemans from Darby?"

And I said, "I am."

He said, "Are you the football player?"

I said, well, I played football. He sat me down to take the test . . . and he did the background check, and everything was good. And I remember going home, a couple days later, going home, and my wife was in the kitchen fixing dinner, and . . . she said, "How did your day go?"

And I said, "Well, I joined the Marine Corps."

And she said, "What?"

I said, "I joined the Marine Corps . . . and in September, I'll go to Quantico for three months, and if that all works out, then we'll come home and we'll go back to Quantico, we'll all go to Quantico for six months, and then we'll go somewhere else." She didn't take it all that well.

Wheeler: So you head down to Camp Lejeune. What did you experience in that first duty station?

Coleman: Camp Lejeune was different. I'd never been down there, and just before, we were at TBS, and I remember one of the instructors talking about the Marine Corps, and the ethos, and how we take care of each other, and all those sorts of things. And there are things you can do and things you can't do, but the Marine Corps takes care of you. Whoever the instructor was talked about Camp Lejeune. He said there was an apartment complex down there, I'll never forget, Beacham's Apartments. And Mr. Beacham would not allow black people to rent from him. So the Marine Corps said, okay, no Marine can rent from you. So immediately, the rule got changed.

And I thought, okay, that's pretty good, I mean, the Marine Corps is progressive, so you thought. . . .

I ended up being a regimental supply officer, so the regimental commander knew me. . . . This is where the mentorship came in. My regimental commander was a person by the name of Colonel [Harold L.] Cy Blanton [Jr.]. And you would've thought I was Cy Blanton's son. He was from Plains, Georgia, or somewhere down there where [James Earl] Jimmy Carter [Jr.] was from. But he was just as honest and clear as you possibly could be, treated me like a son, he really did. Then the next one was [Gerald H.] Jerry Turley, from the Easter Offensive [in Vietnam in 1972]. Jerry Turley, whenever he would go anywhere, he would say, "Come here, son." And he called me "son," he always called me "son," knew me by name, whereas Colonel Blanton the regimental commander—and Turley was the XO [executive officer]—but Colonel Blanton would say, "Ron, I want you to do this." But the other lieutenants were "Lieutenant Smith" and "Lieutenant Jones," or whatever. I moved to be the regimental Four Alpha [S4A, assistant logistics officer], and the regimental Four [S4] was [James L.] Jim McClung, then-Captain McClung. And those three gentlemen—all three white—they mentored me as well as you could be mentored.

The amazing [thing] was that I didn't know very many black officers in the 2d Marine Regiment . . . senior officers; I don't know that I saw a captain, major, or anything like that—a higher ranking officer—in the 2d Marine Regiment. But the mentorship was there. We lived in "TT" [Tarawa Terrace housing area], and there was a captain by the name [Willie J.] Will Oler. He had been prior enlisted, a really good Marine. I was a second lieutenant, and Captain Oler (I always called him "Captain") invited me to his house, set me down, and he said, "I'm going to mentor you, and we're going to map out your career, and this is what you need to do." So as a second lieutenant, Will Oler mentored me. It was a great upbringing, and the Turleys, and the Blantons, and the McClungs, and [John B.] "Black Jack" Matthews, and folks like that just took great care. You say, "Ah, that's hogwash when you say there's no color." If there was color in those folks, they didn't show it. They just treated me like Ron Coleman. It was one of those things where you say, okay, these gentlemen have such respect for you, you can't let 'em down. My tour as a second and first lieutenant was a great tour.

Wheeler: What happened next? You were planning on a short career. Obviously, at some point that changed.

Coleman: I remember I got augmented, which was good. . . . Once you got augmented, you knew you were going to Okinawa. So I go to Okinawa, started off as the supply officer for 3d Med [medical] Battalion, and then we got to deploy as a battalion supply officer [with] LSU [logistics support unit] Foxtrot. That was my first real deployment. I had fun with that.

At this point I'm still thinking, "I've augmented now, so now I can get out when I want to get out. I want to make captain and then I'll get out when I make captain." Well, while I was over there, I got selected for captain, so that was good. I'm about to come home, and Captain Oler was in Okinawa, and so was Captain Cliff Stanley. We bumped into each other in the airport in San Diego. I was at Camp Hansen, and I was walking up to the officers' club, and I saw this black, obviously Marine, but not in uniform. . . . He introduced himself and said he was Captain Cliff Stanley. And he's black, and I said, "Wow." I said I was going to go by and look for Captain Oler. And he said, "Oh, he's my best friend." So Cliff Stanley and Will Oler were best friends. So now Cliff Stanley mentors me, and he takes me around and introduces me to people and tells me what the do's and don'ts of being a young black officer are. So I'm really impressed now.

I'm in Okinawa, I know I'm coming back to Quantico, and Captain Oler says, "You need to go to The Basic School, and I want you to write a note to [Dennis] D. J. Murphy, Colonel Murphy," [who] at this point is the CO of The Basic School. He said, "We're going to try and get you there."

And I said, "Okay." So I think, "Okay, that's a done deal." So I come in, I report to Quantico, and the personnel officer . . . says, "You're going to go to Officer Candidates School." I said, "Captain Oler told me I was going to go to The Basic School."

He said, "Who's Captain Oler?"

I said, "He's in Okinawa."

And he said, "Yes, he's in Okinawa, he's not here, you're going to Officer Candidates School." So he gets on the phone, and he says, "Hey, I've got this young captain here." I'm reporting in my Alphas [service uniform], and he said, "You've got a couple rows of ribbons, and I think you'd be a good person [to have at Officers Candidate School]."

So I said, "Okay." So I go out to Officer Candidates School, and Lieutenant Colonel Solomon [P.] Hill, black officer, first black lieutenant colonel I'd ever seen, is there. And I report in, and I tell him about Captain Oler, and he knows Captain Oler, and he says, "Well, The Basic School has all they need, and we need you here." Lieutenant Colonel Hill and Captain [Henry] Napoleon [Jr.] were the only black officers [at Officer Candidates School] at the time, Hank Napoleon. He said, "No, we need you here."

Colonel Alphonse G. Davis

Allison: You attended Officer Candidates School in 1973. What stands out in your mind as you recall that experience?[30]

Davis: In OCS, there was a big guy, looked like a cross between a cowboy and a football player, a blond-headed guy with a buzz cut. We were eating in the mess hall at OCS, and I'll never forget, another black candidate, [Theodore] Ted Lambert—he eventually became a helicopter pilot—he's in the line, and this guy, for whatever reason, he hits Lambert in the head with a cup, and blood's spewing. All of a sudden, the 11 or 13 of us that were black candidates, we didn't start a fight or anything, but we kind of banded together to say something's got to be done about it.

Allison: Because you thought it was a white on black attack?

Davis: That's right. We took it as a racial incident. The benefit that we had back then was that in the OCS company was a black, First Lieutenant Cliff Stanley, who eventually became a general, and our company first sergeant, named Rogers, and the company gunny [gunnery sergeant], Crawford, also. They were black. They called us in one weekend, all the black candidates, and just read us the riot act and said, "Remember why you're here. Remember why

Commanding officer Col Alphonse G. Davis inspects Marines at the Officer Candidates School at Quantico.
Courtesy of Col Alphonse G. Davis

[30] This oral history interview was completed by Dr. Allison on 16 February 2012.

you're here. Focus on why you're here." Their words weren't so kind and choice, but they invoked the Montford Point Marine folks. They talked about that. That was my first time hearing of them, about what they went through. So I said, "Okay, great." Then we focused on graduating. We did that.

I would see Lieutenant Stanley on the weekends out in town, Quantico. He'd see us in the bowling alley, and he'd just come over there quietly, he and his wife Roz, and they'd be bowling, and he'd say, "You candidates studying this weekend? Getting your laundry done?"

"Yes sir."

He was good at that. He was the guy who was first in this, first in that, first in what have you. But if you check his academic record, the guy just worked hard. And he was giving us those hints. So I said, "Okay, man, this is serious stuff."

Allison: How did you come to select infantry for a military occupational specialty?

Davis: When it came to picking an MOS, a lot of young black lieutenants were pushed into those we call now combat service support or supply. That's when I think the institutional racism comes in, similar to the situation with the Montford Point Marines. Those guys were in support companies, transport companies, truck companies, supply companies, and longshoremen. I think it's an institutional thing because that's where the Marine Corps was comfortable having people of color. Now whether the folks in The Basic School Class 3-73 really had it in their hearts that these guys are not as good, or is it in their heads because that's what they're accustomed to seeing?

When it came time to select the MOSs, I talked to Captain Stanley. He actually was a supply officer but later changed to infantry. He told me to choose infantry. I said, "Why would I want to sleep out with the bugs? Why in the hell would I want to do that instead of just kicking back behind a desk?"

He said, "Well, you're going to see a lot of young black Marines there that need your leadership."

And I said, "Okay." So I selected infantry. Most of the other guys are selecting supply and all that. I think I was the only black guy in our company at TBS [The Basic School] who went infantry.

Allison: It sounds like you benefited from mentoring early on.

Davis: Yes, but it was unofficial mentoring. There's a difference. They had "official" mentoring that I didn't think was so effective. It was like "All black officers report to room . . ." For what?! There was something, a concept I came up with called the "Godfather concept." The Godfather at that time was Frank Petersen; they called him the Godfather. When I first met him, I was a first lieutenant in Puerto Rico, 1975–76, at Roosevelt Roads. I had heard about this guy. He was in a jeep, just sitting behind the wheel licking an ice cream cone. I was like, "Damn, I've seen his picture." So I go up to him, I salute, "Sir, how you doing? My name is First Lieutenant Davis. I saw your pictures and read about you."

I extended my hand, and he said, "Hey, how you doing, brother? Are things going okay?" Again, a very forthcoming guy. He was an aviator, and aviators are laid back and cool. . . .

Going back to the mentoring thing, when I talk about this Godfather concept thing, I was thinking, you've got to have a way that if a young black officer reports in to a new command, then there's a senior black officer, captain or better, hopefully, there. And they contact you, invite

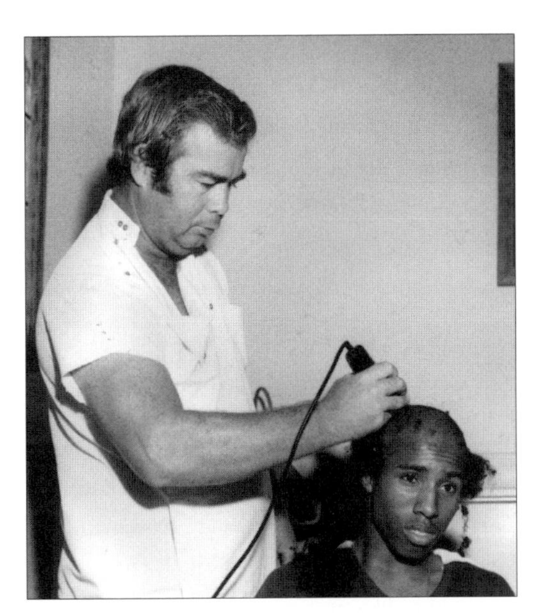

A Marine recruit receives a haircut after arriving at Marine Corps Recruit Depot Parris Island, SC.
Richard Spencer Papers (COLL/5233), Archives Branch, Marine Corps History Division

you to dinner at their quarters or what have you, talk to you, tell you what you're getting ready to embark on in your career because the transition is a little bit different. For example, I would see young black lieutenants wanting to date enlisted Marines. Not smart, dude. Or young black guys back in the early 1970s wanting to date white women. Not smart, dude. The times are not ready for that yet. Like the indelible impact that Cliff Stanley or General Petersen made—it was because of their personalities. . . .

That shaped me on how I became as a leader, to not be a traditional, textbook type of leader. For example later, when I commanded OCS—the first and the only black officer as far as I know to command OCS—I had several black candidates come to my home for dinner one weekend. One of them is a lieutenant colonel selectee now.

Allison: After The Basic School, you went to Camp Lejeune I believe, and took command of a platoon of Marines. What was that experience like?

Davis: I go in as an infantry platoon leader to 2d Marine Division, 3d Battalion, 6th Marines. My first company commander is a Naval Academy graduate, blond-haired guy, [James L.] Jim Clark [Jr.]. I had my hair cut short. I realized I'd be setting an example to black Marines because they put the stocking caps on and all of that.

Allison: Was that allowed, the stocking caps?

Davis: It wasn't.

Allison: This was a big issue in those days, the three inches for hair length and problems with the afro haircuts.

Davis: That's right, the three inches. But I asked Captain Clark if my hair was short enough. The key was, most of the leaders were white, so they were afraid to address that. You know how troops are—if you give them an inch, they're going to take six inches. So I asked him if my hair was cut short enough. His eyes lit up. He said, "Yes, thanks for asking, because you are going to be an example."

Another pivotal thing happened. I discovered that the regimental commander's driver was a guy that grew up across the street from me by the name of George Stewart, a sergeant. I asked the company commander where I might find Sergeant Stewart. So again, going untextbook. "What are you looking for this sergeant for?"

I said, "Well, sir, I think I grew up with him."

He said, "Well, you'll eventually find him." Like that. He never told me where he lived. So I kind of do my own thing and find out where he is. At the end of the day, I get in my civvies, I drive over to the barracks where the enlisted are. I ask Marines, and they said, "Hey, Stew, you got some 'butter bar' out here looking for you." When I go in his room he says, "Damn, boy, what are you doing here?"

I said, "Man, how are you doing?" We just started talking. Again, we grew up across the street, and we used to fight together, we played the dozens together and all that stuff, played football together. And he told me, he says, "Two things. Young black Marines are not going to want to salute you because they're going to think you're their 'brother.' The first one that does that, you grab their ass." That's what he said, "You grab them by the damn stacking swivel."

And I said, "But man, you can't do that; they told me in Basic School."

He said, "Forget what they taught in Basic School." Because we're just coming out of Vietnam and these are bad times. "Forget what they told you. That's what you need to do."

"Okay, man."

So one day I'm walking from my company office, India Company, and I passed H&S [headquarters and service company], which was called "hide and slide" back then, or "heat and steam," or what have you, and they had the casual company. There were four black Marines on the steps out there. They had the black power bracelets on and their covers, they didn't starch them, they'd be flat across. The hair was packed down, and you could tell where the stocking cap went around. They were unshaven, they had the no-shaving chit thing. And that's another thing—I would show black Marines how to shave. I said, "You want to be pretty like me? Let me show you how you shave." I would teach them this stuff. But these four guys, I passed them, and they're kneeling down, but they don't salute. And so I walked two steps past the steps of H&S. Then I remembered what George Stewart told me. I turned around, and I said, "Gentlemen, we don't salute officers?"

One of them looked at me and said, "We don't need to salute you. You're our brother."

And I stepped to the tallest one, and I grabbed him by the collar, and I said, "What's your last name?" He told me. I said, "Mine is Davis. I'm not your brother. Salute." I let him go. They all stood up at this time and gave me one of those really slow salutes. My knees were shaking. And then I just went about my business. Then the word became, "Don't screw with that lieutenant, man, he's crazy." So that was really just kind of like proving myself, not whether or not you know me enough to respect me as a person—that will come later—but I'm a Marine officer. Respect that.

Allison: This was a turbulent time for racial issues. What was the situation in your platoon?

Davis: With my Marines, I had Puerto Rican kids out of New York, I had a lot of black kids out of North Carolina. I had white kids out of North Carolina; the drugs, the alcohol, all that. I just remembered that I needed to be everybody's lieutenant, period—the need to be fair, balanced. That doesn't mean that there wasn't racism or that there weren't hostilities, or that you didn't hear "Uncle Tom" or you didn't hear the "N" word. That was going on. But I would use those instances to say, "Okay, we're a team." So I became a beacon, and they would watch whether or not I would be more favored towards this one or that one. So that was important. I remember this kid, Joe Jefferson. I'll never forget this kid. He always wanted to wear an afro, and I used to always tell him, "Go get a flipping haircut." When I did my inspections, I used to keep a comb in my pocket, and I'd tell the black kid to take off his hat and comb it out. That way you could see how long it was. So there were certain things like that that I had to do to send the message that I'm not going to have double standards.

Lieutenant General Walter E. Gaskin Sr.

Allison: What do you recall of your first assignment as a Marine infantry platoon commander at Camp Lejeune at a time when racial tensions were high?[31]

Gaskin: The night before graduation from The Basic School, we were just sitting around talking to the platoon commander, Captain [Edward F.] "Fast Eddie" McCann. I had just gotten assigned to Camp Lejeune. I'm anxious, I'm ready to go. He said, "I want you to remember this. If your white Marines can't come to you and talk to you about what black Marines are doing or not doing, then you have failed as a lieutenant and a Marine, and I want that title back if you can't do that." He said, "The second thing is that you're going to have tremendous pressure from your African American Marines when you get there for special favors, to see things their way, knuckle knocking and all," which was very prevalent at the time.

Allison: Are you talking about the dap?

Gaskin: The dap, exactly. He said, "All of that will happen to you when you get there. But what you should say is that you are here for them to have equal opportunity at proving that they are good Marines; nothing more, nothing less. Are you proud of your heritage? Absolutely. And they should be proud that you are there, but that's all. They're all Marines. If you can't handle that, don't go." And I always remember that, because sure as hell when I got there, the first thing I had was OOD [officer of the day] duty, and I

MajGen Walter E. Gaskin, commanding general, Marine Corps Recruiting Command, received his two-star rank insignia on 7 October 2005. Gen Michael W. Hagee, then-Commandant of the Marine Corps, oversaw the frocking and reaffirmed Gaskin's commitment as a leader of Marines.
Official U.S. Marine Corps photo

go to the chow hall, and the whole damn line is held up because you've got those Marines going through the dap, you know, it takes them two or three minutes. The law was no dapping in the chow hall. I engage. And there are a few Marines that went to the brig that day because I was an "Uncle Tom," or "I didn't understand," you know, "you think you're white," or "this is bull s——t." They were right. I called the MPs [military police], they're gone. But also what that said to my white Marines is that I am not a black lieutenant, I am a lieutenant and a Marine, period. If you want to impress me, perform.

Everybody is going to apply to the rules. I am not going to shortchange you because you're black, but I'm not going to give you special fa-

[31] This oral history interview was completed by Dr. Allison on 10 May 2012.

vors because you're black, either. If you perform, you'll get just that. And that has been my philosophy in the Marine Corps. What got me there were my experiences, and I am a firm believer that performance trumps everything. I am a part of the performance trumps everything. I remember my first fitness report. I was scared s——tless that day. Lieutenant Colonel Richard [C.] Raines was my battalion commander. He had 19 lieutenants in the battalion and did handwritten fitness reports. We had to go in there and sit in front of him like the Spanish Inquisition. He would sit down, and he would give us counseling on our fitness reports. Colonel Raines said to me, "I didn't think blacks could be officers. I had some damn good black staff NCOs in Vietnam." He said, "You know, I just didn't think they had the mental ability to be officers." I'm sitting there thinking, "This is not going good." I mean, I haven't even seen the report, but I'm afraid. So then he hands me my report to read, and I looked down there, and [it said]: "My number-one lieutenant. The best lieutenant I've seen in 15, 17 years," whatever time was he had in the Marine Corps. "I like his leadership style," you know, "command potential;" "I'm considering him being a company commander as a first lieutenant." It was unbelievable. And I looked at him, and he said, "You changed my mind. Don't you change. You just keep doing what you're doing." And I walked out of there saying that performance counts.

I tell all the young officers. They always ask you when they come up to you, "What did you do?" I always say, "Performance." If I have my job down cold, it relieves all the other issues and thoughts and stereotyping and everything else that comes with any prejudice or bias that they may have. But I can tell you this, I can almost guarantee if you don't perform, if you are average, you can't break out of the pack, you are just barely making it, [then] everything else wrong, or every other bias they have will suddenly surface. They don't like the way you dress. They don't like the fact you have that loud-ass car, you got crazy music, you didn't come to country-western night. Everything else that was there would then fall on the fact that you were not ready because you did not perform. So you have got to be good. You have no option in this.

Colonel Gail E. Jennings

Jennings: One day in 1973, I was walking through the student union at University of Dayton, and they had the contact booth set up for the armed forces.[32] They had the Air Force, Marine Corps; I don't remember the Army being there. I went over to the Air Force, and I got their little contact card and filled out everything. I went to the Marine Corps, liked the uniform, filled out my little contact card. Well, Air Force never got in touch with me, but Marine Corps did. And not only did Marine Corps get in touch with me, Marine Corps stayed in touch with me. About 10 days before I graduated from University of Dayton, I had to let them know one way or the other [whether] I was going with the Corps or not. I nodded my head and said, "Yes, I'll go with the Corps." I didn't get to go to OCS right away because there weren't enough women to make up a class. So I worked during the summer at one of the factories, the auto factories, on one of the assembly lines, and I made my money to buy my uniforms. So I started at OCS, well, in fact . . . I started at TBS after that, but during my junior and senior year of college, I went to OCS . . . and I had an eight-week session that

[32] This oral history interview was completed by Maj Beth M. Wolny on 7 April 2012.

Senior Marine female African American officers gathered at Headquarters Marine Corps on 7 April 2012. From left: Col Sheila Bryant, LtCol Doris A. Daniels, Col Stephanie C. Smith, Col Gail E. Jennings, LtCol Denise T. Williams, Col Adele E. Hodges, LtCol Reina M. Du Val, and LtCol Debra W. Deloney
Courtesy of LtCol Melissa D. Mihocko

I went to Quantico. OCS was a really positive experience for me.

Wolny: In what sense?

Jennings: In that you get indoctrinated into all the tradition of the Marine Corps, all the history of the Marine Corps. That's at a time when, for the women, we even had makeup classes. We had an instructor, some highfalutin someone that had the nice cosmetic bag, out of New York. We had our application. I went home with this big leather kind of suitcase thing with all this makeup in it and whatnot. It was just how it was ingrained in that short period of time, sort of a teamwork and camaraderie that the Marine Corps was all about. I definitely took that back with me. I thought I was leaving a small family in that short eight-week period, and I liked the idea of being associated with something greater than me. I had that experience as I played sports in college, but it was a much more significant kind of attachment, I felt, in that eight-week period. So I had that to carry me through that last 10 months of school before I made the decision that I was going to go with the Marine Corps or not.

Wolny: So that was a very positive experience for you?

Jennings: Very positive experience, even though OCS was not physically challenging for me at all. . . . The women's physical program I don't think was nearly as challenging. I was an athlete going in, but I liked the other things that were involved in the training environment for that eight-week period.

Wolny: How about The Basic School. You went through in 1976. Did that go well also?

Jennings: I didn't enjoy The Basic School nearly as much as I did OCS. When I came through, we were the first company that was integrated with women. We were administratively segregated, like we had a women's platoon, but tactically, we were integrated with the men.

Wolny: Can you explain a little bit more what that looked like?

Jennings: Administratively, we were segregated. We had a women's platoon, so to speak. All the women were set up in Graves Hall. Our platoon commander was female, [Beverly A.] Bev Short. I'll never forget who she was. So in that sense, we were all women a far as that platoon was concerned. But when we went into a tactical environment, when we did all of our field ops [operations], when we did our land nav [navigation], when we did squad tactics, [for] all of those things we were integrated with the men. You'd have women sprinkled in with the men. So when we're coming in from the field and they do port arms, we knew the next thing we were going to be double timing. If it was one of those really long ones, you might have someone behind you that might lift your pack a little bit for you so it wasn't quite as heavy on your hips when you're running in, that kind of thing. But again, you were given tactical assignments out in the field just like the males were. . . .

Wolny: Had you noticed that at OCS or TBS, were there other African American women, or were there other African Americans?

Jennings: At OCS, there was only one other black female officer.

Wolny: Candidates?

Jennings: At OCS . . . [Denise T.] Williams was in my class as well. That's when I first met her. So I think it was just the two of us.

Wolny: But it didn't strike you as odd at the time?

Jennings: No. I mean, I didn't think about it one way or the other. I really didn't. That wasn't my focus, that wasn't my concern.

Wolny: Did you experience situations that were unique to you as a black female officer, discrimination or whatnot?

Jennings: When I got to Cherry Point [North Carolina], I don't think me being a minority or being a female, early in my career—and I qualify that, early in my career—worked to my disadvantage at all. I don't think it was a detriment; it probably was a more positive thing than not, to be perfectly honest. I feel that I got opportunity, especially when I was at Cherry Point, that I didn't see other people getting, maybe because there weren't a lot of women, but I definitely was the only black face running around as a female officer. And I happened to be good at what I did. So again, like I said, that worked well.

I think as I got to be more senior, and again, when I was at Camp Lejeune, I was the disbursing officer, and then I ended up being the comptroller for the FSSG [force service support group]. [This was a] great experience with not only leading within the MOS, but also the credibility I had with the command because I was one of the special officers to the CG [commanding general]. And not only was my expertise for financial management key, but also being a senior female for the command was very key.

We happened to have had a couple of challenges with two different female officers who happened to be minority. Because people sometimes get afraid to misstep, they did ask my opinion in regards to certain things. And I had to call a spade a spade with the CG [commanding general] and the chief of staff one day. Loved them both, thought they were great officers, but I said, "Hey, you put these two officers in the same basket and they're totally different. One individual does have challenges; one individual probably should not be in the Marine Corps now because of this, this, this, and this and has not done those things that she needs to do." And I had no qualms about saying, "This is an individual that is deficient and is trying to use race in order to throw that as a distracter for her lack of competency." This other individual, however, they had two different stories. But they happened to both be female single parents. I said, "But you didn't ask the question on this individual. You just see that that's a single female black officer, and you automatically thought that she had not been married before. I've got a senior colonel over here, colonel, battalion commander—he's a single parent, he's divorced, has a child. But when you look at her, you don't see a single parent that is divorced and, oh by the way, sir, you didn't know she was divorced, did you? You just thought she had a child that happens to be four years old. She's divorced just like that colonel is over there."

They turned red in the face because they knew I had them dead to rights because that stereotype of these two officers. On the surface, their conditions seemed similar, but they weren't. And I knew both of their stories because I had taken the time and talked to both of those officers, not because they told me to, meaning my leadership told me to, but because I was a leader and that's what I needed to do.

TRULY A "NEW CORPS"

by Larry James
Leatherneck, April 1975

You know it is a whole new ball game when all Marines no longer receive advanced infantry training before going to their first duty station or being assigned to a school.[33]

Marine Corps training has not been revamped. It has become more of a polishing process. Corps old-timers—and today that probably includes Marines who enlisted anytime from '74 on back—will find that some dramatic changes are under way. The end result is still the same professional Marine. The means to achieve that goal merely have been altered.

An indication of the shift in training emphasis includes the fact that Marines in the future will not all be required to spend two weeks annually on "range details," with one week devoted to "snapping in."[34]

In further marksmanship training developments, the Marines assigned to aviation units will only need to sight-in from the 500-yard line every three years. In other years, they fire the "B" course from the 200[-yard line].[35]

And, do not be surprised the next time you spot a woman Marine sporting a qualification badge. Certain women Marines will be required to requalify annually.

Today's truly "New Corps" training also deemphasizes rigid classroom hours for essential subjects—and less importance will be placed on written tests.

A further revision cuts the number of essential subjects from 12 to 10. At the same time, the mandatory requirement to be retested annually on all subjects is gone. However, performance "tests" will continue in marksmanship and physical fitness.

With all of the cuts and deemphasizing, the question has to be: Just when are Marines trained? There are two keys to the Marine Corps' updated overall training picture. Considering the needs of the Corps, the individual Marine receives recruit training as either a "basic Marine rifleman" or "basic woman Marine." Following the selection of a military occupational specialty [MOS] for which the Marine is suited, the next step beyond the recruit depot is Skill

[33] The original article came from Larry James, "Truly a 'New Corps'," *Leatherneck*, April 1975. Minor revisions were made to the text based on current standards for style, grammar, punctuation, and spelling.

[34] In military parlance, the term *snapping in* refers to dry-firing a weapon.

[35] This refers to a modified course of fire for select MOS.

Sgt Jeanne E. Jacko fires a Remington 870 pump action shotgun equipped with a riot control canister, 16 July 1979. LCpl Julie Ann Williams (left) waits to fire the shotgun. Shotgun familiarization was part of the training for Marine security guard students at the Marine Corps Development and Education Command in Quantico.
Official U.S. Marine Corps photo DM-SN- 82-03056, courtesy of PFC E. Marshall

Qualification Training. In other words, MOS training.

This may sound similar to the old procedure, but the big difference is that, after boot camp, our 1975 Marine goes directly to a formal school or to a unit. The Marine may receive that skill training to be MOS-qualified from either an on-the-job training program or from a unit training setup. This actually pulls the Marine off the "training line" (the length of time spent in initial or entry-level training) much earlier and makes him available to a unit sooner.

This also saves money by eliminating instructor billets. The in-unit skill training is accomplished by T/O [table of organization] assigned Marines.

The second updated key to training concerns the Marines' instruction after his assignment to a unit. Headquarters Marine Corps has divided this post entry-level training into four main areas: career, mission-oriented, essential subjects testing/training, and related training.

Under old training schedules, units sometimes found it difficult to accomplish all phases. Now, unit commanders have been given priorities for training. The unit's mission emphasized in the mission-oriented phase has top priority. If other training cannot be reasonably accomplished, it can be delayed until time is available.

Marksmanship, too, has a new look. If a Marine has been a qualified shooter several times, does he really need to spend two full weeks each year in the requalification phase? In fact, is it necessary for an accomplished shooter to fire for five days? Not any more. A shooter who has established a reputation for high scores can arrange ahead of time to fire for record on the second or third day.

Instead of annual essential subjects testing for all 10 subjects, time for training and testing is devoted only to those who have problems mastering certain subjects. These individuals will be tested on only the subjects with which they have had difficulty.

While written exams are to be used if the commander feels they are necessary, the real test in all categories is performance.

Marine Corps Order 15110.2H (Individual Training of Enlisted Marines—the new ITEM order) stresses: "Performance tests and observation will be the primary means of determining individual proficiency, supplemented when necessary by written or oral tests."

The new training concept is a far cry from the days when Marines crammed for their TT and GMS tests, which required passing for promotion. The old technical test [TT] (an indication of a Marines' MOS knowledge) and the GMS (the general military subjects test, a forerunner of today's EST [enlisted screening test]) were all geared to reading and comprehending the questions. Now there is peer instruction—innovative or imaginative training—and job/performance aids. They may be new descriptions, but they are all factors in the improved and updated training.

How do individual Marines "see" the new training trends?

While Brigadier General Maurice C. Ashley Jr. served as director of Training and Education at Headquarters Marine Corps, he spoke of training as "the father of each generation of Marines."

In *Parris Island*, artist Thomas O'Hara shows recruits at Parris Island, SC, training with a pugil stick or padded pole to simulate rifle combat.
Art Collection, National Museum of the Marine Corps

"The recruits definitely heard about boot camp before arriving," the general said. "The image is there. He knows recruit training is a tough test of the individual. And, understandably. When the young Marine gets to a depot, recruit training lives up to everything he believes about that image. And the training is set up to ensure there is no letdown."

Even during recruit training, there is a new look about measuring a boot's advancement. Recognizing that written tests place huge demands on reading, writing, and interpreting the questions, performance-oriented training (now referred to as "hands-on" training throughout the Marine Corps) and peer instruction are taking over.

Simply, peer instruction employs students to train other students in a controlled environment. Broken into four steps, the student (recruit) observes a skill being performed and then receives instruction in this skill. In the third step, the Marine is required to demonstrate the skill. At this point, another Marine entering the first phase may observe. During the fourth step, our

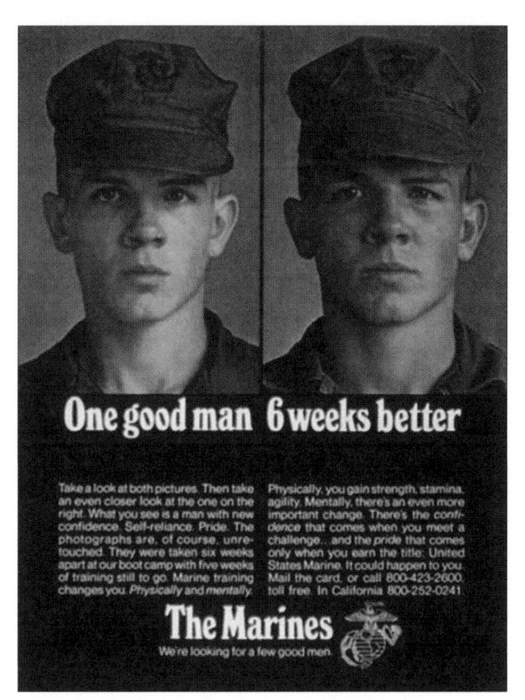

With a caption that reads "One good man 6 weeks better," this Marine Corps recruitment poster purports to show the physical transformation of a recruit from arrival at boot camp to its successful completion six weeks later.
Official U.S. Marine Recruiting poster

original Marine can instruct, which provides the second step for the follow-on Marine.

General Ashley saw this approach as both interesting and challenging. Additionally, it supports "a hallmark of Marines," he said, "assuming he can do it after instruction." A written test did not necessarily prove that.

Recently, Staff Sergeant Michael E. Moro, a drill instructor at Parris Island, told *Leatherneck*'s Tom Bartlett, "This new program is better than the old package. It teaches a recruit quickly. With the new program, the privates pay closer attention, knowing that they're going to be doing your act next."

However, Staff Sergeant Moro believes that the success of the new instruction can be attributed to the actual handling of the training aids by the recruits. The training session becomes little more than a lecture.

Gunnery Sergeant Joseph Gates, academics chief of the Parris Island Recruit Training Regiment, had been a drill instructor for a year and a half before joining the academics section.

"The new training program really impressed me," he said, running a tanned hand through a graying crew cut. "I'm a grunt, and . . . I learn by doing.

"The new program means fewer written tests. We've only been working with it here for about six months, but we find that it permits us to go into more detail."

Another Parris Island Marine, Gunnery Sergeant Bobby Dixon, NCOIC [noncommissioned officer in charge] of the Field Training Unit, which instructs and supervises the training of recruits at Elliot's Beach, [Parris Island] has 17 years in the Marine Corps, and two tours as a drill instructor.

"I've seen many programs tested and tried," he grinned, "but I believe in this one. They come out here, and they either know or they don't know. There's no way they can fake an envelopment or first aid problem.

"As a result of this new application program, the recruits know more about what's going on, and they know how to react."

The training change is probably more noticeable in the period following boot camp. Many Marines wonder whatever happened to ITR. The infantry training requirement for all Marines after boot camp is gone, but for some there is a new training program called field skills training.

On the surface, Marines at Camp Lejeune [North Carolina] seem to be getting the same infantry indoctrination as before. They fire crew-served weapons, receive instructions on tactics, and begin to learn the fine points of being a grunt. But a closer check reveals that only four MOSs are involved. There are Marines being trained as 0311 (riflemen), 0331 (machine gunners), 0341 (mortarmen), and 0351 (antitank assaultmen). The field skills training (FST) schedule resembles the ITR package, but it applies to only the four MOSs.

And, the Marines already belong to a unit. Using the 2d Marine Regiment as an example, the FST instruction unit consists of one officer and 12 enlisted Marines. But, in place of the cadre of special instructors and troop handlers that was necessary for ITR, all of the guidance is provided by these members of the 2d Marines.

The regimental commander, Colonel John E. Greenwood, believes his [MOS] 03 Marines are "better trained than when there was ITR." And he thinks the sergeants and staff sergeants involved in the instruction are also benefiting. The enlisted leaders come from his infantry units. After instructing with the FST for a planned 12–18 months, they return to a company. At this point, the colonel believes he has a better leader.

The evolution of today's FST in the 2d Marines has been polished and changed slightly, since the last ITR in mid-1972.

Initially, a division-level infantry training program was tried. But, it was decided a stateside regiment was a better answer for training the Corps' grunts. The 2d Marines began the field skills training in late 1972. At that time, the young Marines reporting in from boot camp were assigned to a company, went through training in the daytime, and were billeted with the company at night.

Colonel Greenwood said this proved to be a little too much for the young troops to handle. "Things got a little out of limbo for some of the Marines right out of recruit training," he said, "and then listening to those 'old salts' of six months in the barracks at night didn't help.

"Now that we have gone to centralized billeting for the FST Marines, we believe the problem has been solved," the colonel commented.

Captain John Gaieski, a former company commander in the regiment, is in charge of the 2d Marines' FST unit. Of the six weeks—30 training days—about 70 percent is practical application instruction, he stated. Most classes are geared for 25–40 Marines but entire companies do go through, particularly when a company with a deployment scheduled in the near future is being formed. The equivalent of one battalion a year goes through the second FST.

The first phase of FST includes five days of assembly, welcoming aboard, unit assignments, personnel processing, equipment issue, and an actual base orientation. The basic Marine infantryman is shown where Disbursing, Special Services, Red Cross, and other important offices are located. It is not the case of a check-in sheet with a list of building numbers.

And, from the beginning, there are troop handlers. These are not T/O assignments like the old ITR. They are squad leaders from units to which the Marines are assigned. This way, a corporal from a line company gets six weeks to know his men before they become an intricate part of the company.

"We started with only one troop handler," Captain Gaieski said. "Now we have up to 25, particularly when we have men from the 2d Recon Battalion going through our FST."

Corporal Walter Haas of I/3/2 [Company I, 3d Battalion, 2d Marines] is typical of the caliber of Marine assigned as a troop handler by the 2d Marines. Last year, he was the second rifle-

Air traffic controllers at both El Toro and Cherry Point Marine Corps Air Stations guide Marine aviators to landings and takeoffs, 1965.
Defense Department photo (Marine Corps) A412173, Marine Corps History Division

man for the third fire team in K/3/2 [Company K, 3d Battalion, 2d Marines]. This unit was selected as tops in the annual Corps-wide squad competition. Haas, who has three years in the Marine Corps and had gone through ITR about two months before it was phased out, said the FST Marines are really motivated.

"This hands-on training really appeals," he said.

Staff Sergeant Charles Dedmond, an 0311 tactics instructor from C/1/2 [Company C, 1st Battalion, 2d Marines], believes the FST is "just outstanding."

"Personally, I believe the experience the instructor gets here is valuable when he goes back to the company," Dedmond said.

Staff Sergeant Kenneth Browne, a 38-year-old mortar instructor with 19 years in the Corps, definitely has the perspective to discuss FST. Browne has been a member of the FST unit since March 1973. A couple years before that, he spent 18 months as an instructor at Lejeune's Camp Geiger. Prior to that duty he spent a year as a troop handler. He claims the specialized FST instruction is not as thorough as when he was at ITR.

"But one advantage is that I work with smaller groups than at Geiger," Browne said. "The largest number I have had was 43 Marines versus 150 to 200 in ITR.

"The biggest problem the FST instructor has is to counter the scuttlebutt the young Marines pick up at night in the area. Some of the six-month 'salts' pass out bum dope. At Geiger, they were isolated."

Although most of the 12 instructors favored

the old ITR since it could accommodate more Marines for the advanced infantry training, much of the original ITR material is now taught during recruit training. FST is not intended to replace ITR. It is strictly meant to qualify Marines in the four infantry MOS's.

The one thing that all of the 2d [Marines] FSTU [field skills training unit] instructors agree about is the leadership value for the enlisted Marines, who will be rotated as instructors or troop handlers.

The Marines have 10 days of MOS qualification in the second phase [of] basic skill training. Following this, everything is put together in organizational/mission training. The individual Marines discover how their particular skill blends into the overall infantry picture.

Duty and training with the 2d Marines' FSTU is not an eight-to-four proposition.

"All of my instructors pull duty with the night study halls," Captain Gaieski said. And the troop handlers have a lot more to do than merely seeing that their men move from class to class. Corporal Haas said his day starts at 0500 with the Marines in the barracks. At the end of the training day at 1700, the troop handler often spends until 2100 or 2200 talking and reviewing with the troops.

This personal interest, along with the centralized billeting, has greatly lessened disciplinary problems for the 2d Marines.

This relationship of small unit leaders and the new Marines was one of the points General Ashley stressed from the Headquarters Marine Corps level. It ensures there is no letdown of that Marine image with which the recruit arrived at boot camp.

While still director of Training and Education, the general cited the 2d Marines as "biting the bullet" to accomplish the needed training. At the same time, the regiment was able to maintain its combat readiness.

General Ashley also said that he believes the Marine Corps concept in training new Marines will carry its advantages over to career men. Beyond the improved leadership derived by Marine NCOs involved with FST, the possibility of establishing "adventure training" is being considered.

As part of the "no-letdown" idea, a career Marine who is in supply or administration might be able to put in for something like jump training. This would apply to a Marine who desires to continue in his MOS, but has always wanted to attend what he envisions [as] more glamorous training. The general felt that if the Corps makes "adventure training" available, the individual will probably have to contribute his share of off-duty time while undergoing the training.

Someday, the clerk who always wondered what it would be like to drive an M60 [Patton] tank may get the chance.

It is also believed that MOS training for many fields does not have to take a formal school approach. Job/performance aids will allow precise step-by-step directions for Marines to work on equipment or weapons. With very little experience or knowledge of the item, Marines can efficiently perform both simple and complex maintenance and repair activities. Simply, a Marine will be able to accomplish work while actually learning a skill.

Today's Marine training still produces the same professionals who were on Hills 861 and 881 at Khe Sanh eight years ago. But the new look on individual training is now giving the Marine unit's primary mission even greater consideration and significance.

YOUR FOREVER EXPERIENCE

by Robert Church
Leatherneck, November 1975

On this 200th anniversary of our beloved Marine Corps, I would like to write a letter and (if I could) send it back through time to a certain young Marine who marched in the green legions of the Corps more than 30 years ago.[36] Strangely enough, his name happens to be the same as the writer's . . .

Private Robert Church, 10 November 1975
Platoon 400, 7th Reconnaissance Training Battalion
Marine Corps Recruit Depot
Parris Island, South Carolina

Dear Private Church:
It is May 1943, and you are in boot camp, the threshold to glory! For you (my younger self) and your buddies of Platoon 400, this is where it all starts—your training, your salty pride, your service to Corps and country, your Marine memories. Here in the sweat and toil of boot camp, you begin the greatest adventure of your life. This is the birthplace of your forever experience.

[36] The original article came from Robert Church, "Your Forever Experience," *Leatherneck*, November 1975. Minor revisions were made to the text based on current standards for style, grammar, punctuation, and spelling.

Your forever experience! You sail away from the monotonous gray of commonplace existence into the exciting blue of life extraordinary! Your odyssey will span all the years, places, men, women, and events of your life in the Marine Corps. You would not trade it for a king's crown, and you will never forget it. It will be your forever experience, and it started the moment you stepped aboard boot camp.

Of course, right now, in your first few days of training, your experience consists of trying to keep up with everything your drill instructor keeps throwing at you: push-ups, double-time, drill, and all the other little exertions that leave you with quivering muscles, heaving lungs, and rivers of sweat! There may be times when you wonder if you will make it—and you want with all your heart to make it! Many Marine boots know that feeling. Well, do not worry, you will make it. You will go all the way to graduation, and you will come out grinning with the joy and pride of knowing you have earned your place among the elite!

Yes, Private Church, I know just how tough boot camp is. In memory, I can still feel the blisters that made you limp in your stiff new field shoes so many years ago! (You were a mighty hard DI [drill instructor] back then, Platoon Sergeant Koons . . . and a first-rate Marine!)

It is hard training, indeed, but it is all part of a carefully planned and tested program that is transforming you into a highly efficient fighting machine. No, not a robot. A man. But a very special kind of a man with a toughened body and a trained, alert mind, able to function equally well with others under command, or on your own initiative—like a single bolt of lightning!

At the same time, you are learning the techniques of survival, so you will have the best chance possible of coming safely through whatever battles may lie ahead. You are learning to hit the deck with enthusiasm. You are learning to "keep that big, bouncing butt down, Church!" You are learning to go over an obstacle faster than a man can aim and fire a rifle at you. You are also learning that often the best defense is a swift, overwhelming offense. And you are learning that, in battle, the Marines take care of their own, by covering fire and other tactics, and by buddy helping buddy.

No, not all Marines survive their battles. Many fall—but not without meaning or purpose, no matter what the armchair critics may say! They fall doing their duty for their country—and with the greatest of honor! And they lie asleep with their fallen buddies until the final reveille calls them forth forever.

Is that such a bad way to go?

Think about it.

Meanwhile, Private Church, you are learning a lot, and you are learning it well. And, as time goes by, you will also discover that, along with everything else—or perhaps because of everything else—something unique is happening inside you. It began your first day in boot camp, and it will grow as you grow in the Corps. It is the very heartbeat of your forever experience, and you will have it all the days of your life. They call it esprit de Corps.

U.S. Marine Corps recruitment poster *U.S. Marines, Uncle Sam's Right Hand*, designed by R. McBride, ca. 1917.
Willard and Dorothy Straight Collection, Library of Congress Prints and Photographs Division

Esprit de Corps. The Marine mystique—the intangible but very real spirit that makes the qualitative difference between a U.S. Marine and any other military man in the world. It is love. It is pride. It is devotion. It is all that and more too. You cannot fully explain it. To understand it, you have to experience it yourself. And to do that, you have to be a Marine.

Look at any Marine when he hears a band playing his own hymn. Look at any Marine when he is saluting his country's flag. Look at his face, and you know he will go all the way for his country and his Corps. All the way to kingdom come if he has to!

Featured on the cover of *Marine Corps Gazette* in November 1953, Cpl Tony Kokinos's painting *Heritage* was later discovered hanging in the supply office at the Marine Barracks and brought up to record in May 1975. *Art Collection, National Museum of the Marine Corps*

That is just a hint of what esprit de Corps is, Private Church. It is why you will honor your Corps—second only to the flag you defend.

Which brings us to the moment of truth. The defense of that flag, and the country it represents, has been the primary mission of the

Sgt Tom Lovell, *Flag Raising, Mt. Suribachi Iwo Jima*, oil on canvas.
Art Collection, National Museum of the Marine Corps

Marine Corps for 200 years of magnificent heroism, and it is still our primary mission!

And what of the flag? What is there about it that makes men willing to lay down their lives to keep it flying? What makes young men actually eager to climb a hell on earth like Mount Suribachi to plant the Stars and Stripes at the summit?

Simply this: that flag, that Old Glory, that beautiful Star-Spangled Banner, symbolizes the very highest principles of human decency ever conceived in the hearts and minds of men!

And your primary mission, Private Church, is to help your Corps defend that flag with all your might! It is as simple as that. And as worthy!

You see, Uncle Sam and his "few Marines" have come down a long road of years shoulder to shoulder. The road began as a narrow, cobblestone street in Philadelphia; it extends down to today; and stretches on into all our tomorrows.

The milestones along this road have the names of places that are shrines to the courage of the Marines who fought there: the Bahamas (where Marines made their first amphibious landing, led by none other than our old friend, Captain Samuel Nicholas, formerly of Tun's Tavern); Tripoli (source of the Mameluke sword); Chapultepec (and the halls of the Montezuma); Guantánamo Bay; Cavite; Samar; Peking and Tientsin (Boxer Rebellion); Belleau Wood; Mont Blanc Ridge; Meuse-Argonne; Guadalcanal; Tarawa; Guam; Iwo Jima; Okinawa; Tokyo; Chosin Reservoir; Con Thien; Chu Lai; Hue—and many, many other milestones, including the decks and rigging of countless ships, in naval engagements starting with the American Revolution and continuing today.

And the 200-year road is paved with the valor of men whose names, coming one after the other, sound like the measured thunder of a mighty bell, tolling liberty through the centuries: Samuel Nicholas, Presley N. O'Bannon, Archibald Henderson, Archibald Summers, Smedley D. Butler, Daniel J. Daly, John Quick, John A. Lejeune, Alexander A. Vandegrift, Evans F. Carlson, Merritt A. Edson, Gregory Boyington, Leland Diamond, and Lewis B. Puller . . .

And so many others their names would fill an honor roll reaching from here to eternity. And in a very special place on the honor roll would go the legendary Private First Class G. I. Grunt and his buddies, those heroes of heroes, the tired, dirty, sweating, shooting, unsung sons of the rifle companies!

We salute them, every one!

Marines like these know what every thoughtful American knows, that freedom is not free. The price is high, but we Americans pay it because we happen to think freedom is worth it. And the Marines always pay a lion's share of the bill. Gladly. Proudly. The first Marines thought America had something worth fighting for. Today's Marines still think so.

Nobody likes to fight a war. But until somebody figures out how to establish a permanent peace with freedom for all, somebody has to be able and ready to guard the rights and lives of free people. The Marines have that capability, and use it whenever Uncle Sam calls them into action.

Private Church, the Marines of your generation won your war in the islands of the Pacific, and you won it heroically. But there have been other costly battles between your point in time and the present—Korea, Vietnam—and special missions such as the Cuban missile incident, the Lebanon landing, and the Mayaguez rescue.

Now, in the year 1975, 200 years since the Marine Corps was established, we live in a world that seems more troubled than ever. There are those who would take away our freedom today, if they could. (But they damned well cannot!)

And there are others, sitting on a so-called "neutral" fence, who snipe at us verbally. We Americans are "decadent," they say. "Immoral."

It seems they can see the speck in our eye, without being aware of the plank in their own.

Sure, we have our "lunatic fringe." What country has not?

But when there is disaster anywhere in the world, who is the first in with help? Who is the first to land with food, medical supplies, blankets, shelters—planeload after planeload of help—without ever counting the cost? The decadent Americans, that is who!

And when innocent people in danger of mass slaughter call out for help—who is the first (and sometimes the only) to go in and guard them, feed them, evacuate them without ever counting the cost? The "immoral" Americans, that is who! With the United States Marines leading the way in!

Uncle Sam and his "few Marines" are ready—as they have been for 200 years—to go anywhere, anytime, in the cause of justice, honor, and simple human mercy.

These missions have the wholehearted approval and support of the American people. Not the oddballs, of course, who would not support anything, including themselves. Not the cynics who look on from afar with delicately raised eyebrows. No, our strength comes from the vast center of our population, which is as solidly American as it ever was.

We have many millions of decent, patriotic citizens who love our country as much as ever. Gutsy Americans who are not ashamed of a teardrop in the eye when they see Old Glory still proudly waving. They know what she stands for, and what it has cost to keep her flying.

The Prayer at Valley Forge depicts Gen George Washington kneeling in prayer under trees at Valley Forge during the American Revolutionary War. Engraver John C. McRae and publisher Joseph Laing generated this popular print after an original painting by Henry Brueckner, ca. 1889.
Library of Congress Prints and Photographs Division

And we have millions of strong, reverent Americans who are not ashamed to ask God's blessing and guidance for our country, our leaders, our men and women in uniform.

There is a famous painting of General George Washington before a battle, kneeling on one knee in the snow, praying for his men and his country.

That spirit of faith and reverence still prevails in the great heart of America. We are still "one nation under God."

Thank God that we are!

No godless nation or group of nations in the world can prevail against that spirit! And as for the godless cynics in our midst—they can go plumb to hell, which no doubt they will!

Well, Private Church, as you can see, you have got a lot going for you, and a lot to live up to, as you join the elite in the front ranks of our country's defense forces. Your life as a U.S. Marine stretches like a shining road before you, and whether you stay on that road for one tour of

SSgt Ermelinda Salazar, a woman Marine, was nominated for the 1970 Unsung Heroine Award presented annually by the Ladies Auxiliary to the Veterans of Foreign Wars. SSgt Salazar, determined to help the children of the St. Vincent de Paul Orphanage in Saigon, Vietnam, in her off-duty hours, holds two of the youngsters in this painting by artist Alex Young.
Art Collection, National Museum of the Marine Corps

duty, or 30 years, you will find it the high road of your whole life. You will never have a greater opportunity to reach so high a level of service and personal fulfillment.

So there is something you should know, something you should be aware of at all times. It is one of the most important things anyone could be aware of—but very few are! It is something I should have been aware of, but was not.

This is it: be aware that you are always building memories! Wherever you may be, whatever you may be doing, you're also building memories.

Now you will be building memories of men, women, duty stations, voyages, islands, the rattle of gunfire, the coppery taste of fear, the sound of laughter, the many details of life in the Corps.

But most importantly, you will be building memories of your own performance from moment to moment, and day to day. Some of these will be the big memories that will come winging home to you . . . either like golden eagles or like vultures. They will bring you satisfaction, or regret, as long as you live.

Once an action has been completed, once a moment has passed, once today has become yesterday—you cannot call it back and change it, or erase a line of it, no matter how much you wish you could! Each moment is a segment of eternity that becomes part of your forever experience.

That is why it is so important—to both you and the Corps—to live the kind of life that produces good memories, ones that will bring joy and pride whenever they come to mind. Private Church, live so the Corps will always say, "We are glad he was one of us." And you will always say, "Me too!"

There are two simple guidelines that can help you build good memories. One is this: do not take anything or anyone for granted! Everything, and especially every person—and your personal involvement—is a potential memory, treasured or tragic.

And the second guideline: when you are going to do something, always ask yourself, "Will this deed be worthy of 200 years of Marine Corps honor? Would a really first-rate Marine do this?"

If the answer is "no"—then by all means do not do it! Because if you do, the memory of it will cast a shadow on your forever experience that will haunt you all the days of your life!

To sum up: the quality of your own service to your Corps and country will determine the quality of your memories, and thereby the quality of your forever experience.

So you see, Private Church, you will want to be the very best Marine you possibly can so that, 32 years from now (way up in 1975), you will be able to look back at a forever experience that gleams like purest gold in your memories—and in your Marine Corps record!

Well, Private Church, this letter will not reach you in boot camp because you graduated many years ago, and marched away into the mists of time . . .

. . . and now you are an old-timer who has lived his forever experience. An old-timer full of memories, of both sunlight and shadows. But an old-timer with the fierce love and pride of the Marine Corps still surging through his veins!

I am wearing civvies now—but just inside lives a young Marine in dress blues . . . forever!

As for this letter, I guess the best thing we can do is make it an open letter to all Marines, everywhere. And we salute you all!

We often think of our own old buddies—the Marines of the World War II generation—and to them we say, "Thank you for the unforgettable camaraderie."

We think of the generations of Marines who served before our time, and we say, "Thanks for what you built, and guarded through 200 years, and passed on to us."

We think of all the Marines of the future, and we say, "You will have a lot to live up to. But you will do it. Marines always have."

We look at the magnificent Marines of today, and we say, "You are doing just fine! You are as good as we were, and better! We wish you Godspeed, a happy voyage, and a treasured forever experience!"

And you women Marines of yesterday and today—did you think we would forget you? So trim and chic? The prettiest, proudest sight in all of God's green earth! Forget? Ah, no. Love is the word! And respect! We wish you a golden shower of bright memories always!

And now to our beloved Marine Corps in your 200th year of glory—we are overflowing with pride and devotion as we present to you the smartest, most heartfelt salute a Marine can give!

Finally, to the flag of our country—the basic reason for our existence as Marines—we pledge our allegiance, and our lives, forever!

Semper Fidelis!

EMPHASIS ON PROFESSIONALISM FOR A NEW GENERATION OF MARINES

by General Louis H. Wilson
Sea Power, January 1976

We in the naval Service often talk about the uses of sea power in matter-of-fact terms because we accept without question the need for strong naval forces.[37] We understand that to keep our place in the international order means we must maintain the means to protect our sea lines of communication. Occasionally we forget that, in today's society, virtually nothing is accepted without question. We discuss maritime defense issues in journals such as this, but very often we fail to examine the assumptions and premises on which the need for such defense is based. It is important that we do so from time to time, for others will certainly do it for us if we fail.

To that end, I would like to review some of the philosophy upon which our plans are based. With that background, we can then discuss some of the details of our ideas for 1976 and on into the future.

The world is changing rapidly, and the pervasiveness of change is apparent to most observers of the international scene. At the strategic level, there is, to be sure, a stability based upon essential equilibrium among the nuclear nations. But, beneath the umbrella of the nuclear standoff, a broad spectrum of behavior is possible, not all of it friendly to the United States. Provided that we maintain the necessary strategic deterrence, the most likely challenges that we shall have to face in the future will be on the level of conventional military confrontation.

In order to meet those challenges, the Marine Corps must remain ready, strong, and flexible.

A REVIEW OF BASIC CONCEPTS

In examining our nation's defense needs, we should reexamine our basic concepts to rid ourselves of that which is outmoded and revitalize that which remains valid. Doubts have been raised in the past about the viability of amphibious warfare, and those doubts tend to resurrect themselves from time to time. In order to ensure the soundness of our future planning, we might, then, dwell for a moment on the basis for our amphibious structure.

Future employment of the Marine Corps could take place in a wide range of circum-

[37] The original article came from Gen Louis H. Wilson, "Emphasis on Professionalism for a New Generation of Marines," *Sea Power*, January 1976. Minor revisions were made to the text based on current standards for style, grammar, punctuation, and spelling.

stances in any theatre on the globe, from full-scale war, with or without nuclear weapons, to making a show of force at some remote spot. The toughest scenario is, of course, the former. Whatever the scenario, the Navy-Marine Corps team will have a vital role to play. Those who doubt the efficacy of an amphibious landing in a total war environment may have a point: such a landing would be extremely difficult. The most likely circumstances, however, would probably involve U.S. and NATO or other allied forces attempting to hold the line against an advancing aggressor somewhere on the Eurasian landmass.

In that case, or in other similar circumstances, friendly forces holding a line or conducting a delaying action might need reinforcement. A mobile, ready, general purpose force integrated with air is ready-made for such a task. If the line were holding, or if a counteroffensive were planned, an action on the flanks of an advancing or retreating enemy might have the same effect as at Inchon [South Korea], though probably less dramatic.

There is no way to execute such a flanking move unless a forcible entry into enemy controlled territory is possible. It would most certainly be conducted from the sea, if one imagines the battle lines reaching from coast to coast. If the enemy withdrew forces to resist such a landing, his front line would be weakened accordingly. Regardless of his reaction, the threat of amphibious attack would seriously limit his options. Given a powerful amphibious force, we would have several viable options, which in turn would seriously hamper the enemy's freedom of action. The crucial question is whether a large successful amphibious landing would be possible under the circumstances. If not, then it would be postponed until conditions favored the amphibious attack.

But, whether or not such an attack would be possible at a given point in time, the possession of a major amphibious capability is an extremely valuable asset. An enemy could never be sure whether it would come or not—we have been known to attempt the "impossible" before. In other words, in the most difficult of all possible circumstances, there is every reason to have an amphibious capability, and no reason for not having it.

Training—of the body, of the mind, of the total man—was a hallmark of the Wilson Commandancy (1975–79). *Gen Louis H. Wilson* by Albert K. Murray, oil on canvas, 1976.
Art Collection, National Museum of the Marine Corps

Outside the circumstances described above —and they can be translated into any locale or time frame—other possible actions by those hostile to U.S. interests are forever limited by our ability to put a major force ashore on short notice and sustain it there for as long as it is necessary to prepare for additional options. The U.S. amphibious capability is unique, it is respected by friend and potential foe, it remains a viable, important form of response to a broad

"A forcible entry into enemy-controlled territory..." in any future conflict, Wilson points out, "would most certainly be conducted from the sea," and that is the reason, he notes, for the Corps' continuing emphasis on amphibious training and doctrine. Col Peter Michael Gish, *Onslow Beach*, watercolor on paper.
Art Collection, National Museum of the Marine Corps

spectrum of actions. And the potential for employment of amphibious forces is worldwide; our focus remains similarly wide.

Because of the broad-based nature of our possible time, place, and degree of deployment, we have to maintain maximum flexibility and maximum readiness. All the good things one may say about the Marine Corps fades to platitudes if we cannot instantly react as we are paid to do. We emphasize our readiness over and over, for it is the sine qua non of our existence. And our focus in the Marine Corps will not waver from that lasting goal of complete readiness.

To that end, we are constantly examining our policies, personnel, and equipment, to ensure that they meet the needs of the mobility necessary for our amphibious mission.

One of the vehicles for testing ourselves is training. There is no guarantee that methods proven successful in training will stand up in combat, but it is obvious that those who fail in the training environment will not survive on the battlefield. And, although proof of guaranteed success is elusive, history shows that well-trained units have a much higher probability of success than others, regardless of circumstances. So we are placing great importance on training at all levels, being aware that it serves two ends—testing and preparation.

NUCLEAR AND HIGH-ARMOR THREATS

Our training must reflect the most difficult circumstances we might encounter. We must train for defense against nuclear weapons, for we dare not assume they will not be used.

The next most critical problem is the high-armor threat. We are planning to meet this threat by expanding our deployment of the [M47] Dragon and TOW [tube-launched, optically tracked, wire-guided] missile systems and by adding the M60A1 tank to our arsenal as a replacement for older models.

It used to be a cliché that "the best weapon against a tank is a tank," the implication being that it was the only weapon. Recent advances in hand-held infantry antitank weapons have modified that idea. Better tanks are vital for our armored support needs, but the rifleman is no longer helpless against the tank. Further, we have no plans to build "heavy" divisions; such would be inconsistent with our mission. We are uniquely dedicated to the concept that integrated air and ground weapons are our best means of defeating enemy armor.

A very important step in improving our combat readiness has been the establishment of an air-ground combat training program at the Marine Corps Base Twentynine Palms, California. A certain amount of time will be spent in preparing to move entire units through a combat arms training program. The concept of such training is very important for two reasons:

- First, the training will, as far as consistent with safety, be realistic. Ground unit commanders and support weapons controllers will be using live ordnance to refine their skills in the coordination of all the assets that an infantry commander has at his disposal to support him in the execution of his mission. For the artillery, tank, aircraft, and other support units, the training will be realistic since they will actually be conducting live-fire missions. The infantry units will get an opportunity to observe the effects of supporting arms on the ground and will gain greater confidence in their use. By conducting live training in direct support of the infantry unit, the support units will become more proficient in their primary task.
- Second, the training program will allow for innovative thinking and improvement on existing techniques. The program cycle will be repeated for different units that pass through, so that those conducting the program will be able to make refinements and try new ideas incrementally, without jeopardizing an entire training exercise. New weapons and equipment items will be introduced as they appear, at first in the form of demonstrations and later as an integral part of the exercises. In keeping with our policy of concurrent testing and training, commanders and their staffs will be expected during the full unit phase of each exercise to demonstrate that they can effectively coordinate and use supporting arms. The possibilities of the program are exciting, and the benefits that will accrue enormous. Our combat power will be substantially improved. By coordinating the entire program at Twentynine Palms, we will be able to support the program from our existing training budget. We will train and test both Regular and Reserve units, and we will include electronic warfare as part of the exercises we conduct to create as realistic an atmosphere as possible.

Two of the new weapons systems that figured prominently in the Marine Corps' plans for the future were the F-18 and the LHA. Col H. Avery Chenoweth, *F/A-18 Hornet*, oil on canvas.
Art Collection, National Museum of the Marine Corps

AIR SUPPORT FOR THE FRONT LINE

To further enhance our frontline Marines' ability to do his job, we are looking hard at our aviation requirements for the future. Whatever form our air support takes, it has but one purpose: to support the ground trooper. All our aviation problems must be addressed from that perspective. One of the biggest problems is money, and financial constraints will be a permanent reality in procurement of all items. Aircraft are particularly vulnerable to monetary limitations because of, among other things, long lead time and the need for sophisticated equipment.

Aircraft also make considerable demands on our personnel. It takes highly qualified men and women to maintain and support modern airplanes. They require schooling of weeks or months duration as well as on-the-job training before they are fully effective. Those factors lead inevitably to the fact that we must seek the simplest aircraft consistent with our needs, and we must try to procure as few different aircraft as we can.

Consistent with those goals, we anticipate that by the 1980s our aviation assets will be a combination of V/STOL [vertical/short takeoff and landing] craft for light attack and the [McDonnell Douglas] F18 [Hornet] for fighter/attack missions, as a gradual replacement for the

[McDonnell Douglas] F4 [Phantom II]. We are monitoring all aviation developments closely and are working in close harmony with the Navy in the aircraft field.

We are looking as well at rotary wing requirements, and [we] are planning now for aircraft to replace those in our present inventory. Of course, the new [Sikorsky] CH-53E [Super Stallion] will enhance our lift capability. Beyond that, we must realize that we cannot always afford to design a new aircraft from scratch when an existing model becomes obsolete. On the other hand, we are not prepared to completely reorganize our ground elements to maintain the principle of tactical integrity, if such reorganization is necessary solely to take advantage of "on-the-shelf" items.

In all our aviation requirements, both rotary and fixed wing, we intend to take advantage of modern electro-optical systems, sophisticated weapons, and other technological advances to provide the best possible support for the Marine in the rifle squad on the ground.

NEW TECHNOLOGY FOR AMPHIBIOUS OPERATIONS

Because of our unique seaborne role, we are constantly concerned with developments in amphibious shipping. The LHAs [landing helicopter assault, or multipurpose amphibious assault ships] that are coming along will enhance our lift capability, and we are vitally interested in follow-on technology to the work, which has already been done in hydrofoils and surface effect vehicles.

Much of the doubt that some observers have recently expressed about our capacity to conduct an amphibious landing is based on the assumption that 30-year old techniques and equipment will be used indefinitely into the future. Although we will still land from ships onto a hostile shore, there will be great differences in landings of the future from those of Tarawa and Okinawa. The basic problems will remain the same; methods of solving them will continue to change dramatically.

There also will be occasions when Marines are needed very rapidly in areas that cannot be reached by ship, or at least not fast enough. We have worked with the Air Force in the past to help move Marines to trouble spots. We will continue to do so, and our cooperative efforts with the Air Force will be part of our training plans as well. We must ensure that we are prepared to meet our legal mission of "such other duties as the President may direct."[38]

To be ready to fulfill our primary and secondary missions, our divisions and wings must be adapted to the most likely place and method of employment for each. By law, we have three active divisions and aircraft wings, and one of each in reserve. They do not necessarily all have to be the same size and shape, as long as they are flexible, ready, and mobile. We will make any necessary shifts of personnel or equipment to ensure the constant readiness of our FMF [Fleet Marine Force] elements for the most likely contingencies.

We will continue to keep smaller units from the three active divisions positioned where they are most needed generally, afloat with Navy fleets. Circumstances may arise, however, where the prepositioning of an amphibious force in a threatened area might give a potential aggressor pause, and reduce his range of options considerably. But, however our divisions, wings, Marine Amphibious Brigades, and Marine Amphibious

[38] Title 10, U.S. Code § 5063, Pub. L. No. 99-433, 100 Stat. 1043 (1986).

The slogan "If everybody could get in the Marines, it wouldn't be the Marines" played heavily in the recruiting campaign that ran from 1971 to 1984. *Richard Spencer Papers (COLL/5233), Archives Branch, Marine Corps History Division*

Units are structured, and wherever they are located, they will be specifically designed for their amphibious role. If we are again committed to sustained land combat, we will make whatever adjustments are necessary to carry out that assignment.

Within the Department of Defense, we are fitted under the category "General Purpose Forces." Considering the missions assigned to the Corps, we think of ourselves as a ready, mobile general purpose force with amphibious expertise.

"WE WILL NOT SACRIFICE QUALITY"

Our personnel are, of course, our most important asset. Whatever plans and concepts for the future, they will have to begin with the individual Marine. A division of 18,000 men is built one man at a time, both on paper and in practice. And we look to every man in each division to carry his share of the responsibility for that division's readiness, from the commanding general to the rifleman.

Our recruiting initiatives are an attempt to assure that every man who comes into the Corps will be able to fill a position in that division or wing and have the potential to move up through the ranks and assume greater responsibilities as his knowledge and experience grow. To that end, we will continue to have a goal of 75 percent high school graduates among our enlistees. We will not sacrifice quality for quantity.

We will continue to recruit women for our ranks, although we do not contemplate any combat role for them, in accordance with existing law. We now have approximately 350 women officers and about 3,000 women in the enlisted ranks. Current plans are to raise those figures to about 480 officers and 3,700 enlisted by 1980.

Those figures are not very large, but, as long as the legal restrictions against placing women in combat units exist, we are limited in the billets in which we can use women. We have women in many varied assignments, and all but the combat military occupational specialties are open to women Marines.

Regardless of how women serve in our ranks or in what capacity, they are Marines, and the standards we impose on Marines apply to all. Naturally, in such areas as physical fitness, we use different methods of measurement, but no one is exempt from being fit and ready. We have always believed that Marines are Marines and, except for the differences over which we have no control, we do not try to differentiate between the sexes.

LEADERSHIP: THE MOST IMPORTANT FUNCTION

If people are our most important asset, then leadership must be our most important function. We are constantly evaluating our leadership programs, recognizing that, as our society changes, our leadership needs change. Young men and women who enter the Marine Corps today are experiencing pressures that did not exist 20 years ago. On the other hand, they are better educated and more sophisticated than the 18- and 19-year olds of the 1950s.

Society is emphasizing individual awareness, and our programs emphasize self-respect and respect for others as part of our fundamental approach. Old leadership ideas have not disappeared; they are merely taking a new form to meet the needs of a new generation of Marines. I emphasize that they are still Marines in every sense of the word. They respond to good leadership as always.

Wilson identified leadership as the most important function of the Marine Corps. His philosophy is reflected in the recruiting slogan: "Marines. To lead is our tradition."
Official U.S. Marine Recruiting poster

In return for the favorable reaction we are getting from our Marines on our recent initiatives to emphasize personal responsibility and individual quality, we are trying to pay more attention to our good Marines—the overwhelming majority—and provide them with additional reasons to stay on the team. We are reducing personnel turbulence by stabilizing the length of tours and eliminating unnecessary permanent moves. We are looking for ways to provide better educational opportunities for our Marines who want to improve themselves and increase their chances of promotion. There are certain expenses involved, but the rewards to the Marine Corps and to society as a whole are manifold, whether the Marine stays on in the Corps or returns to the civilian world. In summary, we are emphasizing professionalism in the Marine Corps for the coming years, and our emphasis is on the individual Marine as part of our air-ground team. I have reiterated here our philosophy is based upon our maritime role and our all-important partnership with the Navy.

The vitality of our nation is linked to the sea, and our focus will continue to be there, as our name implies. We perceive that we have a job to do, that the American people trust in us to be prepared to do that job, and that we are and will continue to be ready. We shall not betray that trust.

THE PERSONNEL CAMPAIGN ISSUE IS NO LONGER IN DOUBT

by Lieutenant General Bernard E. Trainor
Marine Corps Gazette, January 1978

Few campaigns in Marine Corps history have been as difficult and critical.[39] Few campaigns have been so dramatically marked by defeats and victories. In no other campaign was the future of our Corps so threatened.

The campaign was not fought on some far offshore, but in the cities and towns of America and on the drill fields of Parris Island and San Diego. The campaign can be called the Personnel Campaign, 1973–77. It dealt with recruiting and recruit training. While there are still pockets of resistance, it is probably safe to say that the campaign has ended, and the issue is no longer in doubt.

The precarious struggle began sometime after the public attitude turned against the war in Vietnam. Coincidentally, a social upheaval had taken place in the United States, which was particularly manifest among the youth of the land. Its most negative aspects were marked by a drug culture, a climate of permissiveness, racial discord, a rejection of authority and established values, cynicism, and a philosophical commitment to rights without responsibilities.

In this climate, the Marine Corps came home from Vietnam, its banner held high, proud of the professional manner in which it had acquitted itself in that unpopular war. The Corps settled down in a business-as-usual fashion and resumed peacetime training as though nothing unusual had happened or was happening. The order of the day was continue the march, no compromises, no short cuts. In this regard, we can be thankful for our steadfastness, because in the difficult days that followed, the discipline of the Corps sustained it. But to think that the Corps would be unaffected by the changes in society was unrealistic. We were drawing our recruits from that society. Good and the bad aspects would be reflected. Though reluctant to admit it, the Marine Corps, like the other Services in the early '70s, was having serious disciplinary problems.

On top of this, the draft ended. The era of the all-volunteer force was ushered in. In the following pages, I would like to tell the story that ensued. One aspect deals with recruiting,

[39] The original article came from LtGen Bernard E. Trainor, "The Personnel Campaign Issue Is No Longer in Doubt," *Marine Corps Gazette* 62, no. 1 (January 1978). Minor revisions were made to the text based on current standards for style, grammar, punctuation, and spelling.

the other recruit training, but at all times the two were inextricably mixed.

RECRUITING

There were three basic errors made in the post-Vietnam recruiting situation, which bore bitter fruit for the Corps for years thereafter. The first error dealt with the subject of education. The Marine Corps had always encouraged young men to remain in high school before enlisting. A standing recruiting goal of 65 percent high school graduates existed. From a practical standpoint, however, it did not appear that the market could support this goal. As a result, greater dependence was being placed on mental testing of the IQ variety, the sort of test that measures the trainability of an enlistee. To many, this index of quality was superior to the mere possession of a high school diploma.

I'm Looking for a Few Good Marines to Help Me Find a Few Good Recruits, Marine Corps Recruiting poster. *Art Collection, National Museum of the Marine Corps*

There was a deceptive logic behind this thinking, which recognized that the nationwide variance in educational quality made it almost impossible to use a high school diploma as an accurate measure of intelligence and ability on the part of the recipient. As some were quick to point out, frequently diplomas were nothing more than social certificates, indicating that a student had vegetated in a school for four years. It seemed to make sense that a test for trainability was a more useful instrument to measure an enlistee's potential for useful service, regardless of whether the enlistee had finished high school or not. So testing soon became a primary index for quality measurement. The 65 percent high school graduate goal remained on the books, but it also remained unachieved, as the actual numbers enlisted dipped for a time below 50 percent.[40] Further, given the recruiting conditions at the time, a mandatory high school percentage would sorely narrow the recruiting market and could threaten the Corps' total personnel strength.

For a period of years, the Marine Corps had been filling its ranks largely with high school dropouts. In retrospect, it is clear that we had erred. It is one thing for a young man to be trainable, as measured by a validated testing system. It is quite another for him to be receptive to training. The very factors that prompted many enlistees to drop out of high school also

[40] The issue of high school diplomas was raised in January 1974 when the House Appropriations Committee report pressed for a retroactive 55 percent high school graduate requirement on the Marine Corps for that fiscal year. As the year was half over, this was a clearly impossible task. It did, however, signal a concern over the number of non-high school graduates coming into the Corps.

Photographic print after the classic Marine Corps Recruiting poster *We Don't Promise You a Rose Garden*. *Richard Spencer Papers (COLL/5233), Archives Branch, Marine Corps History Division*

compelled many of them to resist training both at the recruit depots and in subsequent Service schools and assignments.

In short, the Marine Corps, by taking a large number of dropouts, had been taking a large percentage of losers. Our ranks were being infiltrated by too many nonachievers whose only real potential was trouble. What the Corps had failed to fully appreciate was that the high school diploma, notwithstanding its value as an index of mental capability, was indeed a social document. It told us some things about the young man. It told us that he had the perseverance to stick to a job for four years, and it also told us that he exhibited acceptable social behavior, or else he would have been invited to leave school before completion. Most importantly, it gave us a clue as to the influence of his home life. The high school diploma with the testing system was the proper combination to ensure quality, but we suffered through a trying period before this was properly recognized. In the meantime, the problems appeared to grow almost geometrically.

The second recruiting error was an error of assumption. Throughout the draft years, the Marine Corps did reasonably well in recruiting. Prospects were not knocking down the doors to the recruiting office, but on the whole, adequate numbers of applicants came to enlist. The assumption was made, therefore, that because we were a volunteer organization, the end of the draft would not impact adversely on our recruiting effort. As it turned out, the recruiting service soon came to the chilly realization that a large percentage of our so-called volunteers were volunteering to avoid being drafted into the Army, or as the result of an undefined awareness of some form of military obligation to the country. Our enlistment success was, in fact, draft-motivated. Without that motivation, success soon went out the window.

Not having foreseen that recruiting in an all-volunteer environment was a brand new ball game, the recruiting service was ill-prepared to cope with selling the Corps in a competitive market. The results were dramatic and bad. Recruiters were still held to quotas, but they had not been given the skills, tools, or management techniques necessary to survive in what can be best described as a sales world. On top of this, Marine Corps publicity was almost wholly dependent upon public service advertising. Only after a long struggle was money authorized for paid-print advertising to make the Corps competitive in the manpower market.

While "management" at district and Headquarters Marine Corps levels desperately sought to gear up for the new situation, the "salesmen,"

Life in the Marine Corps, Marine Corps recruiting booklet, ca. 1978.
Marine Corps Recruiting Collection (COLL/636), Archives Branch, Marine Corps History Division

like good Marines, did the best they could. They were mission-oriented. For each of them, the mission was numbers or quota. In retrospect, it is a marvel that the recruiters did as good a job as they in fact did. Many a marriage went on the rocks, and many a nervous stomach disorder developed, as recruiting sergeants expended superhuman efforts and man-killing hours trying to find "a few good men"—particularly when they could not promise a prospect a "rose garden." Nonetheless, given the lack of draft pressure, the unpreparedness of the recruiting service to compete in the recruiting market place and the propensity to enlist high school dropouts who scored acceptably on the trainability tests, it was inevitable that the quality of the recruits would drop. Add to this the pressure of monthly quotas, and it was equally inevitable that the percentage of recruiting errors and downright cheating would rise.

A final error in the recruiting equation resulted from another false assumption, that drill instructors were miracle workers and could make a Marine out of anyone. This belief was a convenient one for the recruiter who was having difficulty making quota. It was enhanced by a wartime Department of Defense requirement that called for the enlistment of 20 percent Mental Group IVs. The recruiter could rationalize that he should not stand in judgment over what seemed to be a poorly qualified prospect's potential for success in training. Making Marines,

Pvt Leroy D. Curry of Dayton, ND, scores his shot on prequalification day during week five of the nine-week training cycle at Edson Range, Camp Pendleton, CA, on 12 April 1978. His final qualification score was 190.
Defense Department photo (Marine Corps) 03510878, courtesy of MSgt C.H. McCormic

after all, was the job of the people at the recruit depots. Besides, if the prospect had any potential at all, the drill instructor could bring it out and make him a good Marine. Thus evolved the insidious principle of "when in doubt, ship him."

This combination of factors soon pumped far too many poor quality recruits into the recruit depots on either coast. The ingredients for mischief were present and working.

RECRUIT TRAINING

Marine Corps boot camp has always been a mysterious thing. Raw, callow youth, fresh from city, town, and farm by some almost magical feat are turned into smartly disciplined Marines capable of smiling in the face of death, and prepared to storm the "Halls of Montezuma and the Shores of Tripoli." All this is done in less than 90 days. The master craftsmen who turn out these spirited wonders are the legendary drill instructors [DIs]. But like the recruiters, fate conspired against the DIs in the early '70s. Seeds of later trouble were sown at the recruit depots.

Again, trouble stemmed from related causes. In the first instance, the DI was being overworked. He worked on a rigid and tight training schedule. To aggravate matters, the recruit training schedule had grown a bit like Topsy.[41] A new requirement added here and there seemed insignificant, but the cumulative effect was to put the drill instructor on a treadmill where he had to run just to keep from losing ground. On top of this, recruit platoons had grown to unmanageable size. Given the task facing the DI, it was an exceptional NCO, indeed, who could maintain the training pace and his composure at all times.

A primary cause for later anguish, however, was the introduction of the declining quality recruits. Overwork and a percentage of inferior quality material spelled trouble for the DI and the Corps. Being proud Marines, the DIs, again like the recruiters, were mission-oriented. If the goal was to turn out Marines, that is exactly what they would do, regardless of the incoming quality. The DI would force them into a mold, push and pull them through training, pop them out of the mold at the other end, and hope that they would have benefited from the experience and perform as acceptable Marines. In too many instances though, it was a case of trying to make something out of nothing. Anybody who served in the operating forces during the dark ages of

[41] Topsy was a trained circus elephant who made headlines in 1903. See Kat Eschner, "Topsy the Elephant Was a Victim of Her Captors, Not Thomas Edison," *Smithsonian*, 4 January 2017.

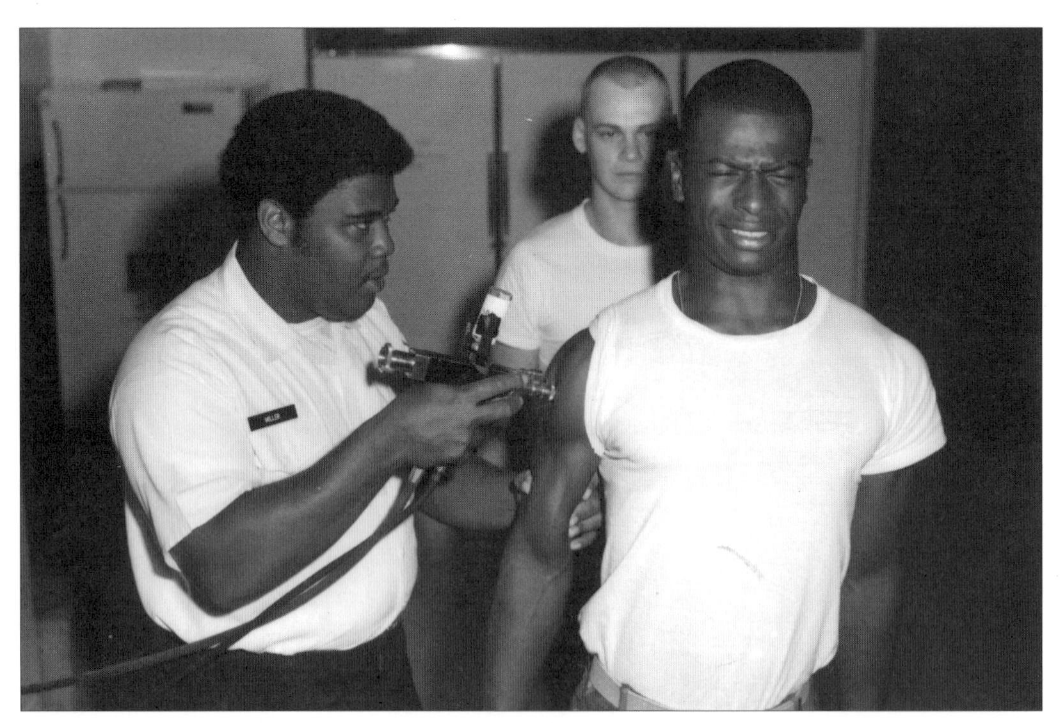

A recruit receives mandatory vaccination at Marine Corps Recruit Depot, Parris Island, SC.
Official U.S. Marine Corps photo 016033722

the early '70s knows that, far too frequently, it did not work. There were no miracle workers. The computer adage of "garbage in–garbage out" applied to the recruiting and recruit training processes. Many ill-adjusted, antisocial young men ended up in our ranks. They caused great damage to our Corps before the pendulum began its return swing to quality.

In the process of meeting a demanding schedule, trying to train recruits while coping with a percentage of misfits found in every platoon, the frustration factor for the DI began to tell. Certain improper practices began to creep onto the drill field in the name of discipline and motivation. At best, these could be described as petty harassment: the screaming, the gratuitous profanity, the order/counterorder. At worst, they can be defined as debasement, maltreatment, and abuse, to include the laying-on of the hands. The roots of these unhappy practices went well back into recruit training history, but for the most part, past incidents of real abuse had been aberrations. Unfortunately, in the '70s, the process became institutionalized. Supervisory safeguards against it, while present, proved inadequate. They were inadequate because they were geared to what could be considered a normal cycle of training with a normal body of recruits. The conditions of the '70s were neither.

By 1974, we hit our low point. The Fleet Marine Forces were complaining about the product given them by the recruit depots. Drill instructors were complaining about the poor quality of recruits they were receiving, and recruiters were complaining about quotas and the quality of the recruit market place. There was

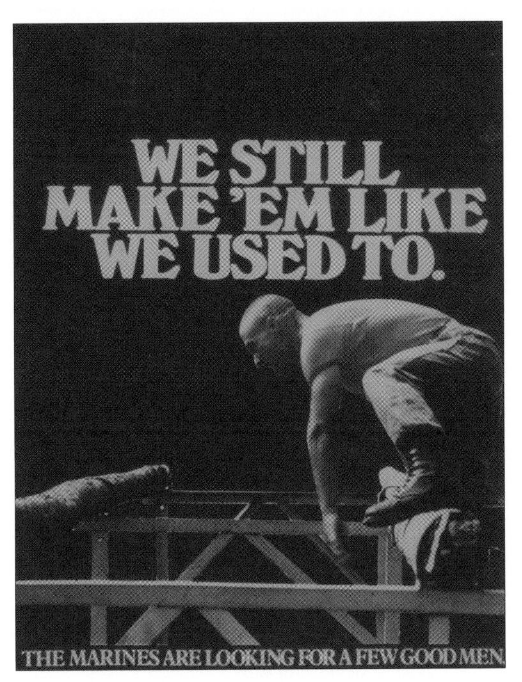

As part of the "Few Good Men" recruitment campaign, the Marine Corps issued a poster that assured potential recruits with the slogan: "We still make 'em like we used to."
Richard Spencer Papers (COLL/5233), Archives Branch, Marine Corps History Division

a certain degree of finger-pointing, as all Marines experienced the exasperation of a decline in quality in our Corps. Fortunately, a new day was about to dawn. Marine leadership faced the problem squarely and started actions in the recruiting service, at the recruit depots, and in the operating forces to set the situation straight.

Unfortunately, however, time had run out on the Corps. The public eye focused on the Marines shortly after the renaissance began, but before it became visible. One has only to look at media coverage of recruiting and recruit training abuse and the commentaries upon the disciplinary statistics within the Corps to appreciate how badly shaken public confidence became. It was an experience that scarred those involved, and one that no Marine should ever forget.

EMERGING FROM THE DARKNESS

The first and most essential step out of the darkness had to start with the recruiting service. Only by shutting off the alarmingly high input of inferior recruits could the Corps return to normal. The damage done by these poor performers cannot be overemphasized. Not only did they create problems at the recruit depots, but those who made it through boot camp created such trouble in the operating forces that many young Marines of admirable quality became disillusioned with the Corps. Reenlistment rates dropped. More than a few good Marines went UA [unauthorized absence] just to get out of the growing unwholesomeness of garrison life. The crisis of public confidence in the Corps that was later to emerge in 1976 was preceded by a crisis of spirit in the Corps itself. It permeated every level. Duty with the FMF had its appeal replaced with apprehension for officer and enlisted alike.

It was obvious, therefore, that action had to be taken to ensure that only quality recruits were enlisted. To do this, drastic action had to be taken in three areas:

- Criteria for enlistment
- Quality and training of recruiters
- Recruiting management and quality control measures

Upgrading in these areas was neither easy nor amenable to overnight achievement, if for no other reason than all were interrelated. For example, if the enlistment criteria were raised, it meant that the recruiting market was narrowed, which meant that it was more difficult to recruit, a fundamental problem in the original equation; that is, when quantity became hard to get, cheating became an attractive near-term solution. If this was to be avoided, it called for recruiters who could recruit effectively in a narrow market. In turn, those recruiters had to be trained to a level commensurate with the job. On the management level, it called for developing these resources and controlling them—no easy task. Nonetheless, the recruiting comeback was undertaken with a vengeance with actions in the three areas occurring unevenly, but concurrently.

A general officer was put in charge of recruiting. Recruiting guidebooks were developed. The selection process for recruiters was tightened up. The syllabus at recruiter school was revamped. Industry was searched for sales and market management techniques. Sales and self-confidence programs ranging from Dale Carnegie to Xerox courses were introduced in search of means to improve recruiter capabilities. Statistics, reports, and data of all sorts were analyzed to improve performance in what was to become a tight market, because it was to be a quality market. The Marine Corps' advertising program took on a new look with reemphasis on a "few good men." Enlistment criteria were reexamined, and standards for enlistments were raised. As already noted, the House Appropriations Committee had sought a mandatory high school percentage in 1974. By the end of the fiscal year, the Marine Corps had no doubt about the efficacy of that congressional move, but more importantly, it had also become convinced that quality recruiting goals could be achieved and maintained by skilled Marine recruiters in the all-volunteer environment. Thereafter, the requirement for high school graduates and the qualifying scores on the enlistment test batteries were incrementally raised.

As standards were raised and the ability of the recruiters improved, various recruiting systems began to emerge. These were the results of trial and error and adaptation from business and sales-world expertise. While each varied in one or more respects, there were certain common elements, which were in large measure designed to make maximum use of minimum resources. The systems permitted the recruiter to analyze his market and to program his time and activities profitably. It showed him how to make use of natural recruiting vehicles within his locale (e.g., radio, TV, newspapers, patriotic and Marine-affiliated organizations, etc.). It prompted him to establish required standards of performance to achieve specific goals. For example, in using the telephone, the recruiter could formulate how many calls he had to make to get the requisite number of appointments to achieve the necessary number of applicants to result in the number of enlistees desired.

The drill instructor has been called "a Marine's Marine." The DI instructs, guides, and molds recruits into basic Marines, where discipline is the difference between success and failure, whether on the battlefield or in garrison.
Richard Spencer Papers (COLL/5233), Archives Branch, Marine Corps History Division

The disorganization and the sense of futility that plagued the recruiter at the onset of the all-volunteer era and frequently drove the weak recruiter to recruiting malpractice slowly disappeared. What began to emerge was a truly professional recruiting force which, given the training and assets, could do the job regardless of a narrowing market occasioned by tougher enlistment standards and fluctuating unemployment.

There were many setbacks in the process. It is a truism within the recruiting service that recruiters will recruit to the lowest common denominator. This tendency required that strict and stern controls be implemented and enforced to ensure that recruiters were adhering to the quality standards set down by the Commandant. Corollary to this is the fact that a recruiter's productivity will suffer whenever a "change" takes place. This is an interesting phenomenon that, while purely psychological, is nevertheless real in its impact. So it was that every time criteria changes were made, recruiter performance would go into temporary decline. At times, also, during the get-well period, other events, not psychological, put a real crimp on recruiting. Two of the most damaging were the Roth Amendment at the beginning of 1975, which put a freeze on travel funds, sorely restricting the mobility of recruiters, and a congressionally directed reduction in recruiters and recruit advertising funds in the spring of 1976.

It is ironic that the Congress, which has been so concerned about the quality of Marine Corps recruiting, has the propensity to occasionally make that job all the more difficult. The cuts have been substantially restored for FY 1978, although the desired monies for recruit advertising are still below adequate levels.

As recruiting was revitalized, it was often a case of two steps forward, one step back. But it became apparent that progress was being made. Quotas were being met and, more importantly, quality was on the rise. During this ongoing revitalization period when it was still uncertain whether the recruiting service was going to achieve both quality and quantity, the Commandant made it clear to all involved in the recruiting business that quotas were to be considered goals but that quality was a requirement. The message of quality over quantity came through loudly and clearly.

In 1975, the Commandant made another decision, which was to enhance recruiting management, but more importantly, assured a quantum jump in quality control. Effective in June 1976, operational control for recruiting shifted from Headquarters Marine Corps to the commanding generals of the two recruit depots: San Diego and Parris Island. Henceforth, recruiting west of the Mississippi (8th, 9th, and 12th Marine Corps Districts) with roughly 50 percent of the annual quota, became the responsibility of [the] CG [commanding general], MCRD San Diego, while the other half from the eastern United States (1st, 4th, and 6th Districts) became Parris Island's responsibility. In effect, the Commandant charged the two depot commanders with sole responsibility for providing the operating forces with basic Marines in requisite numbers and of requisite quality. If the operating forces were not happy with the product, one did not have to look far to fix blame. If the field was happy with what they were getting, credit could also be quickly acknowledged. This action, together with judicious use of the expeditious discharge program throughout the Corps, crowned with success the long, tough fight along the comeback trail. The bad blood was being pumped out and new, good blood was being pumped in.

By the end of calendar year 1976, it was clear that the Corps was well along the way in its quest for quality. All the indices of performance reflected improved health. Recruit attrition was down; physical and mental categories and high school graduates were up. In the field, expeditious discharges began tapering off. UA, desertion, and crime went down dramatically. A sense of spiritual rejuvenation was evident throughout the Corps.

In spite of clear improvement in recruiting, the task of getting good recruits remained difficult. The shocking fact was that only two out of every five young men within the target population of 17–21 years of age met the qualifications for service. If a recruiter was to be successful he had to work hard for those two in competition with the other Services, the private and public sectors of the economy, all of whom were looking for the same caliber person. The task proved manageable, however, in proportion to the skills, organization, and determination of the recruiter and the support he received from the recruiting hierarchy.

As indicated earlier, a recruiter has to approach his task systematically to achieve success. As high school graduates prove to be the best bet for successful service, it is obvious that the successful recruiter targets this population and places high value on lists of high school seniors and graduates within his area. This allows him to concentrate his efforts on the quality market. Armed with such lists, the recruiter systematically attempts to contact and gain an appointment with as many prospects as he can. In this endeavor, he will use the mail, telephone, home visits, personal referrals, and high school visits to make initial contact. The key axiom in this process is the golden rule of the sales world: activity equals productivity, which translates as the more you do, the greater your success.

To understand the dimensions of activity involved in recruiting just one qualified enlistee, the following statistics are helpful. The recruiter, using the means cited above, must generate sufficient activity to contact 181 potential enlistees in order to secure 18 appointments to realize a show rate of 9 of whom 4 will actually apply for enlistment, 2 of whom will qualify for enlistment yielding the 1 who will actually enlist. This effort means 108 working hours for the recruiter and a cost of approximately $1,500 per recruit.

It might be appropriate to note at this point that the recruiting service, like its business world counterparts, is constantly striving to increase productivity while lowering costs. Aside from internal management improvements, one valuable technique to recruit economically is to make use of new enlistees to generate help in the recruiting business by providing referrals to the recruiter. This is done while the enlistee is waiting to ship to boot camp and again when he is home on leave. Command recruiting is also vital and cost-effective in the recruiting effort, although it has not yet been exploited to its full potential. Consider this: 192,000 Marines take 30 days annual leave, which equates to 5,760,000 man-days of exclusive Marine contact with the civilian community. If only 1 percent of those on leave subscribed to the adage "every Marine gets a Marine," the Corps would realize a free recruiting gain of almost 2,000. A fledgling pilot program conducted this past year actually demonstrated that, with appropriate command attention and emphasis, the command recruiting rate can be in excess of 2 percent. Corps-wide this means that we have a built-in capability to achieve 10 percent of 1978's recruiting goals by use of fellow Marines on leave. The message to commanders should be obvious:

New arrivals at Parris Island, SC, await classroom instruction on their first night of basic training.
Richard Spencer Papers (COLL/5233), Archives Branch, Marine Corps History Division

Marines can play a powerful part in regenerating their Marine Corps.

As the consolidation of recruiting and recruit training under the depot commanders went forward, quality control measures were being refined at each recruiting level from the substation up. The underlying philosophy of quality control was in keeping with the "whole man" concept and helped to provide an answer to the question: "Will this man make a good Marine?" In other words, the prospective enlistee not only had to meet the stiffening standards for enlistment as they existed in regulations, but also had to pass muster as the type of man mature Marine NCO's and officers on recruiting duty wanted to see in the Corps. This meant delving into the background of the man to include work record, character, and reputation.[42] Frequently, this was difficult to do because of community institutions' misconception of the intent of the Privacy Act. Nonetheless, if we were to ensure quality, the recruiter had to do his best to overcome local obstacles so he could look at the "whole man" and learn as much about him as possible before he enlisted him. Where feasible, this evaluation process was extended to include interviews by an officer or senior NCO at the recruiting-station level.

To ensure that recruiters did not take the requirement for quality in the "whole man" lightly, the recruiter's social security number was included on the enlistment papers of each man he brought into the Corps. Thereafter, throughout a man's first enlistment, we had immediate knowledge of who enlisted him. The recruiter's SSN [Social Security number] became, in effect, a personal stamp of approval and attestation to the quality he was putting into the Corps. Additional quality control measures were introduced to ensure that recruiters who were doing a good job could be quickly identified and recognized. Needless to say, it also illuminated the marginal or poor recruiter and immediately identified any recruiter who was foolish enough to cut corners. Some of these measures included a printout comparing the success of a local recruiter, in both quality and quantity, with his Army, Navy, and Air Force competitors. Another reflected the record of a recruiter's input at the depots and identified his attrition and the reasons therefof.

Other quality control programs identified variances in such things as enlistment test scores, numbers of dependents, and educational background. Unusual or unexplained variances could

[42] The whole man concept has been particularly valuable in assessing the potential for success of the 25 percent non-high school graduate enlistees. Such an applicant needs a minimum of 10 years of school, higher test scores than the high school graduate, and demonstrable potential for achievement.

then be double-checked. With the establishment of an assistant chief of staff for recruiting at each depot, any aberration, administrative or otherwise, could be spotted and checked out almost as soon as a recruit arrived for training. The system identified the good recruiter and the poor recruiter and permitted each to enjoy or regret the consequences of his actions. The "when in doubt, ship" syndrome became a thing of the past.

The Armed Forces Entrance and Examining Stations (AFEES), through which all Service enlistees process, also have contributed to improved quality by tightening their procedures thereby screening out the unqualified.[43] Various versions of the basic entrance test have been introduced for random use in testing enlistees. Identification procedures for the test applicant have also been made stricter. This minimizes the likelihood of test cheating and use of substitutes on tests. Likewise, a new system called exclusive jurisdiction limits an applicant to a single AFEES, thus preventing an unqualified applicant from shopping around from AFEES to AFEES until he found one where he could beat the system and enlist. A third improvement has done much to counteract the difficulty in obtaining background information as the result of the Privacy Act. This has been the introduction of trained interviewers to screen the candidate in a one-on-one interview during the process of initiating a National Agency Check.[44]

All of these quality control measures are paying off. Like any system, however, somebody who has the determination to do so can beat it. But the fact remains, it is tough to beat this system and, more importantly, if somebody does beat it, the chances of being caught are high indeed. In short, the present quality control procedures keep the honest, honest and go far in ensuring that only quality enlistees join our ranks.

At this writing, Marine Corps recruiting in the all-volunteer environment is working. The goals for FY 1977 have been met. We move into FY 1978 with confidence that we can get our share of the quality market. But make no mistake, it is hard work. It is also resource-sensitive, in that we can meet whatever goals are set for quality and quantity, but only as long as we have a realistic level of assets to do the job.

It is no secret, however, that as the national demographic trend turns downward, recruiting will become increasingly difficult. Theoretically, at least, we can reach a point in the mid-1980s where the market will no longer produce results as a function of assets employed. It is too early to determine how we will fare as we move toward the low demographic era. But in the meantime, by ensuring a quality input and by ensuring an interesting and challenging lifestyle for our new Marines, we will reduce attrition and increase first-term reenlistments. The two achievements, in turn, will lower the manpower demands and make the recruiter's task both manageable and attainable.

REFORMS ON THE DRILL FIELD

In 1974, as the improvements were getting underway in recruiting, it became apparent at the two recruit depots that major readjustments in recruit training were also in order. It was obvious that it was going to take awhile before the effects of higher recruiting standards were felt on the drill field. In the meantime, however, steps

[43] In January 1982, AFEES became the Military Enlistment Processing Station (MEPS).

[44] National Agency Check (NAC) is the minimum investigative requirement for final clearance up to secret and for interim clearance up to top secret for certain categories of personnel. A NAC is also an integral part of a background investigation.

A platoon marches in formation during basic training at Marine Corps Recruit Depot Parris Island, SC, 23 November 1979.
Official U.S. Marine Corps photo 016354179

had to be taken at the depots to reverse the soaring attrition rates and to minimize the potential for abuse that, in large part, were by-products of low quality recruit input. Positive leadership and command attention were the two most readily available prescriptions for action. Both depots tightened supervision and reemphasized positive leadership at all levels.

However, a formidable, unique, and most unusual problem complicating this prescription came onto the drill field: attitude. As a group, drill instructors believed that given their rigid schedule and the questionable caliber of some of their recruits, a heavy-handed, high-stress approach was the only way to ensure that the system produced a good Marine. The DIs, as good and as dedicated as they were, failed to see that some of the practices that had crept onto the drill field were not only counterproductive, but also dangerous to the well-being of the Corps.

Too many potentially good Marines were being turned off by the periodic absurdities that they saw in training and ended up being discharged in the process. On the other hand, many who should have been sent home made it through training. By looking at the disciplinary problems plaguing the Corps, it was obvious that a great number who did graduate from boot camp were not necessarily the stellar leathernecks their DIs thought them to be. The robot-like response to authority on the drill field did not extend beyond the main gate. Conformity was being confused with discipline and respect for authority.

Unfortunately, the DI's conviction as to how good Marines were made was reinforced by a perception that the practices followed on the field were time-honored and an integral part of "old Corps" success. Translated, this meant that the DI perceived that what he was practicing was practiced successfully on him as a recruit. In some cases this was true, unfortunately. But for the most part, it was myth caused by recollections colored by the passage of time. Prompted by an absolute conviction that they were the only ones who knew how to make Marines, the DIs as a group resisted, and in some measure, defied the early actions aimed at eliminating the artificial stress and downright foolishness that had become part of the recruit training process.

Drill instructor attitudes, therefore, constituted a distinct obstacle to the reforms that were getting underway. The problem was exacerbated by the fact that a lot of Marines throughout the Corps shared the views of the DIs. To this collective group, boot camp was viewed as an initiation rite rather than as a training, testing, and development process.

Aside from ensuring an input of better quality recruits, three major tasks had to be initially accomplished, if corrective action was to be successful at the recruit depots:

- Reduce the institutional potential for abuse
- Provide for adequate supervision and enforcement of command policy
- Change the attitude of drill instructors

The first essential step in reducing the potential for abuse was to reduce some of the pressure on the DI and on the recruits. The DI was operating in a tense, shrill environment. Everything and everybody connected with the drill field was wound tight. If there was any humor at Parris Island and San Diego, it was well hidden. Everyone went about his tasks tensely and grim of visage.

As for the recruits, it was an atmosphere of fear compounded by terror from the moment they stepped off the bus at the depots. The cultural shock of passing from permissive middle-class America into a highly structured and demanding military environment is traumatic in itself. But this was not considered anywhere near sufficient stress by many DIs. So screaming, shouting, and institutional hysteria enveloped the Marine aspirants from the moment of their arrival.

Common sense and our own personal experiences tell us that one cannot learn in a condition of stark fear. Yet that was the situation into which we thrust the new recruits. Then when they reacted poorly, indignity was heaped upon them, which made the condition worse, which in turn invited more indignity ad nauseam. All this was supposedly designed to condition the recruit to the stress of combat, instill in him immediate response to orders (often contradictory), and to cull out the weak. In fact, it was asinine on all counts, totally contrary to the philosophy and principles of traditional Marine Corps leadership in combat and in garrison.

As bad as this business was for the recruit, it had an even worse impact on the DI. There was no such thing as a relaxed moment. He had to keep the recruits under strain at all times. In the process, he succeeded in winding himself tighter and tighter. This situation was compounded by the length and busyness of the DI's day, which permitted virtually no breathing room or flexibility. It was full speed ahead from 0500 to 2100

Arthur J. Barbour, *USMC Instructor on Firing Line*, ink wash.
Art Collection, U.S. Navy

for 11 weeks. Woe unto the recruit who accidentally or deliberately disturbed the headlong rush of events, for he was the stuff of which maltreatment was made.

A series of steps were taken, over time, to reduce the surrealistic pressure cooker aspect of training while retaining that which has been time-tested and proven good in the recruit training process. Training was to remain personally demanding, but it was to be conducted with firmness, fairness and dignity and, when necessary, with compassion. The traditional leadership philosophy of the Marine Corps was to be reaffirmed: teacher to student, father to son. The recruit would have to give 100 percent of himself daily in his quest to be a Marine. The drill instructor, however, was there to help him in this quest, not to harass him.

One of the early steps taken to ease the pressure on the DIs was to reduce the number of recruits they had to train. Platoons that had been running as high as 90 men were limited to a more manageable 75. Concomitant was a decision to eliminate the artificial stress; that is, the screaming, shouting, and exaggerated convict-like conformity being enforced between reveille and taps. A buddy system within platoons was also introduced so that the slow could learn from the fast, thus easing some of the DI's burden. Recruit leaders were designated, through whom the DI could work, to assist in the accomplishment of simple and mundane tasks. This also tended to identify recruits with leadership potential. Recruits were allowed to talk in the mess halls. One hour of free time was allotted to recruits at the end of the day. Limited to his squad bay, the recruit was permitted during this period to use his own time constructively. Letters could be written, equipment prepared for the next day, instructional TV could be viewed, etc.

Although the reduction in platoon size was universally endorsed, the DI's reaction to these other changes varied. Many applauded what they viewed as a return to reality. Others viewed it as the beginning of the end for the Marine Corps. The majority, however, suspended judgment and did what they were told to do.

In March 1976, Parris Island hosted a recruit training conference with representatives from San Diego and Headquarters Marine Corps attending. There was a free exchange of views to include those of experienced DIs and sergeants major aimed at validating what had been done thus far and to set the course for the future. It was clear at this conference that additional changes were needed. Among other things, the conferees recognized that the DI was still over-committed in his daily tasks and, in large measure, was still on a treadmill.

As a result of the conference, some 68 hours of less-essential training were removed from the program of instruction. While reveille and taps would still go at 0500 and 2100, respectively, scheduled training was set for 0700–1700, six days a week. A significant increase in commander's time was thus made available. This gave both

Recruits stand at attention at Marine Corps Recruit Depot San Diego, CA, in 1983.
Defense Department photo (Marine Corps) 017109883, Marine Corps History Division

drill instructors and series commanders breathing room to accomplish the many nonscheduled tasks attendant to running a unit. It also provided flexibility in adjusting to unforeseen schedule changes and like vagaries. Sundays, for the most part, were to be reserved for divine services in the morning and organized athletics in the afternoon.

Another major step resulting from the conference was the addition of a second officer to the series team, which had the actual job of training recruits. Until then, one lieutenant commanded a series of four recruit platoons. He, like the DIs, was clearly overworked. The addition of an assistant series commander to the team eased the burden of the series commander and allowed him to exercise command over the training of the platoons.

The additional officer was also of value to the drill instructor. The assistant series commander could handle much of the administrative and coordination responsibilities of training. Likewise, problem recruits need not plague the DI. They could be more readily referred to the officers for counseling or other action deemed appropriate. The additional officer, therefore, signaled a reemphasis of the officer's role in the recruit training process. The business of making a Marine became less the personal preserve of the DI and more the responsibility of the series team under the positive command of an experienced officer.[45]

The addition of an executive officer at company level likewise enhanced training, for it enabled company commanders to free themselves

[45] The current series team consists of a series commander, assistant series commander, a series gunnery sergeant, a senior DI, and the two DIs for each of the four platoons in a series.

from their desks to take a more active command role on the company level. This increase in officers at the company and series level will probably prove to be the greatest single assurance over the long haul for continued firm, fair, and dignified treatment of the recruit. It provides for clear, adequate, and positive command at the working level.

A variety of other changes and reforms took place as the result of the conference. All of them were designed to eliminate the hysterical aspects of recruit training and stress positive leadership based upon the fact that incoming recruits were volunteers to our ranks who genuinely wanted to be Marines. They would be given the opportunity to become Marines with all the help necessary; but if they were not willing or able to measure up, they would be discharged. This, however, could and would be done without a lot of superfluous nonsense. The days of trying to make a Marine out of just anybody through force, fear, and humiliation were over. Gone also were the various "motivation" techniques, which may have motivated a recruit to get out of boot camp as quickly as possible, but did not necessarily motivate him to be a good and proud Marine.

The elimination of the negative aspects of recruit training was not universally welcomed by the DIs. The concept of positive leadership and the development of self-discipline as both the means and the end of recruit training were viewed in some cases with suspicion and submerged hostility. Some continued to believe that high stress and abusively imposed discipline were the only ways to make a Marine. Those who felt that way were given a single opportunity to leave the drill field without prejudice. After that, they would be expected to carry out their orders to the letter.

To add unequivocal emphasis to the permanency of the changes taking place, the Commandant made it clear that he was the senior DI in the Marine Corps. If any Marine, officer or enlisted, could not adjust to and support the changes taking place, it was time he looked outside of the Corps for gainful employment. The Commandant further expressed his personal command over the recruit training process by directing that the two recruit depots become mirror images, excepting those circumstances occasioned by geographic, facility, or acute climatological differences. In line with this order, the two recruit depots developed common SOPs [standard operating procedures], POIs [programs of instruction] for recruits, a DI School, and a Series Officer School. Near mirror T/Os were developed.

The Commandant also directed maximum supervision of the recruit training process by depot officers; a capability that was greatly enhanced by the doubling of series officers, the addition of company executive officers, and by the establishment of a billet for an assistant depot commander incident to the recruiting-recruit training consolidation. In addition, the Commandant directed that individual, confidential counseling sessions with recruits be conducted by a series officer to identify unspoken recruit problems at home or in training, including any perceived maltreatment. These sessions also served to apprise the recruit of his performance in training and to motivate him to improve that performance.

It should be remembered that, while the recruiting and recruit training improvements were taking hold, the Marine Corps was under heavy fire from the news media, the Congress, and vocal elements of the public. The [Private Lynn] McClure case in San Diego and the [Private Harry W.] Hiscock incident at Parris Island had become the embodiment of the problems that had come to plague the Corps during the dark period of the early '70s.[46] But, it is at least comforting to know that the improvements in both recruiting and recruit training were initiated by the Marine Corps prior to the public outcry about the legacy of past errors. The problem, therefore, was one of convincing an outraged public that we were truthful when we said that the weaknesses had already been spotted and the necessary corrective action had been initiated on both fronts.

The year that followed the drill field reforms was an incubation period. The DIs adjusted to the changes and carried out their orders, but for months people appeared to be working with extreme caution. There was great anxiety over doing something wrong no matter how innocent of intent. This uncertainty was in large measure due to the increased supervision and the many detailed proscriptions contained in the revised recruit training SOP, which governed every facet of recruit and drill instructor existence. Initially, morale on the field took a nosedive. But during the summer of 1976, the irrepressible optimism of the Marine NCO began to assert itself. Happily, at that point, the improved quality of the recruit also began to show clearly.

There were many statistical and intuitive indicators of the rise in quality. The newly arriving recruits were brighter and in better physical shape. Their attitude was one of eagerness. They displayed more common sense. Even the dental clinic reported evidence of better preservice dental care, from a dentist's point of view, a sure sign of quality. As quality improved, so did morale.

A crop of Marine recruits arrives at the Recruit Receiving Section at Marine Corps Recruit Depot Parris Island, SC, 22 October 1961.
Richard Spencer Papers (COLL/5233), Archives Branch, Marine Corps History Division

Coincidentally, some of the dire predictions that reduced discipline would result from change were proven false. For example, one simple change in the new regimen permitted recruits, on the Sunday before graduation, to have base liberty in service uniform until 1800. The Cassandras predicted all manner of dire consequences attendant to unsupervised recruits wandering around the depots.[47] They were proven wrong. The recruits wore their uniforms proudly. For the most part, they savored their mini-independence in the company of their equally proud families who had come for graduation.

As the year progressed, no indications of a decline in the equality of the training or of the product of the training surfaced. On the con-

[46] Everett R. Holles, "Marines Continue Abuse of Recruits," *New York Times*, 18 July 1976; and James P. Sterba, "Marine Recruit Abuse Continues," *New York Times*, 7 March 1976.

[47] The term *Cassandra* refers to Greek mythology, where the daughter of Priam was endowed with the gift of prophecy but fated never to be believed, or a harbinger of misfortune and disaster.

trary, all the local indicators were positive. Attrition was down. So were disciplinary problems. Scores in academics went up as did rifle range qualifications. Medical problems in the three bugaboos of recruit training—stress fractures, cellulitis, and heat casualties—took a decided drop. Post-recruit training behavior and performance were also encouraging. Close monitoring of field commands indicated increased satisfaction with the product received from the depots.

During the year, one additional event took place, which also marked a significant milestone in the evolution within recruit training. This was the introduction of transition training. It was a well-known fact that a recruit experienced cultural shock upon entry into the Service. It was a new and strangely structured life. It took some getting used to. What was not fully appreciated was the fact that a graduating recruit also experienced cultural shock when he left training for his new duty station. After 11 weeks of rigid regimentation and supervision over most of his waking hours, he had to readjust mentally to the less-structured existence of normal garrison and field life. In short, he had to cope with being his own master in a communal living setting where there was no DI to govern his every move.

To prepare recruits for this transition, a test program was introduced to reduce drill instructor supervision during the nontraining hours of the last week of boot camp. Recruits holding supervisory billets were given responsibility for unit movement and activities. The individual recruit was responsible for planning and using the remainder of nontraining time in the barracks. In the evening of the last week of boot camp, the DI virtually disappeared from the scene by going into what can best be described as a duty NCO status. The magnitude of this step, in the context of recruit training, can be seen in the fact that unit and individual preparation for the final and all important command inspection was left entirely up to the individual recruit and the recruit billet holders. The test program was a huge success. The recruits cooperated with each other and respected the authority of their peer leaders who, in turn, exercised their authority with a mature sense of responsibility. Recruit performance was exemplary. There was a clear determination not to betray a trust, not to let the DIs down by anything less than perfect performance. The test program was validated and finally adopted by both depots.

While this program did much to better prepare a recruit for life beyond the depot, it is particularly noteworthy for another reason. Until it was instituted, all other changes in recruit training emanated from the top. The idea of transition training originated on the drill field. It was supported by the DIs and battalion officers before it ever reached the command level as a recommendation. In short, it was a sign that the drill field community was accepting the fact that recruit training procedures were not set forth in holy writ, that improvements could, in fact, be made.

As the trauma of change faded into the past and life on the drill field stabilized, the two depot commanders undertook another important project. They constituted a joint task force headed by the assistant depot commander at San Diego to evaluate the entire recruit training process. The task force's objectives were to find out what the field wanted the depots to produce by way of a basic Marine, to research the latest techniques and aides to training, and to come up with a validated program of training, which would be both responsive to the field and accomplished in the most effective manner.

This unprecedented project took the task

force members throughout the Marine Corps to commands of every size and variety. Marines at every level were systematically questioned and interviewed. In all, there were almost 4,000 questionnaires administered and 450 in-depth interviews conducted. The task force also included in its labors discussion at the Marine Corps Development and Education Command concerning the future environment and requirements of the Corps in the context of current projections. In addition, use was made of studies and projections of sister Services on a wide range of topics, which might have a bearing on the Corps' recruits and their training. Teams of the task force also visited major commands and recruit training centers of the other Services to study their concepts, techniques, and hardware. When the strings were all pulled together, both commanding generals sat down with the team to examine the results and to develop a recommended course of action for future training.

In this classic recruiting poster, the Marine Corps suggests that the Service is an exclusive club that only a select few can enter. *Special Duty . . . Can You Measure Up?*, Marine Corps Recruiting poster
Art Collection, National Museum of the Marine Corps

It was clear from the task force's findings that the current recruit training program was almost fully responsive to the needs of the field. No dramatic changes or departures in philosophy or content appeared necessary. In general, the operating forces were getting what they wanted from the recruit depots: a new Marine who was self-reliant, responsible, physically fit, proud, and competent in basic military skills. Generic requirements, which did emerge from the task force work, were broken down into specifics and matched against a matrix of training activities and techniques to finely tune the existing program for both efficiency and effectiveness. The report of the task force was then presented at the General Officers Symposium in July 1977. The next step in the days to come is to translate the recommendations of the task force into specific packages of instruction to test and validate the revised elements.

With the completion of the task force's effort, the story of the Personnel Campaign 1973–77 comes to a logical conclusion. It might be useful to sum up exactly where we stand at this juncture.

For recruiter and trainer alike, the bitter memories of the past are ever-present. So they must always remain, if we are to prevent a repetition of past misfortunes. Institutionalization of safeguards against backsliding in either recruiting or recruit training have been established. They should work as well in the future as they appear to be working now. With this in mind, recruiting and recruit training are in good health.

On the recruiting side, a skilled sales force is hard at work using the most advanced advertising, sales, and management techniques to get the "few good men" the Corps needs. Monitoring the effort is a quality control system designed to identify and filter out the unqualified. The market is small. Continued attainment of 75 percent high school graduates and 25 percent high-caliber nonhigh school graduates during the coming year will challenge the recruiter. The techniques that have evolved in the past few years, however, are geared to meet this challenge. It can be met.

On the drill field, the process of making a young man into a Marine remains impressive. At the two recruit depots, the incoming recruit is indoctrinated to Service life, taught basic skills, and made physically fit. But these achievements the Marine Corps shares in common with the other Services. Where the difference comes is in the spirit. Marine Corps recruit training is extremely tough, challenging, and demanding.

While each recruit is treated with the respect due as a human being and citizen, he must work hard to earn the title of Marine. Each day, he must excel. Each day, he must accomplish that which he felt himself incapable of the day before. Each day, demands are made of him, which in their doing, add a level of self-confidence and pride. He grows stronger with a sense of invincibility based not upon bravado but upon real achievement. The imposed discipline of his novitiate days are replaced by the end of training with self-discipline and a sense of loyalty to his fellow Marines. He knows he has earned his emblem and would rather die than let the Corps down.

Those who are intimately involved in the recruiting and training process know that a spirited, self-disciplined, physically fit, and highly motivated Marine is being shipped to the operating forces. What happens when he leaves the main gate is a continuing function of leadership at all levels. If he joins a well-led unit, his performance will be superb. If he joins a poorly led one, he may become disillusioned and view his achievement as a mockery and the Corps as a sham. His performance and the future well-being of the Corps, therefore, depends, as it always has, on the leadership he receives as the newest member of a very old Corps.

FRANK E. PETERSEN JR. BIOGRAPHY

"Once I found out what being a United States Marine was all about, jumping into the tiger's jaw was just something to do. We'd been trained for combat. That's our reason for being. When the time comes, hell, stick out your can. Let's go. Let's see what that old tiger's got. Let's jump right into his big, old jaw."

~Lieutenant General Frank E. Petersen Jr.
Into the Tiger's Jaws

Lieutenant General Frank E. Petersen was born in 1932, and he enlisted in the Navy in 1950. Inspired by the death of Navy Ensign Jesse L. Brown, the first African American naval aviator, in Korea, Petersen entered the Naval Aviation Cadet Program in 1951 and was commissioned in the Marine Corps in 1952. He was the first African American Marine aviator.

He flew Vought F4U Corsairs with Marine Fighter Squadron 212 (VMF-212) during the Korean War and F4 Phantoms as the commanding officer of Marine Fighter Attack Squadron 314 (VMFA-314) during the Vietnam War. He was shot down by ground fire in 1968 but quickly returned to duty. He received numerous awards, including the Legion of Merit with combat "V", the Distinguished Flying Cross, and Purple Heart.

When he took command of VMFA-314 in 1968, Petersen was the first African American to command a Marine squadron. He reflected on his philosophy of command in his autobiography:

I was not unaware of my new role as the commander of a Marine fighter squadron. Of first import, of course, was the squadron itself. It was a live entity, overpoweringly energized. It was equipment, skills, goals, assignments, personalities—all to be melded into a writhing, searing thing with a constant edge that could slash and cut as the ultimate arm that accomplished an assigned mission, whatever the call. It was noise, confusion, smoke, sorted into the attainment of air superiority and support of troops in desperate firefights down there on the ground or the accomplishment of any other kind of missions that Marine aviators may be asked to perform. And all of it was ultimately my call. As commander, I set the tone. How I acted and the rules I decreed help define how all of it would jell so that we'd be a living, effective part of the air group. A kind of pinnacle had been reached, one I consider a definite correct step in an overall career as a Marine Corps officer and aviator.

I knew that I would be running a tight

LtGen Frank E. Petersen commanded the Marine Corps Combat Development Command at Quantico, VA, the "Crossroads of the Corps."
Official U.S. Marine Corps photo, courtesy of Sgt M. G. Lindee

Then-2dLt Petersen climbs from his Corsair fighter bomber at a base of the 1st Marine Aircraft Wing in Korea, 19 April 1953.
Defense Department photo (Marine Corps) A347177, courtesy of SSgt Slatto

ship. One reason was obvious. The first black commander of a Marine fighter squadron had better run a tight ship. I shouldn't have had to think this way, but I did. I felt that I always had to look over my shoulder, to be certain that something or someone wasn't creeping up to snatch what I'd attained. In the final analysis, though, the color of my skin and my concern about what was about to happen "over my shoulder" had to be shelved. I was the commander of a Marine fighter squadron flying 20 fighter airplanes that were technologically light years away from those in which I first began flying. I was the kind of skipper who would never ask subordinates to do something I wouldn't, so early on, I resolved to set the example by flying missions, leading my men in combat, on regular strikes and pulling my time on hot pad as well.[48]

After the Vietnam War, Lieutenant General Petersen commanded a Marine Aircraft Group, a Marine Amphibious Brigade, and a Marine Aircraft Wing. When he retired in 1988, he was commanding general, Marine Corps Combat Development Command. He was the "Silver Hawk" of Marine aviation and the "Grey Eagle" of naval aviation as the senior designated aviator.[49] His designation as an aviator preceded all

[48] Petersen and Phelps, *Into the Tiger's Jaw*, 156–57.

[49] The terms *Silver Hawk* and *Grey Eagle* refer to the Marine and naval aviator, respectively, whose date of designation as an aviator predates that of all other aviators on active duty in the respective Services. Marine Aviators are also considered naval aviators.

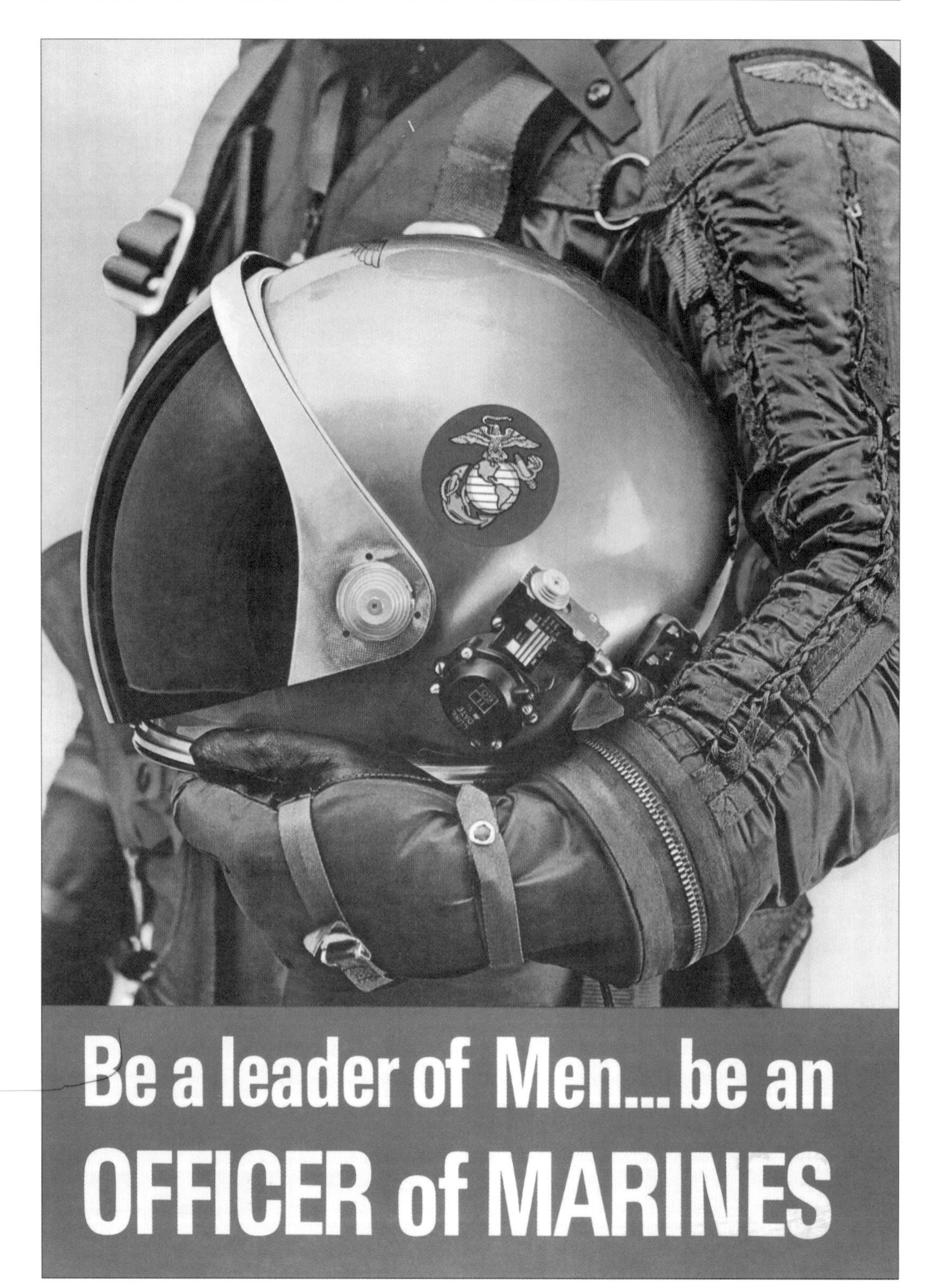

This aviation-focused recruitment poster, *Be a Leader of Men . . . Be an Officer of Marines*, encouraged recruits to consider becoming an officer.
Official U.S. Marine Recruiting poster

aviators in the Army and Air Force then as well.

A true pathbreaker, Lieutenant General Petersen's career as a pilot progressed from piston-engine, propeller-driven aircraft to jet-propelled, supersonic aircraft. As a leader, he eased the transformation of the Marine Corps into an integrated force that maintained its elite status and preserved the legacy of the Corps.

22d MAU Assault by LtCol Albert Michael Leahy. Marines land just south of Pearls Airport, Grenada, during Operation Urgent Fury, the 1983 invasion of Grenada.
Art Collection, National Museum of the Marine Corps

CHAPTER FOUR

The Operational and Educational Renaissance

by Paul Westermeyer

In the 1980s, the Marine Corps continued the process of reinventing itself begun in the previous decade, bringing new weapons systems and doctrines into operation while continuing to demonstrate the Corps' continued relevance to the national command authority. The Marine Air-Ground Task Force (MAGTF) was fully integrated into American strategic thinking, while Marine Expeditionary Units serving in the Pacific and the Mediterranean aboard Amphibious Ready Groups served as a strategic reserve that the president could call on quickly to respond to various types of international crises. MAGTFs were called upon to intervene militarily in Panama and Grenada, participated in numerous humanitarian missions, and acted as peacekeepers in Beirut and Somalia. They proved particularly adept at one of the Corps' oldest responsibilities: protecting Americans caught overseas during a disaster.

The Goldwater-Nichols Act of 1986 combined the warfighting forces of the United States more closely than ever before, creating new theater combatant commands alongside the traditional European and Pacific theater commands.[1]

[1] Goldwater-Nichols Department of Defense Reorganization Act of 1986, H.R. 3662, 99th Cong. (1986).

At the same time, increased emphasis on the dangers of Soviet aggression against the North Atlantic Treaty Organization under President Ronald W. Reagan pushed the Corps to training in mountain and arctic environments and joint training with the Norwegians. The creation of the Maritime Prepositioning Program expanded the Corps' ability to project power globally more quickly than ever before.

In 1991, the Corps mobilized more than two-thirds of its forces for the Gulf War under the umbrella of the I Marine Expeditionary Force. The new weapons systems and doctrine were tried under battle conditions as the new training systems and the Marines of the post-draft Corps were put to the test. The Iraqi military was far from a first-class opponent, but the performance of the Corps in the Gulf remained impressive, and all of the major new systems worked as expected or better.

At the beginning of the twenty-first century, the Corps was better equipped, educated, and trained, riding high on the operational and tactical successes of the Gulf War. Taking inspiration from early Marine thinkers, such as Lieutenant Colonel Earl H. "Pete" Ellis and Major General Smedley D. Butler, visionary Commandant General Alfred M. Gray Jr. created Marine Corps University at Quantico in 1989 as a center for military excellence and education. He also ordered the publication of *Warfighting* (FMFM-1), creating a doctrinal publication that immediately joined the *Small Wars Manual* (1940) and *Advanced Base Operations in Micronesia* (1921) as the "Holy Trinity" of legendary Marine doctrinal works.

The attacks on 9/11 initiated the Global War on Terrorism; a series of conflicts varying in intensity from the massive conventional invasion of Iraq in 2003 to advise-and-assist missions across the globe. Serving alongside the U.S. Army in counterinsurgency campaigns in Afghanistan and Iraq, the Marine Corps faced a myriad and ever-changing roster of evolving foes alongside the other Services in an increasingly joint operational environment.

Throughout these conflicts, the senior leadership of the Marine Corps thrived. Five Marine generals have commanded U.S. Central Command (the most active theater of the last three decades) since it was founded in 1983. One of those Marines, General James N. Mattis, became secretary of defense in 2017. For the first time since the Joint Staff was created, two Marines served as chairman of the Joint Chiefs of Staff, General Peter Pace (2005–7) and General Joseph F. Dunford Jr. (2015–present). The Marine Corps' influence on the making of grand strategy of the United States has never been greater.

A hundred years after the Battle of Belleau Wood, the Corps continues to benefit from and honor the legacy of the Marines who fought there. Those Marines would recognize the *esprit de corps* and determination of Marines today. The technology has changed dramatically, and the missions have only become more complicated during the past century. But across the years, the Belleau Wood Marines can see the Corps has stayed true to its motto, *Semper Fidelis.*

GENERAL ALFRED M. GRAY'S Training and Education Letter on Professional Military Education

by General Alfred M. Gray Jr.

From: Commandant of the Marine Corps

To: Commanding General, Marine Corps Combat Development Command [MCCDC], Quantico, VA

Subj: Training and Education[2]

1. The full establishment at Quantico of our Marine Air-Ground Training and Education Center last fall marked the completion of the center's reorganization and relocation efforts. This particular reorganization and relocation was designed to achieve our overall objective of:
   a. **Improved operational capability and warfighting effectiveness.**[3]
   b. **Upgrade our high standards of excellence in training and education.**
   c. **Focus and revitalize the training of our trainers.**
   d. **Develop and implement our training and education process throughout the Corps.**
2. We are off and running—you and the members of your command have made significant progress and—we have momentum! **Now, let us make it happen and institutionalize as we implement!**
3. My HASC [House Armed Services Committee] testimony on 12 July 1988 regarding professional military education [PME] was **designed to conceptualize our training and education process of the future. As my commander for Combat Development, I want you to develop and implement a concept for PME within the total Marine Corps training and education process.** My intentions were clearly stated during my testimony and during numerous meetings with you and others during the past year. This further **amplifying guidance** includes a **philosophical direction** for schools,

[2] The original letter was written by Gen Alfred M. Gray Jr. to Commanding General, Marine Corps Combat Command, "Training and Education," 1 July 1989. Minor revisions were made to the text based on current standards for style, grammar, punctuation, and spelling.

[3] Emphasis (bold) found in the original document.

On 1 August 1989, the 29th Commandant of the Marine Corps (1987–91), Gen Alfred M. Gray Jr., established Marine Corps University in Quantico, VA.
Art Collection, National Museum of the Marine Corps

which are key players in the training and education process, and **specific guidance** regarding the training needs associated with concurrent actions.

4. **My intent in PME is to teach military judgment rather than knowledge.** Knowledge is, of course, important for developing judgment,

Warner Hall, named in honor of Senator John Warner, is home to the Marine Corps Command and Staff College and School of Advance Warfighting, and is connected to the BGen Edwin H. Simmons Marine Corps History Center. It was completed in 2015.
Marine Corps University Foundation

but should be taught in the context of teaching military judgment not as material to be memorized. **I want Marine NCOs and officers who know how to think about and in war,** who know how to conceptualize an engagement, a battle, and a campaign and then execute the concept. **The focus of effort should be teaching through doing,** through case studies, historical and present-day, real and hypothetical, presented in war games, map exercises, sandtable exercises, free-play, force-on-force "three day wars," and the like. As education progresses, from The Basic School through Command and Staff College, the material should grow more complex, but the essence should remain the same: **teaching officers and NCOs how to win in combat by out-thinking as well as out-fighting their opponents.** In order to implement this order, the following should be considered:

a. The selection, preparation, and subsequent professional evolution of instructors, especially at [the] Amphibious Warfare School and Command and Staff College. Consideration should be given to forming a small permanent faculty of perhaps half a dozen world-class scholars on the military art. Incoming instructors would spend a period of time, perhaps as much as one year, studying this faculty in preparation for teaching. This period of time should also be used to draw on intellectual resources available in the Quantico-Washington [DC] area. The objective is to develop instructors who are truly expert in the areas they teach.

b. The prerequisites required of individual officers and SNCOs [senior noncommissioned officers] for selection to the next higher grade and for attendance at appropriate level schools. Such areas as tests, individualized professional development, prior self-study, and experience are some examples of prerequisites.

c. The mission of each Marine Corps school in preparing Marines for leadership demands and for assignments of increasing responsibility.

d. The current evaluation process, based on lesson plans, ELOs [enabling learning objectives] and TLOs [terminal learning objectives], is inappropriate for education, although it may have use for training in techniques. A new evaluation process must be devised that recognizes the inherent impossibility of "objectively" or quantitatively measuring an art.

5. Concurrent actions in the training and education environment may have an impact on the development of a PME concept and present areas, which should be examined and validated. These include:

a. Examine the relationship of the transients and trainees portion of patients, prisoners, transients, and trainees to the training days prescribed in POIs and the efficiency of movement to and from training and education opportunities. Provide a plan to this Headquarters to reduce, streamline, and more efficiently control the process of conducting training and education.

b. **Training must be focused on winning in combat.** This requires recreating the conditions of combat as realistically as possible in peacetime field exercises **at all levels,** from squad through MEF. The uncertainty, confusion, fog, and friction that characterize combat must be essential elements of the training environment. Usually, this can best be done through free-play, force-on-force exercises.

c. We must institutionalize this kind of training by improving our ability to structure, umpire, and critique field exercises. **I consider this a high priority.**

d. Examine the procedures (i.e., command, maintenance, development) employed regarding our training ranges to include range improvements; develop a Marine Corps Training Range Master Plan (to include a Range Improvement Plan); provide the Master Plan to me for approval.

Recruits from Company H, 2d Recruit Training Battalion, read about Col Chesty Puller during a history class aboard Marine Corps Recruit Depot San Diego, CA, on 28 June 2013.
Official U.S. Marine Corps photo, courtesy of Sgt Liz Gleason

e. Examine the education programs associated with command and control (i.e., Communication Officers School, Computer Science School); assess the requirement for a specific, dedicated C4 [command, control, communications, computer] Systems School; and, if appropriate, develop a plan to create a C4 Systems School in lieu of our existing schools. We need to develop a philosophy of command and control, and not simply define it in terms of systems.

f. Examine the current process for the development of MAGTF tactics and techniques; create an organization, procedures, etc., for the development of MAGTF tactics and techniques and the method of introduction into training and education programs.

g. Examine the current process for incorporating doctrine and changes to doctrine into our training and education system; and, if appropriate, institutionalize this process or methodology (i.e., a better understanding of doctrine as it applies to strategic, operational, and tactical level of operations).

h. Develop the plan for a Marine Corps University at MCCDC: a focal point for planning, doctrine, training, education, etc.; it should provide insight and guidance on naval strategy, national strategy, etc.; it should not be a consolidation effort, but a strengthening of our process.

i. Examine the contribution an upgraded library and research facility will have on training, education, and warfighting.

j. Examine the increased incorporation of history in our training and education process. History should be used to teach officers military judgment, not to make academic historians or simply teach facts.

k. Examine our methods of training the trainers; determine what standardization and upgrades are needed and institutionalize the process.

l. Examine marksmanship training from the perspective of warfighting. Determine the changes needed to ensure:
(1) emphasis on basic marksmanship training, and field and requalification training;
(2) enhanced training for mission-unique requirements; and
(3) enhanced MOS training for marksmanship instructors and range officers.

6. **Philosophical direction** regarding our officer/SNCO/NCO schools:

a. Training and education will emphasize the dictums I consider important to win: maneuver warfare; combined arms; deception and surprise; electronic warfare; fully developed communications capability (C3 [command, control, communication] superiority); flexible mobile logistics; stressed NBC [nuclear, biological, chemical] (tactical nuclear planning and offensive/defensive chemical capability); exploitation of existing mobility; active/passive ground-based air defense; and operational security.

b. The education process will **emphasize how to think,** and stress the development of a logical thought process.

c. The training and education environment **should challenge** all officers, SNCOs, and NCOs to bring out their best.

d. Formal process of **feedback from graduates** will be institutionalized and shared by all educational staffs.

e. **The environment of training and education** should emphasize the use of mission-type orders, history, battle studies, and low-intensity conflict.

f. **Emphasize a generalist perspective** among our career force.

g. The **focus** of all training and education for officers should be at least two grades beyond their current grade, deal in a balanced way with joint requirements, and challenge the student.

h. The **focus** of PME for SNCOs will be centered on the preparation of assuming duties of greater responsibility. This will encompass the basic, career, advanced, and senior levels.

Marines conduct a military education exercise at Antietam National Battlefield in Sharpsburg, MD.
Official U.S. Marine Corps photo, courtesy of PFC David Staten

i. **The SNCO Academy will focus** on a continual strengthening of the leadership development/professional warrior environment addressing battle skills in a platoon/section through battalion/squadron level. An added emphasis will be toward professional self-study/reading program.

j. **The Basic School will focus** on "basics" and the emphasis will be on leadership of a rifle platoon. The basics should not be defined simply as techniques. They include the basic concept of a Marine officer as someone who is a vital part of a corporative, not a bureaucratic officer corps and who, therefore, is an active, contributing member of the total team, someone who takes initiative in furthering the goals and values of the Marine Corps and its ability to win in combat. He is a thinker as well as a leader. A professional warrior environment will be the focus with sufficient emphasis to ensure continued influence throughout the company-grade assignment. An important residual will be the continued need and importance to professional development of a self-study/reading program.

k. **Amphibious Warfare School**

should focus on MEB level of operations and the conduct of such operations in a joint environment. An important adjunct to the course will be the development of professional study/reading programs.

1. **Command and Staff College [C&SC] will focus** primarily on MEF-level operations **in the sense of operational art** and the conduct of such operations in a joint environment. Joint/combined operations and exercises, global wargames, national/international strategy, and associated thought processes and orientation will also be emphasized. Self-study/reading programs will be focused on matters pertinent to the high level/joint staffs, international strategy and high-level command. Provide for a portion of C&SC to receive joint accreditation.

7. We are close to getting the education system to where it must be as we round out this decade and enter the new century. To get there, the following must happen:
   a. Do not staff or examine these issues to death; let us get on with it—you have the charter.
   b. Keep me and the force commanders informed.
   c. Not later than 15 September 1989, I will expect the first of what should be a quarterly in-progress review (IPR) on all aspects of our effort.
   d. IPR meetings will be at Quantico. Any roadblocks should be identified with COAs [courses of action] for removal. Decisions will be made!
8. Use the **philosophy** expressed in **FMFM 1, *Warfighting***

A. M. GRAY

WARFIGHTING Philosophy of Command

Warfighting, Fleet Marine Forces Manual 1

It is essential that our philosophy of command support the way we fight.[4] First and foremost, *in order to generate the tempo of operations we desire and to best cope with the uncertainty, disorder, and fluidity of combat, command and control must be decentralized*. That is, subordinate commanders must make decisions on their own initiative, based on their understanding of their senior's intent, rather than passing information up the chain of command and waiting for the decision to be passed down. Further, a competent subordinate commander who is at the point of decision will naturally better appreciate the true situation than a senior commander some distance removed. Individual initiative and responsibility are of paramount importance. The principal means by which we implement decentralized command and control is through the use of mission tactics, which we will discuss in detail later.

Second, since we have concluded that war is a human enterprise and no amount of technology can reduce the human dimension, our philosophy of command must be based on human characteristics rather than on equipment or procedures. Communications equipment and command and staff procedures can enhance our ability to command, but they must not be used to lessen the human element of command. Our philosophy must not only accommodate but must exploit human traits such as boldness, initiative, personality, strength of will, and imagination.

Our philosophy of command must also exploit the human ability to communicate *implicitly*.[5] We believe that *implicit communication*—to communicate through *mutual understanding*, using a minimum of key, well-understood phrases or even *anticipating* each other's thoughts—is a faster, more effective way to communicate than through the use of detailed, explicit instructions. We develop this ability through familiarity and trust, which are based on a shared philosophy and shared experience.

This concept has several practical implications. First, we should establish long-term working relationships to develop the necessary

[4] The original manual came from *Warfighting*, FMFM-1 (Washington, DC: Headquarters Marine Corps, 1989), chapter 4. Minor revisions were made to the text based on current standards for style, grammar, punctuation, and spelling. Emphasis (italics) found in original document.

[5] John Boyd introduces the idea of implicit communication as a command tool in "A Discourse on Winning and Losing: An Organic Design for Command and Control" (briefing, May 1987).

familiarity and trust. Second, key people—"actuals"—should talk directly to one another when possible, rather than through communicators or messengers. Third, we should communicate orally when possible, because we communicate also in *how* we talk—our inflections and tone of voice. Fourth, we should communicate in person when possible because we communicate also through our gestures and bearing.

Commanders should command from where they can best influence the action, normally well forward. This allows them to see and sense firsthand the ebb and flow of combat, to gain an intuitive appreciation for the situation that they cannot obtain from reports. It allows them to exert personal influence at decisive points during the action. It also allows them to locate themselves closer to the events that will influence the situation so that they can observe them directly and circumvent the delays and inaccuracies that result from passing information up and down the chain of command. Finally, we recognize the importance of personal leadership. Only by their physical presence—by demonstrating the willingness to share danger and privation—can commanders fully gain the trust and confidence of subordinates. *We must remember that command from the front should not equate to oversupervision of subordinates.* At the same time, it is important to balance the need for forward situation, which is often best done from a central location such as a combat operation center. Commanders cannot become so focused on one aspect of the situation that they lose overall situational awareness.

A Marine student in the Infantry Officers Course armed with a Colt 5.56mm M16A2 Rifle, waits on one knee during an assault on the Delta Prospect Range at the Marine Air Ground Task Force Combat Center training area at Twentynine Palms, CA. *Official U.S. Marine Corps photo, courtesy of LCpl Joey Chavez*

As part of our philosophy of command, we must recognize that war is inherently disorderly, uncertain, dynamic, and dominated by friction. Moreover, maneuver warfare, with its emphasis on speed and initiative, is by nature a particularly disorderly style of war. The conditions ripe for exploitation are normally also very disorderly. For commanders to try to gain certainty as a basis for actions, maintain positive control of events at all times, or dictate events to fit their plans is to deny the nature of war. We must therefore be prepared to cope—even better, to *thrive*—in an environment of chaos, uncertainty, constant change, and friction. If we can come to terms with those conditions and thereby limit their debilitating effects, we can use them as a weapon against a foe who does not cope as well.

In practical terms, this means that we must not strive for certainty before we act, for in so doing we will surrender the initiative and pass up opportunities. We must not try to maintain excessive control over subordinates, since this

G-1, Task Force Ripper CP reconnoitering at Umm Huul, by Col Avery Chenowith, shows Marines preparing to breach the Saddam line during the 1991 Gulf War.
Art Collection, National Museum of the Marine Corps

will necessarily slow our tempo and inhibit initiative. We must not attempt to impose precise order on the events of combat since this leads to a formulistic approach to war. We must be prepared to adapt to changing circumstances and exploit opportunities as they arise, rather than adhering insistently to predetermined plans that have outlived their usefulness.

There are several points worth remembering about our command philosophy. First, while it is based on our warfighting style, this does not mean it applies only during war. We must put it into practice during the preparation for war as well. We cannot rightly expect our subordinates to exercise boldness and initiative in the field when they are accustomed to being oversupervised in garrison. Whether the mission is training, procuring equipment, administration, or police call, this philosophy should apply.

Next, our philosophy requires competent leadership at all levels. A centralized system theoretically needs only one competent person, the senior commander, who is the sole authority. A decentralized system requires leaders at all levels to demonstrate sound and timely judgment. Initiative becomes an essential condition of competence among commanders.

Our philosophy also requires familiarity among comrades because only through a shared understanding can we develop the implicit communication necessary for unity of effort. Perhaps most important, our philosophy demands confidence among seniors and subordinates.

THE NCO AND MANEUVER WARFARE

by Captains William H. Weber IV and David J. Furness

Marine Corps Gazette, April 1993

"As leaders, we must push power downward to the young Marines who hunger and thirst for more responsibility—and are quite capable of handling it. . . . In combat, NCOs carry battle. Is it fair to expect them to do this in war if we have not trained them in peacetime—by allowing them significant responsibility and authority?"

~General Walter E. Boomer

On the third day of the ground war during Operation Desert Storm, a platoon of Company A, 1st Light Armored Infantry Battalion, moved north toward Kuwait International Airport as the point element of the battalion's drive, conducting an aggressive forward reconnaissance with the battalion moving up quickly behind them to assume blocking positions north of the airport.[6] *At 1700, an Iraqi strongpoint halted the battalion's movement, and the point platoon came under sporadic machine gun fire. Far over on the right flank, a light armored vehicle, commanded by a corporal, pushed out on its own to try and find a way around this obstacle. Unsupported by any friendly forces, the vehicle commander found a route that would allow the battalion to bypass the strongpoint and continue its mission. The corporal called his platoon commander: "Red 1 this is Red 6, I've got a way around over here on the right. We can get on a trail under those power lines on the 68 grid line and continue moving." The platoon commander acknowledged the transmission and immediately changed his direction, moving his platoon and the rest of the battalion around the obstacle and onto the route blazed by the corporal. The battalion accomplished its mission, arriving in its blocking position an hour early, ensuring that other units could seize the airport without concern about possible Iraqi reinforcements descending from the north. A junior NCO, using initiative and making decisions, allowed the battalion to accomplish its mission. His platoon commander listened to him and was willing to commit the battalion to a route chosen by a corporal. This is maneuver warfare.*

Understanding maneuver warfare is difficult. Executing it is even harder. Our current warfighting doctrine, as set forth in [*Warfighting*] FMFM-1 and [*Tactics*] FMFM-1-3, fails to address the key to maneuver warfare execution—creating a command environment that empowers our

[6] The original article came from Capt William H. Weber IV and Capt David J. Furness, "The NCO and Maneuver Warfare," *Marine Corps Gazette* 77, no. 4 (April 1993). Minor revisions were made to the text based on current standards for style, grammar, punctuation, and spelling.

Campaigns to increase NCO retention, such as the poster *You Can Count on the Corps, Stay Marine*, recognized the value of NCOs to maneuver warfare. *Official U.S. Marine Recruiting poster*

NCOs. Without NCOs who are willing and able to make decisions on their own, and without officers who will support those decisions, we will surely fail. Our doctrine of maneuver warfare "applies equally to the Marine expeditionary force commander and the fire team leader." It is a decentralized system that "requires leaders at all levels to display sound and timely judgment." How do we make our Marines into the warfighters our doctrine demands? The answer is simple. If we are going to fight the way our doctrine dictates, we must empower our NCOs so that decisions are made and executed at the lowest level.

Empowering our NCOs is critical for many reasons. In *Men Against Fire*, S. L. A. Marshall observes

> *that no commander is capable of the actual leading of an entire company in combat, that the spread of strength and the great variety of the commander's problems are together beyond any one man's compass, and that therefore a part of the problem in combat is to determine which are the moral leaders among his men when under fire, and having found them, give all support and encouragement to their effort.*[7]

If a commander cannot control all of his people, then how can he ensure unity of effort and reliable action by subordinates? The solution lies in proper leadership, and commanders must take notice. If we, as an institution, fail to develop our NCOs so that they are the tactical equals of junior officers, we risk slowing our physical and mental speed at the point of contact and throwing away many valuable opportunities for battlefield success. Our NCOs must make and execute decisions.

The empowerment of the NCO starts with the leadership of the unit commander and the command environment he fosters. He must personally take an interest in his NCOs' professional development and give them the responsibility and knowledge to train junior Marines. FMFM-1 describes this command environment:

> *All commanders should consider the professional development of their subordinates a principal responsibility of command. Commanders should foster a personal teacher-student relationship with their subordinates. . . . [and] should see the development of their subordinates as a direct reflection on themselves.*

Besides our doctrinal emphasis on junior leaders who make decisions, we also stress the use of mission tactics as the best way to take ad-

[7] S. L. A. Marshall, *Men Against Fire: The Problem of Battle Command* (Norman: University of Oklahoma Press, 2000), 62.

M1A1 Abrams of 1st Tank Battalion, 1st Marine Division, stage for the assault on Red Beach aboard Camp Pendleton, CA.
Official U.S. Marine Corps photo, courtesy of LCpl Giles M. Isham

vantage of these junior decision makers. Mission tactics require units that can quickly execute basic and advanced techniques and are led down to the lowest level, by men who make timely decisions. This decentralized decision authority must include the NCO. In his book *On Infantry*, John A. English concluded his discussion of the subject by saying,

> *the decentralization of tactical control forced on land forces has been one of the most significant features of modern war. In the confused and often chaotic environment of today, only the smallest groups are likely to keep together, particularly during critical moments.*[8]

The noncommissioned officer therefore holds the key to the execution level of maneuver warfare, the level where we translate our doctrine into action. The Marine Corps must acknowledge this. Many company grade officers do not regard the NCO as a leader, trainer, and decision maker. They are not comfortable with young corporals and sergeants training their Marines or having the freedom of action necessary to grow as leaders and decision makers. Hampered by the need to clear all decisions before executing, the junior leaders atrophy. Action at the point of contact slows to a crawl. Opportunities and battles are lost.

But we can solve these problems. Our company grade officers must force our NCOs to accept more responsibility, while simultaneously focusing their junior officers and staff NCOs on improving the abilities of NCOs to lead, think, and fight—added responsibility should be accompanied by the training that will allow NCOs to succeed. This is empowerment.

[8] John A. English, *On Infantry* (New York: Praeger, 1984).

A Marine from 3d Battalion, 2d Marines, fast ropes out of a CH-46E Sea Knight onto a rooftop at the Military Operations in Urban Terrain facility at Camp Lejeune, NC.
Official U.S. Marine Corps photo, courtesy of Sgt Brandon E. Vancise

The command environment we foster and the schools we send our NCOs to must work together toward this end if we expect to create true warfighters.

Conversations we have had with several of our contemporaries convince us that we were truly fortunate as new lieutenants to be sent to units that viewed the tactical expertise of NCOs as their number one priority. One of our company commanders explained his priorities to all of his new lieutenants and staff NCOs in these terms:

- The NCOs would be the principal trainers of Marines.
- Our responsibility was to ensure that the NCOs were tactically proficient and could conduct effective training.
- Our focus of training would be at the individual, team, and squad level.

He understood that expertise at the small unit level is essential to the execution of our warfighting doctrine. He believed that by focusing the efforts of the company on building strong NCOs and strong teams and squads, the company as a whole would succeed. He realized that limiting the NCOs participation in the planning and execution of small unit training negated their responsibility for ensuring that their unit was prepared for combat. It also diminished their credibility in the eyes of their Marines. By clearly defining the NCOs' proper role in the conduct of training, we began to give them the power they needed to succeed.

We believed that, for the NCOs to train their Marines, they had to become unquestioned experts in techniques and procedures, battle drills, and tactics. They needed to understand completely our warfighting doctrine and what their role was in it. To reach these goals, we established a program at the company level that should be a part of every unit in the Marine Corps. The program began with an examination to test the basic skills of the NCOs. Then, using existing publications, such as *The Essential Subjects; The Marine Battle Drill Guide and Command Tasks; Soldier Skill Level 1, 2, 3, 4 for MOS 11B and 11C*, the company began to train the NCOs while simultaneously conducting basic individual training. Once our NCOs mastered the basic skills, we began to teach them advanced warfighting techniques and concepts. We conducted sandtable and map exercises designed to illustrate the fundamental maneuver warfare principles. These exercises forced the NCOs to develop their own courses of action and express them in a standard five-paragraph order format.[9] We stressed the ability of the NCO to communicate his plan to his Marines. There were no right and wrong answers, but unsound tactical thinking was thoroughly critiqued. The intent of this training was to develop the thought process that we could later expect the NCO to employ in combat. We tried to train the NCO to become the tactical equal of the junior officer.

Initially, the transition from troop handler to small unit leader was difficult. Many junior NCOs lacked the technical and tactical expertise to educate and train their Marines, and they were reluctant to assume their new role. They quickly overcame this initial reluctance when they realized that we meant to give them real power. They then became eager to learn about their profession. They understood that their increased role brought additional responsibilities, and they rose to the challenge. After several months of intensive training, augmented by many quotas to NCO school and the division's squad leader's course, the technical and tactical proficiency of the NCOs skyrocketed. These corporals and sergeants were hungry for additional tactical expertise—they were becoming professionals. As they began to instruct their Marines, some astonishing things happened:

- The NCOs' status as leaders increased dramatically.
- A clear teacher-scholar relationship grew between the NCOs and their Marines. The NCOs' focus became training. They became inquisitive and interested in professional reading, Marine Corps Institute courses, and tactical decision games.
- The example set by the NCOs constantly reinforced critical skills and techniques throughout the company. A strong bond developed between the NCO and his men, leading the Marines to an almost unquestioned faith in their junior leaders.
- The entire company's growing competence in battle drills, techniques, and procedures allowed the NCOs to adapt easily to rapidly changing tactical environments. They became much more aggressive and willing to make significant decisions. During free-play exercises, the NCOs' newfound freedom translated to dramatically improved physical and mental speed at the small unit level.

The company became a warfighting machine driven by a desire to learn and become more competent. The worth of an NCO rested not only on how well he could drill or prepare for an inspection, but also on how he could think and execute on the battlefield. The officers were constantly challenged professionally because even the most junior leaders in the company clamored for knowledge. Morale was never a problem because the command environment bred professionalism.

This is what a Marine Corps rifle company should be.

The second element in our drive to create true warfighting NCOs is formal education. To get a better understanding of the importance of education in our present situation, it is wise to look to the experience of the German Army

[9] The five-paragraph order format covers SMEAC: situation, mission, execution, administration (logistics), and command (signal).

Sgt Jeff Seabaugh, squad leader with the 15th MEU(SOC), moves his Marines in Zubayr, Iraq, on 23 March 2003 during Operation Iraqi Freedom.
Official U.S. Marine Corps photo, courtesy of LCpl Brian L. Wickliffe

between the two World Wars. The treaty ending World War I dramatically limited the size of the German officer corps. Given their reduced officer strength, they realized that they would have to rely more heavily on their NCOs. The Germans moved quickly to raise the competence level of their troops. During the 1920s and 1930s—while the American enlisted man was wasting away—the Germans began sending their NCOs to a new school, a school dramatically different from any other in the world.

The Germans designed their NCO schools to create decision makers. They believed that a man could be trained to make decisions quickly by getting him in the habit of making decisions. Once that habit was in place, all that remained was to give him the experience necessary to make good decisions. They were, in essence, attempting to teach intuition, an effort which modern research suggests was not in vain. Recent research also suggests that the best way to do this was embodied by the structure of the German NCO school: force men to make decisions again and again, punishing only timidity, while gradually giving them the experience and knowledge needed for the development of battlefield intuition. They will become battlefield leaders.

This school was very different from our current NCO schools, where much of a Marine's time is focused on garrison skills that, although admittedly important subjects for the NCO, are of limited value in the face of the enemy. Our educational system makes the considered point that drill is but a means to an end, but there is a demonstrably superior way of reaching our goal of NCOs who crave responsibility and consider themselves as elite professionals in the U.S. Marine Corps. Specifically:

- NCO schools should be lengthened from their present 38 days to at least four months. The schools should devote this additional time to two things—decision-making exercises and directed study on the art of war, study aimed at making the NCO both knowledgeable and hungry for more learning.
- Attendance at NCO schools should be an absolute requirement for promotion to sergeant.
- Because NCO schools would be truly professional schools, we would expect to be paid back by the Marines who attend it: any Marine graduating would serve at least two additional years in the Fleet Marine Force before he could leave the Corps. Attendance at an NCO school would therefore be voluntary, and any Marine refusing to attend would not be promoted past the rank of lance corporal. Failure at NCO school would mean reverting to lance corporal for the remainder of the enlistment or immediate release from the Corps.

Making these changes would be costly. The future success of our Corps, however, depends on producing Marines who can fight our doctrine on a decentralized battlefield. Institutionally, we are not doing this, and we must. The money must be spent, and the NCOs must get the power they need. We must force our future company commanders to let their NCOs do their jobs. We must educate our company grade officers at Amphibious Warfare School on the real role of the NCO in warfighting and on how to train him to fill that role. We must also change the expectations of the NCOs by changing our NCO schools—the NCO should leave these vital schools fully expecting to be challenged professionally by his officers. He should return to his company with a burning desire to take charge on the battlefield and, by doing so, force the Marine Corps to execute what we spend so much time talking about—our doctrine.

EXPEDITIONARY AIRFIELDS

by Major General Terrence R. Dake

Marine Corps Gazette, August 1994

"From the Sea" has been the naval strategy for more than a year now.[10] Coincidental with this new focus for naval forces, we have been involved in the fastest decline in defense resources since the end of World War II. This combination of powerful forces has moved the Navy and the Marine Corps to undertake actions to make the most of the resources available. The most visible of these moves to date is the integration of Marine Corps fighter attack squadrons aboard aircraft carriers as integral parts of carrier air wings. Not only do Marine Corps squadrons aid in filling shortfalls in the number of naval squadrons, they bring the combined arms expertise of the Marine air-ground team to the deck of the carrier. The resultant team is very capable of supporting Marines ashore. These shipborne Marines are fully integrated members of the resident carrier air wing. This is an important point to remember. When Marines are in their usual Marine Air-Ground Task Force (MAGTF) configuration, Marine air is in direct support of the ground component. When we are integrated, we are in direct support of the carrier that, if we are true to the focus of littoral warfare, will fly sorties in support of Marines ashore. When the carrier is in a direct-support role, the early sorties flown in front of the ground forces would be flown by both Navy and Marine pilots.

What has been lost in the fanfare of Marines again going aboard ship is the role that expeditionary airfields (EAFs) can and will play in the projection of power by naval expeditionary forces. This capability, which exists only in the Marine Corps, is the natural extension of carrier aviation. The expeditionary airfield is a system that was designed to allow Marine aircraft to phase ashore in areas where a suitable airfield may not be available. Once ashore and unfettered by the limitations that are inherent in carrier operations, Marine aircraft are capable of high sortie rates and continuous operations. The same characteristics that allow Marines to operate from these austere fields also allow common Navy aircraft, such as the [McDonnell Douglas] F/A-18 [Hornet], to do the same. In other words, it is a two-way street from the

[10] The original article came from MajGen Terrence R. Dake, "Expeditionary Airfields," *Marine Corps Gazette* 78, no. 8 (August 1994). Minor revisions were made to the text based on current standards for style, grammar, punctuation, and spelling.

Two Marine AH-1 Cobra helicopters on the ground, being refueled during Operation Restore Hope.
Defense Imagery DD-SD-00-00796

shore to the carrier and from the carrier to the expeditionary airfield.

The expeditionary airfield opens up options for the projection of airpower that do not exist with carrier-based aviation alone. Whether used as a forward arming and refueling point (FARP) for carrier-based aircraft or as an alternate basing site, the expeditionary airfield can increase sortie rates. It can act as a divert field for strike-damaged aircraft that cannot make the trip to the carrier or are unsafe to bring aboard. It provides an intermediate fueling point to extend the combat radius, both range and time-on-station, of carrier aircraft when tankers are not available. Perhaps most important, the expeditionary airfield allows more naval aircraft, either Navy or Marine, to be forward based, which equates to more combat power in the face of the enemy.

In Vietnam and [Operation] Desert Storm, expeditionary airfields were the primary basing sites for the majority of Marine aircraft, both fixed-wing and helicopters. My point on EAFs is this: expeditionary airfields provide a capability that has been used frequently in the past and a capability the nation needs in the future. Even in Desert Storm, where ports and airfields were in better condition than in any other region except Europe, the airfields had good runways but little parking space and usually no parallel taxiways. One of the first tasks of the Marine Corps was to create airfields capable of supporting our aircraft. The rapid flow of tactical aircraft into the theater of operations quickly filled all air bases to overflow capacity. More than 4 million square feet of AM2 mat, with its associated accessories, were used to augment existing airfields in Southwest Asia.

During [Operation] Desert Shield, a sports complex at King Abdulaziz Naval Base was

An AV-8B Harrier jet with VMA-513 takes off from Bagram Air Base, Afghanistan. The Harrier is carrying a Guided Bomb Unit 16 and an AN/AAQ-28(V) Targeting Pod.
Official U.S. Marine Corps photo, courtesy of LCpl Andrew Williams

turned into a full-service airfield from which 66 [McDonnell Douglas] AV-8B Harrier attack aircraft and 12 [North American Rockwell] OV-10 [Bronco] aerial reconnaissance aircraft carried the fight to the Iraqis. These valuable aircraft would not have had a home if it were not for this remarkable capability; there simply was not room anywhere else.

Using the fuel and ordnance from maritime prepositioned ships, these aircraft flew sustained operations during the early phases of Desert Storm, preparing the battlefield for the ground attack. When the ground attack into Kuwait began, these aircraft surged from that base, as well as from forward-operating sites, for quick turnaround of combat aircraft.

The standard mission began at King Abdulaziz, near the port city of Jubail, Saudi Arabia, about 140 miles from the southern border of Kuwait. After dropping bombs on targets in Kuwait, the aircraft rearmed and refueled at Tanajib, a small strip 27 miles south of the Kuwaiti border. After flying a second bombing mission into Kuwait, typically only an hour after the first, the aircraft recovered back at King Abdulaziz. The use of Tanajib cut more than 200 miles from the mission profile, increasing the sortie rate, but most important, increasing the amount of ordnance on target. To put flexible basing into perspective, consider this: in Desert Storm, 86 Harriers generated 3,342 strike sorties while their carrier-based cousins, 209 F/A-18s, [Grumman] A-6s [Intruders], and [LTV] A-7s [Corsair IIs], flew 6,109 strike sorties. The strike sortie "box score" breaks out like this: Harriers, around 39 strike sorties per aircraft; carrier air, about 29 strike sorties per aircraft. The Harriers had one additional advantage in that all of their sorties were accomplished without aerial refueling. This expeditionary effect, which boosted the sortie rate at King Abdulaziz and Tanajib, can also exist between an expeditionary airfield and an aircraft carrier.

The naval Services have used expeditionary airfields often in past conflicts. In Italy, during World War II, expeditionary airfields provided en route staging bases for Navy aircraft. During the Pacific Islands campaign, a Marine aircraft group operated from a planked runway, using a catapult and arresting gear. The constant search for improved flexibility in the rapid deployment of Marine aviation led to the development of the expeditionary airfield concept. Development ensued throughout the postwar years and, in 1956, the Commandant of the Marine Corps formally established an operational requirement

Osprey Number 7 of VMM-263 taxis on flight line at Marine Corps Air Station, New River, North Carolina by Sgt Kristopher Battles.
Art Collection, National Museum of the Marine Corps

for the system. Specific time and space parameters were defined as follows:

- Small, quickly constructed tactical support airfields of a temporary nature to accommodate at least one squadron (24 aircraft).
- Ready to use in the first three to five days of an amphibious assault.
- Usable for 30 days to support the landing force in tactical operations ashore.

In 1958, the expeditionary airfield concept was approved, and the system was designated "short airfield for tactical support" or SATS. While the time and space parameters have been expanded to support the increased performance characteristics of today's aircraft, the premise of "small, quickly constructed tactical support airfields of a temporary nature" is as valid today as it was in 1958.

What an expeditionary airfield actually

consists of is frequently not well understood. Too often, it is thought to be only matting and arresting gear. In fact, it is a system that can be as simple as placing arresting gear on a captured/abandoned runway or as robust as any full system airfield, to include weather services and aircraft maintenance. Full-service EAFs are operated routinely at Marine Corps Base (MCB) Twentynine Palms, California, and Marine Corps Auxiliary Landing Field Bogue, North Carolina.

The flexibility of the expeditionary airfield system allows a variety of airfield configurations to be constructed, tailored to the specific needs of the aircraft and mission. On the upper end of the expeditionary airfield system, 8,000-foot airfield complexes can be built to accommodate heavy transport aircraft. Air Force [Lockheed] C-141s [Starlifters] routinely operate from the Strategic Expeditionary Landing Field at Twentynine Palms. The Naval Air Systems Command is working closely with the Air Force and Marine Corps on EAF compatibility with the [Boeing] C-17A [Globemaster] transport aircraft. On the lower end expeditionary airfields, 96-by-96-foot vertical takeoff and landing (VTOL) pads can be constructed to accommodate rotary- and fixed-wing VTOL aircraft. The most common size is a 3,850-foot runway from which F/A-18s and AV-SBs can operate. The expeditionary airfield system is uniquely suited to this purpose and plays a key role in the close air support mission.

The basic components of the expeditionary airfield system are the AM2 matting and its associated accessories. AM2 matting consists of 2-by-12-foot and 2-by-6-foot aluminum panels coated with an epoxy nonskid surface. These mats have four interlocking edges permitting easy assembly into rectangular patterns of virtually any size and proportion. AM2 mat can be used to form runways, taxiways, parking areas, or ramp space for efficient aircraft operations and maintenance. In addition to AM2 matting, expeditionary airfields provide a variety of lighting components for night operations. Future lighting will include man-portable, infrared-capable, battery-operated systems with remote control, which can be installed in minutes with minimal personnel. Thus, we will be able to rapidly light a facility after an attack or light a forward operating base during covert operations.

An integral concept in the expeditionary airfield program is the aircraft recovery system. The M21 aircraft recovery system will stop a tailhook aircraft in 600 feet of rollout at an expeditionary field or on an existing runway, thereby greatly increasing the basing flexibility of naval aviation. Under development is the M29 transportable system, a more mobile and versatile system, that will allow for expanded operations.

Getting EAF matting and equipment to where it is needed is an important consideration. Weight and cube are critical logistic commodities. Efforts are under way to reduce EAF footprints and to find alternative methods of transporting them. Both Maritime Prepositioning Squadrons 2 and 3 have enough AM2 matting to build either a 1,000-by-72-foot or a 1,500-by-54-foot runway with 11 tactical parking spaces. Concurrent to setting up the matting, the associated Marine Air Traffic Control and Landing System, in this case probably consisting of a tower, tactical air navigation system, radio, and other equipment, would be flown in on two C-141s. Larger EAFs, like the full service 3,850-foot field with 75 tactical parking spaces, are usually shipped via break-bulk vessels and married with more robust air traffic control lo-

A C-130 Hercules practices takeoff and landing from one of three different runways aboard the island of Tinian, Northern Mariana Islands.
Official U.S. Marine Corps photo, courtesy of LCpl Eugene E. Clarke

gistics including radar. Approximately six C-141 loads are required to deliver a full package.

The Navy and the Marine Corps are committed to making the best use of dwindling resources. It is not always obvious how best to invest so that readiness, force structure, and modernization are kept in balance. However, innovative use of existing assets is always a winner. In that regard, the alliance of the expeditionary airfield and carrier aviation will result in mutual benefit. Joint exercises and wargames are opportunities that we should use to explore and to advertise this capability. A good start is already being planned for carrier aircraft to operate from the Strategic Expeditionary Landing Field at Twentynine Palms. I expect the future will bring more operations where Marines fly from carriers and Navy aircraft operate from austere fields operated by Marines. Naval expeditionary forces, with the capability to operate from afloat and ashore, confront the enemy with combat power that extends along a continuum from the early strategic strikes to ground operations ashore. Only the naval Service contains all of these capabilities in one integrated package.

IT'S NOT NICE AND NEAT

by Lieutenant General Anthony C. Zinni

Proceedings, August 1995

Today's military operations are not like those the Services have traditionally trained to execute.[11] *But when "something" has to be done, U.S. Marines, sailors, soldiers, and airmen answer the call. General Zinni, the commanding general, I Marine Expeditionary Force, who led a seven-nation combined task force during Operation United Shield—the withdrawal of UN forces from Somalia—provided insights to this new world in his address at the most recent Naval Institute Annual Meeting and Annapolis Seminar.*

First, I would like to talk a little bit about Operation United Shield, which should interest you because it primarily involved a naval force, and its success stemmed both from the training and the capabilities that naval forces bring to these kinds of operations. Then I would like to expand a bit more into the nature of these operations and what I think that means for the future.

United Shield was born sometime around last August, when the United Nations realized it was going to have a hard time extracting itself from Somalia and asked the United States to help. Several nations agreed to participate—with the largest force contributions coming from Italy—but all wanted U.S. leadership and involvement.

The situation required that the withdrawal be amphibious. There was no other way that the rear-guard forces of the United Nations—the Pakistani and the Bangladeshi units—could have been extracted safely. The threat to the airfield—not only from the militias of the warlords, but also from the bandits who roam the streets of Mogadishu and fight each other and anybody else in their way—was too high.

The planning began in New York City. We sent a planning team from the U.S. Central Command, which worked with the contributing nations, with the United Nations headquarters, and the United Nations forces in the field. A planning team also went into Mogadishu in January 1995 to work directly with General Aboo [Samah bin-Aboo Bakar], the four-star Malaysian commander of the UNOSOM [United Nations Operation in Somalia] forces.

Our task force took station off the Somali coast near Mogadishu on 8 February. We were in position three weeks before we landed to set up the final protective perimeter for the UN

[11] The original article came from LtGen Anthony C. Zinni, "It's Not Nice and Neat," *Proceedings* 121, no. 8 (August 1995). Minor revisions were made to the text based on current standards for style, grammar, punctuation, and spelling.

forces. Before the landing, I went into Mogadishu and met with the warlords—five of them, including [General] Mohammed Farah Hassan Aidid. I think you may have read about him. I have known General Aidid for years.

General Aidid, Ali Madi, and several others owned the militias and the turf around the port and airfield. I was interested in several things:

- To find out what their intentions were.
- To secure their cooperation, if possible.
- To warn them of what would happen if they did not cooperate—making sure they understood that what was coming in was not just another UN force, but this time a U.S.-led Coalition, and that the rules of engagement would allow us to protect ourselves and take care of any threats.
- To make sure that they were not deluded by any notion that anything less than an overpowering force was over the horizon, ready to come in.

Sgt Gardi and Kurdish Elder, by Col Peter M. Gish, shows a young Marine conversing with a Kurdish elder in northern Iraq. Kurdish elders, known as *mullahs*, are revered for their wisdom and their opinions carry much weight.
Art Collection, National Museum of the Marine Corps

In meeting with them, I found mixed feelings about the operation. Many of the faction leaders were feeling good about the U.S. forces returning; some even wanted us to stay longer. Obviously, that was not in the cards. General Aidid, on the other hand, felt this was not good—that the return of U.S. or other foreign troops would not be well received by his people and would present difficulties for him. But he promised me that he would cooperate, and there would not be any interference.

All the time I had known General Aidid, he had never lied to me; my dealings with him always had been straightforward. I was confident that he would not interfere—nor would any of his militia. But I was not as confident that he could prevent bandits and rogue elements from interfering on the turf that he controlled. On the other hand, I was certain that the other faction leaders would police their neighborhoods.

As things turned out, *all* the faction leaders cooperated, even to the point of helping control some of the looters and bandits as best they could while we were ashore. We felt that our worst problem would be looters, demonstrators, and the bandits who roamed the streets—who are much like street gangs here, but far more heavily armed and much more aggressive. And they, in fact, turned out to be the problem.

We spent a lot of time in the first three weeks there preparing the battlefield—not only in the physical sense of the engineers preparing our positions and our barbed wire but also in working with the UN units that I would have under my tactical control. We cooperated in developing a plan for some very tricky tactical movements.

We conducted nine tactical evolutions or operations ashore. These are the most complex of operations—amphibious landings, relief in

Two Marine armored amphibious vehicles from the 15th MEU emerge from the surf onto the beach at Mogadishu Airport, Somalia, as part of Operation Restore Hope.
Defense Imagery DD-SD-00-00717

place of one force by another from a different country, withdrawal under pressure, and amphibious withdrawal under pressure.

We did seven of these totally at night; two were a mix of day and night. We made two landings at night; we withdrew under fire at night back to our ships. The Pakistanis withdrew back through our lines in the middle of the night. We relieved the Bangladeshi force at the port, and we relieved the Italians in place on the next-to-the-last night during the hours of darkness. We conducted a noncombatant evacuation of UNOSOM civilian employees, media personnel, and refugees—some Ethiopian Christians who had stowed away on a ship and were left at the port, living in tetrahedrons at the water's edge, made themselves known to us as we were getting ready to leave. They would have been goners if the Somalis—their ancient enemies—had caught them. So we took them out of there and back to Mombasa.

We also provided a day-and-night defense of the airfield and the port.

Originally, the plan said that we would be ashore for 7 to 10 days, but we completed the job in 73 hours. We could have done it in 48 hours, if the ships that the United Nations contracted had been there in a more timely way and had been operated more professionally. Dealing with a drunk master and his crew when he came in to pick up UN forces made life interesting during the last day and the last night ashore.

In those 73 hours, we experienced 27 firefights, everything from snipers to individuals with rocket-propelled grenades firing at our positions. On the last night on the last beach—a place we called "the Alamo"—we were being hit by heavier fire, and we had small groups of Somalis, 12–15 strong, coming at our wire.

I have never appreciated an amphibious tractor any more than that final night, when we were able to continue fighting on the beach with our troops protected, then turn and hit the water in our armored amphibians to get back to the ship.

My own tractor did not quite make it all the way on its own power. It caught fire about 1,000 yards out and began to drift back toward the beach and take on water. Then the tractor that came out to tow us also lost power and if you are a three-star general, stuck in the bottom of a tractor drifting back toward a beach full of a lot of bad guys, you begin to feel a little queasy. You are trying to act very bravely, and what you need right then and there is just what I got from the "trac" commander, Corporal Deskins. He stuck his head down in the troop compartment and said, "General, our trac is on fire. We've shut down the engine. We're taking on some water. We're drifting back to the beach, and we can see the bad guys. The tractor that's trying to tow us has lost power and is drifting back with us. The ship is getting farther away."

"But," he added, "don't worry—everything's going to be okay."

LAV on Patrol by Col Peter M. Gish shows a Marine LAV patrolling in Mogadishu, Somalia, during Operation Restore Hope.
Art Collection, National Museum of the Marine Corps

And I knew that was the time for a general to be a good PFC [private first class] and listen to the corporal; because with that kind of optimism and professionalism, I could not go wrong. It took five and a half hours to get back to the ship, but Corporal Deskins finally got me there, so I was happy.

Our force consisted of 23 ships. It was made up of 16,500 soldiers, sailors, airmen, and Marines from seven nations. We had 89 aircraft: Harrier jump jets; Italian and U.S. attack helos; and U.S. Air Force AC-130 gunships, flying out of Mombasa. With that kind of force—90 percent of which was naval—we were able to accomplish the mission without taking a single casualty ashore. And, again, that is not for want of the Somalis, especially the bandits, trying to inflict casualties.

This operation typifies my personal life for the past five years. It all started in the hills of Northern Iraq, with the Kurds. Next, I spent time in the 12 republics of the former Soviet Union in an operation called Provide Hope [1992], where we provided humanitarian assistance, bringing food and medicine into those areas. I also was involved in the planning for Operation Provide Promise [1992–96] in Bosnia, and I provided relief for the initial airlift of food and supplies into Somalia in Operation Restore Hope [1993], then went back again with the special envoy to Somalia, Robert Oakley, on 6 October 1993, to get the prisoners out and

A Marine corporal helps a Somali woman on crutches from a clinic in Mogadishu, Somalia, where U.S. Navy doctors conducted a medical civic action program in support of Operation Restore Hope.
Defense Imagery DD-SD-00-00838

rehabilitate General Aidid, and then back again for Operation United Shield [Somalia, 1995].

My command, the I Marine Expeditionary Force, has done some things in the past five years that I am sure you have heard about—things like Desert Storm and Vigilant Warrior [Kuwait, 1994], which was an alert to go back because the Iraqis were moving again. And, of course, we have gotten energized a couple of times over Korea. In the past five years, we have conducted a total of 20 real-world operations.

We have been to places like Bangladesh, Rwanda, and Somalia. We have also been to Ecuador, helping build schoolhouses and providing medical and dental care in some of the remote areas of that country.

And we have not only been doing these kinds of things on foreign shores; we have also been doing them within our own borders. We have fought forest fires in Montana and Washington [State]. I sent 1,000 Marines there last summer to fight the forest fires that were threatening some of our small towns. We sent Marines to Los Angeles during the Rodney King episode, to help quell the riots. We sent Marines for earthquake relief twice and for flood relief in the Southeastern United States. My Marines participate every day in counterdrug operations throughout the Southwestern United States, and we participate in operations dealing with illegal immigrants.

The nature of conflict and the level of commitment have changed drastically since the Berlin Wall came down. We are now immersed in these things called [military] operations other than war [MOOTW]. That is a strange title, because a lot of these require the application of deadly force as a defense against deadly force.

If you talk to my counterpart, Lieutenant General [Robert B.] Johnston on the East Coast, he can give you the same listing of operations. The names may change, and the places may be Guantánamo and Haiti, and it may be hurricanes instead of floods and earthquakes—but he is doing the same kinds of things.

On the plane coming across country, I read something about the nature of conflict and war, and some of the changes we are experiencing. In World War I, only 5 percent of the fatalities were civilian. In World War II, the figure rose to 50 percent. In wars and conflicts today, civilian casualties are moving up to 80–90 percent. We have become a lot better at not killing each other in great numbers as soldiers, but a lot worse in anticipating and dealing with the aftereffects of war.

As seen in the use of land mines and other forms of indiscriminate killing, the targeting of civilians is something I am seeing more and more

LCpl Brandy L. Guerrero, radio operator, Communications Detachment, MEU Service Support Group 11, 11th MEU(SOC), kisses an Iraqi baby waiting to be examined during a humanitarian assistance operation in the village of ash-Shafiyah, Iraq. This operation provided medical and dental treatments to Iraqis during Operation Iraqi Freedom.
Official U.S. Marine Corps photo, courtesy of GySgt Chago Zapata

frequently. War always is messy—even after it is over. When Desert Storm ended, everybody left, and the Shiites and Kurds were left behind in miserable shape. The operation we began in the hills of Kurdistan on 11 April 1991, called Provide Comfort, still goes on today.

These kinds of operations are consuming our armed forces right now. I get called to testify on the [Capitol] Hill every once in a while, and the question I will be asked every single time is: Should the military be doing this? Whether we should or should not, I will tell you this—we are. We recently ran an exercise out my way called Emerald Express [1995]. It searched for ways we could do these new kinds of things better: humanitarian assistance, disaster relief, peacekeeping, peacemaking, peace enforcement, noncombatant evacuation opera-

A lieutenant tests a formation during riot control training. Marines from Echo Company, 2d Battalion, 8th Marines, were part of the U.S. Security Detachment Panama, Marine Forces South, providing security during the turnover of the Panama Canal and U.S. bases to the Panamanian government.
Official U.S. Marine Corps photo, courtesy of LCpl Michael I. Gonzalez

tions. All these sorts of things that my Marine expeditionary units live with, day in and day out.

On the Hill, I was challenged a few times about why we ever get involved in this. Well, we get involved with this because we get asked to do it. Who else could do it? It is nice to say as a Marine or a soldier or a sailor or airman, "We don't want to do it, that's not what we're here for."

But I will tell you what—I have walked the ground and seen a lot of dead children. I have seen a lot of people who have starved to death or have been brutally massacred alongside a road. And something inside me says, "Maybe I shouldn't be doing this, but dammit, I want to do it. I want to change something. I want to be part of making this better or trying to fix the problem."

Now, those kinds of decisions go beyond my pay grade, but this is something we have had to live with for the last five years. The missions we get certainly are nontraditional. I have trained and established police forces, judiciary committees and judges, and prison systems; I have resettled refugees in massive numbers twice; I have negotiated with warlords, tribal leaders, and clan elders; I have distributed food, provided medical assistance, worried about well-baby care, and put in place obstetrical clinics; I have run refugee camps; and I have managed newspapers and run radio stations to counter misinformation attempts.

I am an infantryman of 30 years standing. Nowhere in my infantry training did anybody prepare me for all this. I have been seconded to ambassadors twice in my career: once in the former Soviet Union during Operation Provide Hope and once in Somalia during Operation Continue Hope, where I put on civilian clothes and became an assistant to an ambassador-at-large to promote the policies of our nation in that region.

We can say these things are not the matters that our armed forces should be involved in, but this is the direction the new world disorder is going, and there is not anybody else to call upon for help. And these are the kinds of operations we have to do better. We need to learn the nontraditional tasks required to accomplish our mission.

The problem is that today's operations do not go down like the ones that possibly you and I have been trained to run. It is not nice and neat. For openers, you do not get a clean, hard mission that tells you exactly what you are supposed to do. And you do not always get an ideal enemy, another Saddam Hussein, whom you can go after because he is mean and evil and backs a totally wrong cause. It does not work that way anymore. Usually, you are trying *not* to make enemies today.

And you cannot always go in with a force ideally tailored for this operation. What happens is that everybody comes running to the scene, and not necessarily with the ideal force composition. Coalitions are formed. In Operation Provide Comfort, we had the forces of 13 nations; in Restore Hope in Somalia, the forces of 24 nations made up our combined task force; in United Shield, I had the forces, as I mentioned, of 7 nations.

Always the best? No. Always exactly configured right for the operation? No. Always there to operate with the same objectives as you? No. Always completely interoperable with your command and your way of doing business and your doctrine and your tactics and your techniques? No. Always technically and procedurally the same as you? No.

They come from the Third World; they come from a world that grew up in a different doctrinal system; they come with different political motivations; they come with different rules of engagement, which makes it interesting when the shooting begins. And yet you have got to pull these kinds of forces together and get a mission accomplished and make sure everybody goes home feeling good about what they did.

In Somalia the first time around, we had the forces of eight nations defending the airfield. Was that because that airfield was so big or so threatened? No. It was because the forces of those eight nations could go no farther than that airfield when they got off the airplane. For either political or military reasons, that was about it. But they got participation points; and obviously the sense of international legitimacy that is given to you is important to someone. It should not be discounted. So as a commander you have got to take all that into account.

I think that this is in our future for a while. We are in an era of transition like those after all major wars. In this case, it was a Cold War that ended, and I cannot predict whether this will settle out with other superpowers emerging or with a different kind of world order, but I *can* predict a lengthy stay for this period of disorder. And it is not only overseas. There is disorder on the domestic front too. In this country, we do a lot of things that turn inward. We are very careful, because of the laws—*posse comitatus* and others—that restrict the employment of

Helo Relief by Col Peter M. Gish depicts a CH-46E from the Black Knights of HMM-264 delivering relief supplies to a Kurdish refugee camp in northern Iraq during Operation Provide Comfort. In March 1991, more than 760,000 Kurds fled into the rugged Taurus Mountains of eastern Turkey and northern Iraq to avoid the wrath of Saddam Hussein.
Art Collection, National Museum of the Marine Corps

U.S. military forces in these kinds of things.[12] But when disasters hit or when disorder is afoot, you have got to show up with those who can do the job, and we have spent a lot of time involved in things that might surprise you.

Before going off for Operation United Shield, I asked for nonlethal weapon systems. On my first tour in Somalia, I saw that we had a hard time dealing with orchestrated demonstrations. They were used against us as a tactic to defeat the Western approach to handling disturbances.

The Somalis knew we could not use deadly force; if we were provoked into using deadly force, it would just serve the ends of one or more of the warlords orchestrating the demonstration.

We knew that at times we would be dealing with people who are desperate. Looters, thieves, people fighting at food stations who are afraid that you will not come tomorrow and they will never get another chance in line—they are

[12] The term *posse comitatus* refers to common-law or statute law authority of a local law official.

all trying to survive. And when all you have for response is a rifle and its bayonet—that is not the answer. And I knew we would see that again in Somalia during United Shield. That is why I wanted to do better than we did the first time.

I wanted something that was more compassionate in the way we could handle them, and not lethal—but something that also could protect our forces. And we had a whole array of technologies out there available to us. Never once did the use of crowd-control agents put our Marines in danger, because we had the ability to go right over to deadly force at any time.

This is the way we have to be thinking now. There is much more to it than just rifle platoons taking hills.

I gave a pitch the other day to the Retired Officers Association in Southern California, where I am based. One of the well-decorated World War II and Korean War vets came up to me afterward and said, "You live in a far more complex world than I did. Ours may have been greater and more vast in the combat and conflict, but it was much simpler in understanding who the bad guy was and what we had to do and the job we had to get done."

Well, he was right. We are just as likely to need to negotiate our way through something as to fight our way through something these days. And sometimes one comes right behind the other or right before the other, so the world of military men and women today is extremely complex. Critical to success today are training and education, and the depth of knowledge you have about cultures, about such nonmilitary things as economics and politics and policies and the humanitarian aspects of an operation.

You no longer can be only the pure, narrow, military thinker and just worry about fires and maneuver. Fires and maneuver are just two relatively simple battlefield activities that underlie a vast, ever-increasing number of *other* battlefield activities.

Right now, I worry more about psychological operations and civil affairs than maybe I do about fires and about maneuver. When the siren goes off, I do not know if it is going to be Desert Storm revisited or something entirely different. We answer to the U.S. Central Command [CENTCOM] in going back to a Desert Storm-like engagement but we also answer to CENTCOM to form the joint task force for humanitarian and peacekeeping operations. I do not know if we are going to go back into the hills of Korea—because we answer to that unified commander on call in case there is another conflict in Korea—or whether it is to be another Bangladesh or another Ecuador, or whether we get on buses and go up the road to another Los Angeles riot scene or to a forest fire, or to a flood.

In looking at the full range of operations today, it is too simple to say, "Just don't do it," or "We shouldn't do it." Somebody *has* to do it—that is the problem. And all of today's problems are not outside our borders; sometimes they are simmering inside our borders.

THE CRUCIBLE
Building Warriors for the 21st Century

by General Charles C. Krulak
Marine Corps Gazette, July 1997

On the 26th of May 1997, I delivered a Memorial Day address at a solemn ceremony on one of the Marine Corps' most sacred battlefields: Belleau Wood.[13] The tenacity, valor, and sacrifice displayed by the 4th Brigade during that epic battle forever cemented the Corps' reputation as the world's fighting elite. Since Belleau Wood, Marines have been looked upon as professionals, honed to the highest standard, sharpened for any challenge, warriors without peer.

After the ceremony, I spent the rest of the day walking through the wheat fields, forests, and villages where the 4th Brigade fought. This is hallowed ground. Even to this day, the battlefield bears the scars of vicious combat—fighting positions, trenches, shell holes, and shards of shrapnel are everywhere. It was a wonder to me that anyone could survive, much less prevail, in the cauldron that was Belleau Wood. Survival required much more than just courage and exceptional training. The individual Marine rifleman had to be innovative, resourceful, and capable of making the right decision in extremis—in many ways, a force of one. More important, though, each and every Marine at Belleau Wood had to believe in his heart that, although he might seem alone and on his own in the darkness of the forest, he was actually fighting as part of an inseparable team—his unit—Marines who he could never let down.

While walking in the wheat field through which the Marines attacked on the 6th of June 1918, it dawned on me that the Battle of Belleau Wood was won before it was even joined. On the eve of their trial by fire, the Marines of the 4th Brigade were supremely confident in their personal abilities to carry the day, and more importantly, they felt an incredible allegiance to their unit and to their fellow Marines. It was these attributes that enabled them to prevail in the crucible of Belleau Wood. These same attributes—confidence and allegiance—will be necessary for success in the battles that will confront Marines in the twenty-first century. The Corps' Crucible of today is designed to help Marines prepare for those future battles through the inculcation of these attributes.

[13] The original article came from Gen Charles C. Krulak, "The Crucible: Building Warriors for the 21st Century," *Marine Corps Gazette* 81, no. 7 (July 1997). Minor revisions were made to the text based on current standards for style, grammar, punctuation, and spelling.

Laurel Stern Boeck, *Gen Charles C. Krulak, 31st Commandant of the Marine Corps* (1995–99).
Art Collection, National Museum of the Marine Corps

I know that many of you have already heard of the Crucible. Some of you are even beginning to receive Crucible-trained Marines into your units. Let me share with you our rationale for starting the Crucible, identify what this training evolution entails, and then discuss the opportunities and the challenges that it poses to us as leaders.

The Crucible was part of the Corps' focus on values under Krulak as seen in the recruitement poster *Our Core Values. Honor, Courage, Commitment.* *Official U.S. Marine Corps Recruiting poster*

WHY THE CRUCIBLE?

The Crucible was not implemented because we found our tried-and-true methods of recruit training to be flawed. Nothing could be further from the truth. We developed the Crucible for two major reasons. The first reason is that we saw a change in the operating environment in which our Marines will be employed. Decentralized operations, high technology, increasing weapons lethality, asymmetric threats, the mixing of combatants and noncombatants, and urban combat will be the order of the day vice the exception in the twenty-first century. Our Marines must be good decision makers. They must be trained to the highest standard. They must be self-confident. They must have absolute faith in the members of their unit. This is why we have instituted the values program for all Marines. This is why we have enhanced the way we transform America's sons and daughters into U.S. Marines. This is why we have included the Crucible as part of the transformation process. We must ensure that our newest Marines fully understand and appreciate what the Marine Corps represents, and that, as members of the world's fighting elite, they must uphold the sacred trust we have with our great nation—and the sacred trust that we have with each other. The Crucible is designed specifically to contribute to the making of this kind of Marine. Preparing our young Marines for battle is the genesis for the Crucible.

The second reason for the Crucible was derived from subtle changes in the societal norms and expectations of America's youth. We have all heard the term *generation X*, a term often associated with a negative connotation. Yet, it is from this generation that we recruit the Marines who will be our future. It is, therefore, important for us to understand just how the young people of today view the world, to understand what motivates them. Almost two years ago, we brought in a team of psychologists to tell us about generation X. From them, we learned that young people today are looking for standards, and they want to be held accountable. They, for the most part, do not mind following, but they can lead and want to lead. Most want to be part of something bigger than themselves. They want to be something special. Most believe in God. Many do not fully recognize it as such, but they want to have faith. These traits manifest themselves in a tendency to join—join gangs, join fraternities and clubs, join causes. These are exactly the same attributes and attitudes that offer the Marine Corps a tremendous opportunity. Generation X does not want to be "babied." These young Americans are looking for a real challenge. They desperately want to be part of a win-

Marine recruits drag a fellow recruit through an obstacle during the Crucible at the Marine Corps Recruit Depot Parris Island, SC, on 3 December 2015.
Official U.S. Air Force photo, courtesy of SSgt Kenneth Norman

ning team; they crave the stature associated with being one of the best. These are the Marines of the future, the warriors of the twenty-first century. The Crucible is giving them exactly what they want—and exactly what we need.

WHAT IS THE CRUCIBLE?

Remember that transformation is a four-step process: recruiting, recruit training, cohesion, and sustainment. The Crucible is the centerpiece of the recruit training phase. It is a three-day training evolution that has been added to the end of basic recruit training, designed specifically to make Marines better warriors. It features little food, little sleep, more than 40 miles of forced marches, and 32 stations that test physical toughness and mental agility. The events are designed to focus primarily on two areas: shared hardship and teamwork. We wanted to create a challenge so difficult and arduous that it would be the closest thing possible to actual combat. We wanted to create for the recruits a Crucible that, once experienced, would be a personal touchstone and would demonstrate for each and every recruit and candidate the limitless nature of what they could achieve individually and, more importantly, what they could accomplish when they worked as a team. To accommodate this culminating event, we lengthened recruit training to 12 full weeks. The Crucible has been strategically placed in the 11th week of training, a week we have designated "Transformation Week."

The drill instructor is still the backbone of the recruit training process. The drill instructor's role in the first 10 weeks of recruit training

Marine recruits provide security while another exits a tunnel during the Crucible at Marine Corps Recruit Depot Parris Island, SC, on 14 July 2016.
Official U.S. Marine Corps photo, courtesy of Cpl John-Paul Imbody

remains as it always has been. However, during Transformation Week, the drill instructor trades his or her traditional campaign cover for a soft cover or a helmet, and transitions to the role of a team leader and mentor for the Crucible process. The drill instructor guides the recruits, seeking to build confidence in their individual abilities and to emphasize the importance of the team. The objective is to build a sense of unit cohesion so that, by the end of the Crucible, the individual recruits see the value of working together in a common cause to overcome the most arduous tasks and conditions.

The drill instructor's job is not over, however, when his or her recruits complete the Crucible. There is a week remaining—Transition Week. It is the time when our newest members have the opportunity—and the responsibility—to increase their knowledge and confidence so that they are fully prepared for what lies ahead. It is during this last week that the drill instructors debrief the recruits' Crucible experience, identifying and reinforcing the teamwork and values that allowed them to prevail in times of duress and hardship.

THE OPPORTUNITY AND THE CHALLENGE

The results of the first iterations of the Crucible have been impressive, not only in the increased sense of pride and maturity in our new Marines, but in other, more tangible, ways as well. For example, liberty incidents of the Crucible-trained companies going through infantry training battalions at the schools of infantry have decreased dramatically. Both schools report that companies composed of Marines who have completed the Crucible are performing better than Marines who underwent the syllabus prior to implementation of the Crucible. Recruiters report that these new Marines, when assigned to the Recruiters Assistance Program, are more responsible and more confident. These are preliminary results, but clearly we have hit the mark. We have taken a proven process that produces the finest fighting men and women in the world and actually improved it!

Now as these Marines—tempered in the Crucible—enter our ranks, it is up to every leader in the Corps to combine the strengths of our experienced Marines with the intensity of our new Marines. This amalgamation will increase unit warfighting capabilities. As always, Marine leaders must capitalize on the strength that every Marine brings to the team.

You have great Marines now. Your new Marines will be the same in many ways yet will be different for their Crucible experience. Think about how you will capitalize on that difference. Think about how you will meet this challenge. While it is true that leadership fundamentals are timeless, the method of application varies with every scenario and with each individual. I have complete confidence that in this organization of leaders you will find the methods to maximize this opportunity wherever you are—in your fire team or shop, in your battalion or squadron, in your Marine air-ground task force, or on your staff.

The battles ahead will be violent, chaotic, and lethal. It is our responsibility to prepare our Marines for these future trials. They, like their forefathers at Belleau Wood, must have complete confidence in their individual abilities and in those of their unit. The Crucible helps instill that confidence. But, it only helps. It is up to us to do the rest with good, old-fashioned, Marine Corps leadership.

Semper Fidelis. USMC

PREPARING THE MARINE CORPS FOR WAR

by General Charles C. Krulak
Marine Corps Gazette, September 1997

In 1993, we took a major step forward in ensuring that our Corps would always be "The most ready when the Nation is least ready."[14] In that year, we published *Marine Corps Order (MCO) P3900.15*, which defines the Marine Corps Combat Development Process (CDP), now called the Marine Corps Combat Development System (CDS). In that order, we codified an integrated process by which we identify, obtain, and support necessary combat capabilities for the Marine Corps. The CDS is not about the procurement of things. It is about the procurement of capabilities. Things do not win battles. Marines win battles . . . Marines who can outthink, outmaneuver, and who have the capabilities to overwhelm their foes.

As part of the CDS, the Commandant of the Marine Corps is responsible for publishing a document called the *Commandant's Planning Guidance* (CPG). The CPG is intended to be the foundation of Marine Corps planning, the cornerstone of our efforts to maintain a combat ready Marine Corps. Two years ago, we published the CPG—a comprehensive document that serves as the schematic for how we make Marines and win battles for the nation. Now, at the halfway mark of this commandancy, it is time to revisit the CPG and remind ourselves that our priority must always be maintaining our focus on preparing the Marine Corps for war.

UNIQUE CONTRIBUTIONS

We have made significant contributions to the nation's defense in several new and very unique ways. Identifying a gap in our nation's ability to rapidly respond to chemical and biological attacks, a Chemical Biological Incident Response Force (CBIRF) was created. This unit is ready for use in the fight against those who would attack our nation asymmetrically. It has been deployed to several real-world contingencies, making our population and leadership safer—at the Olympic Games in Atlanta, at the president's inauguration in Washington, and at the economic summit in Denver. It is ready for worldwide deployment and is improving on ways in which it can be used to "train the trainers" in organizations and agencies preparing for similar contingencies.

[14] The original article came from Gen Charles C. Krulak, "Preparing the Marine Corps for War," *Marine Corps Gazette* 81, no. 9 (September 1997). Minor revisions were made to the text based on current standards for style, grammar, punctuation, and spelling.

A Standing Joint Task Force Headquarters (SJTFHQ) has also been created. Recognizing the ad hoc nature in which Joint Task Force (JTF) headquarters are usually created and the inefficiencies incurred in such activations, the SJTFHQ was developed to address those inefficiencies. Resourced by the Marine Corps, it is ready for use by any theater commander. The SJTFHQ has participated in numerous joint exercises and most recently honed its skills in the European Command's Exercise Agile Lion [Lithuania, 1997].

The special operations capable Marine expeditionary unit, or MEU(SOC), deployments offer our nation the quintessential crisis response force. Our country's reliance on their capabilities during the last several years has dramatically increased. But, rare are the occasions when any Service or Service department conducts operations solely with its own resources. Because of this, and because the MEU(SOC)s are often the first on scene, we have increased their command-and-control capabilities so that they might be better prepared to serve as JTF enablers.

The enhanced MEU(SOC) capability to serve as a JTF enabler, the chemical and biological crisis management capability, and the SJTFHQ all provide unique capabilities to our national defense. By anticipating and filling the nation's warfighting requirements, we are preparing the Marine Corps for war.

TRADITIONAL CAPABILITIES

While adding and enhancing some capabilities, our stock-in-trade remains being able to field well-trained and capable Marine Air-Ground Task Forces (MAGTFs). Whether a small, special purpose MAGTF (SPMAGTF) organized for a contained response, or a Marine Expeditionary Force (MEF) employed in a major theater war, the Marines we send to battle must be well trained, properly organized, and ably led. Necessary combat power will be provided to the MAGTF, particularly at the MEF level, through global sourcing from the Total Force: one force consisting of Marines, both Active and Reserve.

Marines receive a class in winter warfare during a snow storm while training at Bridgeport, CA, in the Sierra Nevadas in 1992.
Courtesy of Paul Westermeyer

The Marine Corps Maritime Prepositioning Force (MPF) remains one of the cornerstones of our ability to quickly insert a sustainable and capable force in a time of significant crisis or challenge to our national interests. While offloading operations may take advantage of benign port facilities, it is our ability to offload unassisted by such infrastructure that makes MPF such a versatile means of force introduction. Congress has provided the funding, and we have contracted for an additional ship in each of our three MPF squadrons. This enhanced MPF capability will mean an added expeditionary airfield, field hos-

A Marine rushes toward an objective during a combined arms exercise at Camp Lejeune, NC, on 15 December 2017. The exercise allowed riflemen and machine gunners to work together assaulting objectives while covered by live fire.
Official U.S. Marine Corps photo, courtesy of Sgt James Skelton

pital, and additional sustainment for our committed forces.

But forcible entry from the sea remains the Marines' forte. We continue to work with our Navy shipmates to ensure we reach our resource-constrained, programmatic goal of enough amphibious shipping to lift the equivalent of 2.5 Marine expeditionary brigades. The requirement—the capability which we strive to provide to our nation—remains at 3.0 brigade equivalents. The goal of the naval Services is to ensure a credible amphibious capability is ready when the nation says, "land the Marines."

Once landed, our ability to maneuver effectively is directly tied to our tactical mobility. The [Bell Boeing] V-22 [Osprey], the advanced assault amphibian vehicle (AAAV), and the procurement of the lightweight 155mm howitzer are all part of an overarching architecture designed to make sure we have the mobility to support our doctrine of maneuver warfare. But, we must explore advanced technologies, not just for ship-to-shore movement or for enhanced air and ground mobility, but also for technologies that support the individual Marine's mobility. Their clothing and equipment have a direct and immediate impact on survivability, lethality, and mission accomplishment.

Marine Corps operational forces will continue to be organized as MAGTFs, with the MEF as the principal warfighting organization. We will maintain the amphibious forcible entry option for the National Command Authority.[15] We are enhancing our ability to move significant

[15] *National Command Authority* is a DOD term for the ultimate source for lawful military orders.

warfighting capabilities to a point of crisis, and we are aggressively working at increasing our tactical mobility. We are focused on preparing the Marine Corps for war.

DOCTRINE

We are forging ahead with our doctrine efforts, ensuring that concepts and doctrine are synchronized, covering the gaps, and coordinating materials at every level. New doctrinal publications are coming off the presses. And most important, the doctrine is sound.

To ensure that Marine Corps capabilities are understood and properly employed, we are fully participating in the joint doctrine development process. Having our capabilities fully inculcated in the nation's quiver of warfighting techniques is vital to her defense. Ensuring that our Marines and our fellow joint warriors have a fundamental understanding of warfighting principles is of inestimable importance. We are focused on preparing the Marine Corps for war.

HARNESSING OUR ASSETS

No organization can be truly efficient until it harnesses all of its resources, especially its people. This is particularly true if your business is warfighting. The Corps recognizes that every Marine has something to contribute. We are a diverse institution comprised of men and women representing the cultural and ethnic diversity of our nation. These Marines are our warfighters and race, creed, and gender make no difference. It is paramount that we, as an institution, foster an environment of dignity and respect for all Marines, an environment where all Marines feel proud to be part of something bigger than themselves. Those who cannot act with dignity and respect toward their fellow Marines in garrison certainly have not properly prepared their character for the stresses of war.

Often large organizations fail to take full advantage of their people because there is no mechanism by which good ideas can be surfaced to the top. Recognizing that good ideas come from individuals of all experience levels and from throughout our rank structure, we created Marine Mail.[16] When we started, we asked for answers to three questions: What are we not doing that we should be doing? What are we doing that we should be doing differently? What are we doing that we should not be doing? Since then, we have also come to appreciate just how many good ideas are out there with respect to new concepts, tactics, and equipment that might improve our warfighting capability. The response from Marines across our Corps has been tremendous. We are a stronger warfighting organization for the contributions received through Marine Mail.

Just as we have empowered the ranks of our Corps, we must ensure that we are making maximum use of the talents resident in our most seasoned leaders. An Executive Steering Committee has been created to make better use of the knowledge and experience of our senior leadership at the lieutenant general level. This more formalized process of coalescing ideas and tracking progress has been very helpful in the decision-making process that guides our Corps. Capitalizing on our diversity, emplacing a mechanism to encourage the free flow of new ideas, and maximizing the talents of our senior leaders are all measures designed to prepare the Marine Corps for war.

[16] *ALMAR 001/07, Marine Mail* (Washington, DC: Headquarters Marine Corps, 2007). The Commandant created Marine Mail "to encourage creativity and innovation by providing each Marine, sailor, or civilian Marine, regardless of grade, a method by which their positive and professional ideas could be heard."

LCpl Nathan Long, 7th Engineer Support Battalion, 1st Marine Logistics Group, scouts for enemy contact during a military operation on urban terrain exercise during his Sapper Leaders Course at Camp Pendleton on 2 November 2017.
Official U.S. Marine Corps photo

INNOVATION

Innovation is our key to ensuring that we provide the nation with a Marine Corps that is organized and equipped to fill our role as the nation's expeditionary force-in-readiness . . . ready not just for the battles of today but of tomorrow and the day after tomorrow. The *Quadrennial Defense Review* just finished, and the National Defense Panel underway, seek to define our place in the national defense.[17] It is up to us to develop the operational concepts through which we will affect that role.

Operational Maneuver from the Sea (OMITS) is our operational concept.[18] Using the quantum leap in capabilities of the V-22, the air cushioned landing craft, and the AAAV, we will be able to take maneuver warfare to a new level. We will not be constrained by traditional beach landing sites. We will avoid enemy defenses where he is strong and attack through his weaknesses to destroy his ability and desire to resist. Through an unprecedented ability to generate tempo, we will overwhelm our enemies and protect our force.

We are conducting a series of advanced warfighting experiments to determine, among many things, the best configuration of the force that will execute OMFTS. The Marine Corps Warfighting Laboratory will gather the data from these experiments. Based partly on this data, we will then conduct a comprehensive Force Structure Planning Group to evaluate the structure of our Corps. It is through experimentation that we will we find the recipe for success on tomorrow's battlefield.

Just as we have done in other interwar periods, we are using experimentation and innovation to ensure we are ready for war as it will be, not as it was. Innovation is one of the keys to preparing the Marine Corps for war.

PROFESSIONAL MILITARY EDUCATION

Professional military education (PME) is crucial to our development as warriors. There are few dilemmas that will face our Marines on the field of battle that have not been faced before. Even as the nature of war evolves, the challenges associated with it contain a number of reoccurring themes. The Marine who has not availed himself of the opportunity to learn from the mistakes and successes of others is ill prepared for war. He or she stands a higher chance of needlessly becoming a casualty, endangering other Marines, and failing to accomplish the mission. But, Marines do not fail in battle. We prepare ourselves for it. We ensure we are technically and tactically proficient. We study our trade.

We have placed great emphasis on PME. We want all Marines to receive top-grade ed-

[17] *Quadrennial Defense Review* (Washington, DC: DOD, 1997, 2001, 2006, 2010, 2014).
[18] *Operational Maneuver from the Sea*, MCCP-1 (Washington, DC: Headquarters Marine Corps, 1999).

A McDonnell Douglas F/A-18 Hornet over the South China Sea.
Official U.S. Marine Corps photo, courtesy of LCpl John McGarity

ucation at every level, education that will make them better warfighters. Having said that, we have received considerable input that says some of our correspondence course PME may be too time-consuming, that it is detracting from the accomplishment of our day-to-day mission. We are examining that. The goal is a continuous and incremental increase in the ability and education of every Marine as he or she progresses in rank.

Formal, residence courses are valuable experiences for those Marines who get an opportunity to attend them. We have not always done as good a job as we could have in filling school quotas. This is an area that requires constant monitoring. These courses are where we accrue the skills necessary to allow us to conduct decentralized operations—to fight and win.

The Commandant's Reading Program is designed to help steer our Marines toward books with good lessons.[19] The MAGTF Staff Training Program provides professional education for our staffs. All of our correspondence and resident PME courses cultivate our Marines as warriors and prepare them for additional responsibilities. PME is an essential ingredient in preparing the Marine Corps for war.

DEVELOPING THE WARRIOR

We have already spoken of the preparation of our warriors' minds through education, but there is more to being mentally prepared for combat than being well schooled in the art of war. Our Marines must be mentally tough as well.

[19] The Commandant's Reading Program has since evolved into the Commandants Professional Reading List (CPRL), which is accessible through the Library of the Marine Corps. The Commandant believes the CPRL is necessary for professional development and critical thinking at each level. It is arranged into two sections: Commandants Choice and by grade levels (entry, primary, career, intermediate, and senior). Each Marine is required to read a minimum of five books from the Commandants Choice or grade-level sections each year. These levels coincide with specific ranks and Marines should attempt to read all titles within their level prior to proceeding to a higher level.

The Transformation Process helps make Marines with the depth of character to do the right thing, in the right way, for the right reasons. Marines full of conviction and with strong minds, Marines who have been made to look within for the answers they seek, will be a powerful force on any battlefield to which the nation sends them. The Marine Corps values program is designed to reinforce and help sustain the hardening—the *Transformation*.

Of all the things we do as an institution, none is so crucial as preparing our Marines for the rigors of combat. Tough physical training hardens our warriors, makes them equal to the challenges ahead. The Physical Fitness Test has been made tougher. We will continue to emphasis fitness as a way of life for Marines.

But Marines are more than body and mind. To be a United States Marine, one must prepare the body, the mind, and the spirit. The experiences of the Crucible are the gateway for the development of the Marine spirit. The cohesion-building phase of the Transformation Process is designed to strengthen the bonds between us as warriors. Being a Marine has always been a mystical association of spirit with one's fellow Marines. Capitalizing on this esprit, we have become the band of warriors we are today, feared by our foes, and respected throughout the world.

We make Marines—body, mind, and spirit. Making Marines is all about preparing the Marine Corps for war.

YOU THE MARINE

The last paragraph of the CPG states:

> *In the final analysis, my guidance simply is to be prepared to fight, on the shortest notice, under any circumstances of weather or resistance, in conflicts large or small. Be prepared to integrate Marine combat power smoothly into the overall matrix of other U.S. Services or other nations. Be prepared, in conjunction with the U.S. Navy, to project power from the sea for as far and as long as necessary. Be ever mindful of technological opportunities to enhance combat proficiency and to promote logistic economy. Be also mindful of the deep meaning in Title 10 of the U.S. Code of the requirement that Marines shall be prepared to discharge "such other duties as the President may direct," whatever those duties may be. But, most of all, be prepared to fight and win.*

We, as an institution, are preparing the Marine Corps to fight. This preparation is reflected in everything we do. From revamping the fitness report system so that we are sure to promote our most qualified warriors, to our aggressive efforts in seeking funding for the tools we need to enhance our warfighting capabilities, it is all about preparing the Marine Corps for war. But we cannot achieve our goals as an institution without participation from all Marines—from you. Look for ways to contribute to the readiness of your Corps. Make training for your Marines tough and demanding. Reaffirm your commitment to principle and make a check of your personal character. Conduct or participate in the daily warfighting discussions mandated in the CPG. Send in a Marine Mail, write your ideas in an article for the *Gazette*, or share a lesson learned with a peer. Together, we shoulder the awesome responsibility of *preparing the Marine Corps for war*.

UP TO THE CHALLENGE
Women's Training in Today's Marine Corps

by Colonel Nancy P. Anderson
Leatherneck, February 2002

Like all fulfilling achievements, the acceptance of women into the Marine Corps was hard-won.[20] "The American tradition is that a woman's place is in the home," said Brigadier General Gerald C. Thomas, director, Headquarters Marine Corps Plans and Policies Division in October 1945.

That comment seemed to reflect the attitude of Marine Corps leadership on restricting female accessions. In the years between the end of the Korean and Vietnam wars, the percentage of women serving in the Regular component of the Marine Corps remained fairly constant, approximately 1.3 percent, even though the 1948 law limiting women to 2 percent had been repealed in 1967.

Winds of social change began to blow in the 1970s. Following Vietnam, the military Services actively recruited women to fill anticipated all-volunteer force personnel shortfalls. In early 1975, in response to requests from operational commanders, Marine Corps Commandant, General Louis H. Wilson Jr. authorized the routine assignment of women to 526 nondeployable continental United States operational billets and tasked commanders to review both stateside and overseas billets with a view to substantially expand assignment options for women. Female accessions were monitored to ensure force commanders had sufficient male Marines to preserve a 5-to-1 deployment rotation base and still provide fair career progressions for all Marines. "Fair" would begin at entry-level training.

In 1976, Major Barbara E. Dolyak assumed command of Company L, the all-female officer basic class at Camp Barrett, The Basic School (TBS), Quantico, Virginia. Major Dolyak questioned why it was essential for the 82 percent noninfantry male lieutenants to attend a 26-week TBS while female lieutenants received only 12 weeks of training. The resulting buzz made its way to the halls of Headquarters Marine Corps (HQMC). Separately, plans were underway to shorten TBS to 21 weeks and to create a follow-on Infantry Officer Course for lieutenants assigned the 03 (infantry) military occupational specialty.

The stars—director of Training, command-

[20] The original article came from Col Nancy P. Anderson, "Up To the Challenge: Women's Training in Today's Marine Corps," *Leatherneck*, February 2002. Minor revisions were made to the text based on current standards for style, grammar, punctuation, and spelling.

Women are integrated now more than ever into the Corps.
Official U.S. Marine Recruiting poster

ing general, Marine Corps Development and Education Command, and the Commandant—aligned. White Letter no. 5-76, "Women Marines," was published on 23 June 1976. In it General Wilson stressed the fact that increased opportunities for women demanded positive leadership and management by commanders. He directed that "commanders who are responsible for the conduct of professional schools should review curricula to ensure that the training offered prepares Marines to lead, irrespective of sex."

Twenty-two female and 243 male lieutenants reported to Company C, basic class 3-77, on 4 January 1977. The women formed 2d Platoon, led by Captain Robin L. Austin. That summer, the Corps' leadership realized the importance of physically and mentally preparing women before TBS, and the decision was made to gender-integrate Officer Candidates School (OCS).

By 1979, five consolidated TBS companies had graduated, including some that had men and women assigned to the same platoon. The women received the same tactics training, although they assumed only defensive roles in field problems. The women were assigned to all-female platoons between 1980 and 1992 due to "physiological differences and legal limitations set forth concerning application of phases of offensive combat." The women's training expanded as combat exclusion was less rigidly defined. Male and female TBS lieutenants have been assigned platoons alphabetically by their last name since 1992 and undergo the same training.

For enlisted women, too, change was in

the air. Beginning January 1976, female drill instructors were authorized to attend, and required to complete, Drill Instructor School. The drill instructor is the first Marine a recruit meets when reporting for active duty and the last Marine a recruit will ever forget. Sending to Drill Instructor School all Marines charged with transforming young adults into Marines ensured men and women could be trained to the same high standard.

In November 1980, General Robert H. Barrow, 27th Commandant of the Marine Corps, announced a pilot defensive combat training program for female recruits. Training included field exercises, some combat training, and weapons familiarization. Women would fire with the M16A1 but not for qualification score. However, only three days into the first field exercise, the Parris Island, South Carolina, deputy commanding general, Brigadier General William Weise, observed the training and did not like what he saw. He stated that, while the female recruits were being trained to learn how to defend themselves, "Women do not have the physical or emotional stamina to handle the rigors of the battlefield. I would not want to see my daughters or female friends of mine in a combat situation if I could avoid it." Their training was scaled back.

Teaching field tactics to women was a sensitive issue from Parris Island and Quantico to the halls of HQMC. There was as much concern about public perception of women receiving combat training as with the need to teach all Marines how to stay alive.

Between 1978 and 1981, the number of enlisted female Marines rose from 4,652 to 7,091, and female officers increased from 433 to 526. At this point, in order to achieve training goals, female drill instructors were averaging 340 workdays annually compared to 230 to 250 days by male drill instructors. To better train and lead more female recruits and officer candidates, the number of female drill instructors was slowly increased.

GySgt Watson, USMCR, by Maj Alex Durr.
Art Collection, National Museum of the Marine Corps

Marine Corps Order 1500.24D, Training Policy for Women Marines, published on 20 May 1985, recognized that even combat service support billets removed from the forward line of troops did not guarantee safety. The Commandant, General Paul X. Kelley, directed that female Marines receive the same training as male counterparts in the same units and billet, less offensive combat training.

LCpl Stephanie Robertson, a member of the female engagement team assigned to 2d Battalion, 6th Marines, Regimental Combat Team 7, speaks with locals during an engagement mission in Marjah, Afghanistan, on 18 August 2010.
Official U.S. Marine Corps photo, courtesy of LCpl Marionne T. Mangrum

Recruit training was again modified to include instruction in day and night tactics, rappelling, the confidence course, and defensive field training. Female recruit training was extended from 8 to 11 weeks, mirroring that given to male recruits, and included a three-night field training exercise. Female recruits were tested in close order drill with rifles for the first time in July 1985, while the female drill instructors made their debut with noncommissioned officers' swords.

Private Anita Lobo, in the first female series to fire for score at the rifle range that year, raised a few eyebrows by setting a new Parris Island range record, firing 246 of a possible 250. On 1 November 1986, the commanding general, Parris Island, Major General Harold G. Glasgow, redesignated Woman Recruit Training Command as the 4th Recruit Training Battalion, Recruit Training Regiment. Female drill instructors were qualified as swimming and marksman coaches and drill masters. Senior women were increasingly visible in training and command billets, although male and female recruit training remained separate.

The appointment of General Alfred M. Gray Jr. as 29th Commandant brought the mantra "Basic Warrior" to the Corps. Basic Warrior Training was initiated at the recruit depots and enhanced at the School of Infantry, producing a basically trained, combat-ready Marine, male or female.

The repeal of Title 10 U.S. Code combat restrictions and new secretary of defense policy defining combat risk made Service chiefs responsible for ground combat policies with respect to the employment of women.

In January 1988, training days for female recruits were extended from 57 to 64, while male training remained at 56 days. The increase provided women a condensed version of basic warrior training, as they did not attend the School of Infantry. That December, Company O, Series 4000, became the first female series to execute the entire Combat Assault Course as an offensive exercise. In January 1989, female recruits began to run in boots and utilities and with the rifle.

By 1994, male and female recruit training programs were nearly identical. By October 1996, men and women were undergoing 12 weeks of recruit training. The traditional three phases of recruit training—basic military skills, marksmanship/combat skills and advanced skills—were replaced with a building block approach to forge individuals into an effective team. The 54-hour Crucible would be their toughest challenge.

Female accession quotas continued to rise, and women continued to enlist. A third company was added to 4th Recruit Training Battalion, allowing for a sixth series and putting all series into more efficient training phases.

Following graduation, female recruits were assigned to Marine Combat Training, initially at Weapons Battalion, Parris Island, and then to Camp Lejeune, North Carolina. Male and female recruit training was identical.

The increasing number of women completing recruit training or OCS is more striking when injury attrition numbers are examined. The annual attrition of female recruits and officer candidates for medical reasons is nearly twice as high as among men, more than 12.5 percent in 2000. Musculoskeletal injuries associated with vigorous physical training are the leading cause of female attrition. Commanding officers and staff of the Recruit Training Regiment, OCS, and TBS have focused on this particular gender discriminator during the last decade.

The chief attributing factor is that women arrive for entry-level training in poor physical condition and with weak bones. Neither recruit training nor OCS last long enough to develop the bone density and lean muscle mass that are principal safeguards against these injuries. Strength conditioning programs were initiated at Parris Island and Quantico, using stronger muscles to compensate for weaker bones. Female recruits and officer candidates were taught to run and hike more efficiently and to develop a shorter, faster stride to minimize lower leg injuries. In trying to reduce injuries, female officer candidates would hike different routes, hike different paces, or begin ahead of male candidates. Each commanding officer worked to find a solution.

As perception is reality, it also was necessary to explain the reasons behind the training differences to the male candidates. Men and women worked equally hard, but it is easy to acquire a "second class" moniker when the perception was that women are not working as hard as men. The different physical fitness standards were less a readiness issue than a morale issue. To further manage the perception of fairness, savvy trainers or team leaders would have a female carry the first and last loads during a coed problem if she could not carry as much per load.

Women do not learn in the same way as

Capt Elizabeth A. Okoreeh-Baah, the first female MV-22 Osprey pilot, stands on the flight line after a combat operation on 12 March 2008. Capt Okoreeh-Baah spent five years flying the CH-46E Sea Knight before transitioning to the Osprey.
Official U.S. Marine Corps photo, courtesy of Cpl Jessica Aranda

men. Live grenade toss was stopped between 1986 and 1992 for female recruits when an evaluation showed that 60 percent could not throw a dud grenade the 15 meters required to escape the bursting radius. For the leadership, the solution was to halt training, yet all the women needed was instruction on how to throw. Most women faced the target rather than rotating their shoulders and hips 90 degrees away from it and then swinging around in order to accelerate the grenade. A similar problem occurred at OCS when women were initially not shown how to wrap the rope between their boots to form a step for the rope climb, drawing far less on upper body strength. The women learned, and the instructors found more effective teaching methods.

Personality profiling and educational studies have shown that women are more visual and learn better by seeing than by hearing or telling. At Parris Island, when female recruits were taught about the rifle before learning to fire it, rifle qualification percentages went way up. A change as simple as posting the series training schedule enabled women to see how each class linked into the big picture. Women were not asking to be shown an easier path; rather, they wanted to be taught how to succeed.

Meanwhile, the number of women completing initial training and serving proudly as United States Marines continues to rise and continues to exceed planning estimates. Women in today's Corps represent 6 percent of the active and 4.6 percent of the Reserve force. Women choose to be Marines for the same reasons as men: duty to country, money for education, an escape from a bad situation, growth as a person.

The issue of meaningful military service does not involve gender but, instead, the strength and defense of the United States in general and the most efficient use of personnel in particular. Today's Marines, male or female, know their stuff and have earned the eagle, globe, and anchor.

1ST MARINE DIVISION (REIN) COMMANDING GENERAL'S MESSAGE TO ALL HANDS

March 2003

Just prior to the beginning of Operation Iraqi Freedom, General James N. Mattis, commanding the 1st Marine Division, sent the following message to all of the sailors and Marines of his division.[21]

For decades, Saddam Hussein has tortured, imprisoned, raped, and murdered the Iraqi people; invaded neighboring countries without provocation; and threatened the world with weapons of mass destruction. The time has come to end his reign of terror. On your young shoulders rest the hopes of mankind.

When I give you the word, together we will cross the line of departure, close with those forces that choose to fight, and destroy them. Our fight is [neither] with the Iraqi people, nor is it with members of the Iraqi Army who choose to surrender. While we will move swiftly and aggressively against those who resist, we will treat all others with decency, demonstrating chivalry and soldierly compassion for people who have endured a lifetime under Saddam's oppression.

Chemical attack, treachery, and use of innocent human shields can be expected, as can other unethical tactics. Take it all in stride. Be the hunter, not the hunted; never allow your unit to be caught with its guard down. Use good judgment and act in the best interests of our nation.

You are part of the world's most feared and trusted force. Engage your brain before you engage your weapon. Share your courage with each other as we enter the uncertain terrain north of the line of departure. Keep faith in your comrades on your left and right and Marine Air overhead. Fight with a happy heart and a strong spirit.

For the mission's sake, our country's sake, and the sake of the men who carried the division's colors in past battles—*who fought for life and never lost their nerve*—carry out your mission and *keep your honor clean*. Demonstrate to the world there is "No Better Friend, No Worse Enemy" than a U.S. Marine.

J. N. Mattis
Major General, U.S. Marines
Commanding

[21] The original letter came from Gen James N. Mattis to 1st Marine Division, "Commanding General's Message To All Hands," March 2003. Minor revisions were made to the text based on current standards for style, grammar, punctuation, and spelling. Emphasis (italics) found in original document.

AN OPEN LETTER TO THE "YOUNG TURKS"

by Lieutenant General Robert B. Neller
Marine Corps Gazette, November 2011

I want to take the opportunity to thank the *Gazette* for putting me in contact with Major Peter J. Munson.[22] As a result of his letter in the April issue and my response, we had a conversation on the phone. He also sent me his article, "Back to Our Roots," published in the April 2011 online version of the *Gazette*, and we discussed that as well. As I mentioned in my commentary printed in the June *Gazette* in response to "The Attritionist Letters," I believe it is better to talk and get things out in the open.[23] Consequently, though I have not changed my view as articulated in the "Rebuttal," I have considered the views of the good major, along with the captain (Captain Joseph Steinfels), who responded to me in the August *Gazette*, and many of their peers. I have personally listened to the views of these "Young Turks" in long and sometimes heated discussions during the past few years.[24] The following paragraphs are my view of their views.

Although my initial reaction to both the major's and captain's letter and article was to push back—whiny, do not get it, just pointing out problems without offering concrete solutions, spoiled by a resource rich environment where there is little accountability and a lack of supply discipline, think higher direction means a lack of trust, and on and on. The more I thought about our conversations, Major Munson's article, and the conversations I have had with his peers around the Corps, I came to the conclusion that this officer, like the authors of "The Attritionist Letters," is trying to tell leadership something about where "middle management" is mentally on their perceptions of the current state of the Corps and, more importantly, their expectations for the future. What I think I am hearing them say is:

> *We are tired of trying to fight this war with a supporting establishment—especially manpower and a process for equipping/training being the most consistently named examples—that is not responsive or attuned to the needs of the warfighter. In short, the operating forces are at war and the supporting establishment is not.*
>
> *We are given great freedom of action*

[22] The original letter came from LtGen Robert B. Neller, "An Open Letter to 'Young Turks'," *Marine Corps Gazette* 100, no. 3 (November 2011). Minor revisions were made to the text based on current standards for style, grammar, punctuation, and spelling.
[23] "The Attritionist Letters #1–#14," *Marine Corps Gazette*, 2010–13.
[24] The term *young Turk* refers to those focused on making radical change in the face of tradition.

Commandant of the Marine Corps, Gen Robert B. Neller (2015–present), gets water from the Devil Dog fountain after the American Memorial Day ceremony at the Aisne-Marne American Memorial Cemetery, Belleau Wood, France, on 29 May 2016. U.S. Marines, French servicemembers, family members, and locals gather each year to honor the memory of the Marines killed during the Battle of Belleau Wood. *Official U.S. Marine Corps photo, courtesy of SSgt Gabriela Garcia*

and responsibility for the lives and welfare of those in our charge while deployed to the fight, but when we return to the "world" we are treated like we do not know anything or, worse, like we are not trusted.

You can trust me back in the "world," like you do in combat. Just tell me what you want done, resource me, and let me lead. If I get it wrong then get in my business, but allow leaders to lead and on occasion kick one into the stands. At the same time, stop levying tasks on me that waste the time of the unit and the Marines/sailors.

We "get" the importance of safety and of taking care of the Marines in our charge, but this whole process has gotten out of hand. The great majority are paying for the sins of a very small minority, making all, regardless of rank, experience, and established performance, fill out forms for leave/liberty and be subjected to mandatory and poorly organized group training—suicide, safety, diversity, etc. This is where I feel you do not trust me, and this approach is not going to create the change in behavior and conditions that leadership is looking for. In fact, it may go the other way.

Our inadequate and precious prede-

ployment training program (PTP) time is wasted on noncombat-related training, which—to add insult to injury—is in many cases not well presented and not focused, in addition to being irrelevant. Be more concerned about the quality of the training than about the reporting of the results.

Now these thoughts probably sound like any conversation any group of peer officers has had about their higher during the last 234 years, but the fact that we have been in this war now for almost 10 years makes these concerns, in my mind, both more legitimate and valid. I say this because I believe today's Corps, based on all measures of performance and effectiveness, is a pretty good outfit. Without question, it is exponentially better than the Corps I joined back in the mid-1970s. But we are at a similar point in the cycle of sustained combat—war winding down, the budget knives out, and the nation, although it continues to be supportive, is tired both mentally and fiscally of the cost of war. My own greatest personal concern is that, once this fight ends, with the cuts we know are coming, unless we have a plan to address the issues the future leadership is raising and other long-term problems we know are institutional, this group/generation of officers is not likely to be satisfied (read stick around). I would submit that if we think we can simply go back to the "old Corps" pre-11 September 2001, and the bureaucracy is not tamed/changed/reformed, we will be sadly mistaken and dissatisfied with the results.

Although I think "The Attritionist Letters" and the thoughts of the Major Munsons of the world are a bit overstated, especially the inexplicable correlation between centralized, directed training executed in a decentralized manner equating to a lack of trust, it is done, I believe, for effect. These Marines are trying to tell us what they see and feel after 10 years of war. We now have majors who have never known any other Corps—PTP, deploy, fight, redeploy, PTP, and do it again and again. We have women who have seen more combat than most of us ever did growing up, which is another factor we must consider. The combat exclusion policy for women is insulting to them. I digress.

We would do well to heed and reflect on their "canary in the mineshaft" thoughts, engage them head-on in frank and candid discussion, look for ways to remediate those concerns that are legitimate, and explain our logic for those with which we do not concur. Of critical importance, we must not think that when this fight is over we can/should go back to operations as normal. Fewer and fewer Marines know what that is anymore. We will have to create a "new normal." We all know that coming out of a long conflict is fraught with risk, with historical issues of budget cuts, poor retention, and discipline issues. We have, I believe, begun the process to craft a plan to address these and other issues we have yet to wrestle with in order to keep the Corps the "middleweight force-in-readiness" the nation expects and needs. As important, the Corps must be a place where the best of the best want to stick around to be a part of what lies ahead.

ANTHONY C. ZINNI BIOGRAPHY

"Moral courage is often more difficult than physical courage. There are times when you disagree and you have to suck it in and say, "Yes, sir," and go do what you're told. There are also times when you disagree and you have to speak out, even at the cost of your career. If you're a general, you might have to throw your stars on the table, as they say, and resign for the sake of some principle or truth from which you can't back away."

~General Anthony C. Zinni[25]

General Anthony C. Zinni was born in 1943, and was commissioned a second lieutenant in the Marine Corps in 1965.[25] He served two tours in Vietnam as an advisor to the Vietnamese Marine Corps in 1967, and as a company commander with 1st Battalion, 5th Marines. In between those tours, he served as an instructor at The Basic School. He was wounded in Vietnam during his 1970 tour, and was awarded the Purple Heart.

In the 1970s and 1980s, he commanded the 2d Battalion, 8th Marines, the 9th Marines, and the 35th Marine Expeditionary Unit. In 1991, he served as the chief of staff and deputy commanding general of Combined Task Force Provide Comfort during the Kurdish relief effort in Turkey and Iraq.

In 1992, General Zinni was serving as director of operations for Unified Task Force Somalia. In that capacity, he often met with Mohammed Farah Aidid, then the leading Somali warlord; for these visits his driver, a Corporal Watts, would organize a squad from the headquarters staff for security. In *Battle Ready*, General Zinni described one memorable encounter with Aidid:

Inside the compound, the buildings were all layered with the porches that are typical in tropical countries. Dozens of heavily armed men always swanned about, staring brazenly at my Marines from every level of the buildings. During my meetings, the Marines stood beside the vehicles returning the stares of the cocky Somali gunmen.

Our entrances and exits to and from the compound were normally without incident, which was just as well, considering the possibilities. But one of our entrances proved to be deliciously memorable. As we were stepping out of our Humvees on this particular occasion, I was greeted by shocked faces on Aideed's men. I turned to the second Humvee in line, which seemed to be the source of all the excitement: An African American woman Marine was standing there in her battle

[25] Tom Clancy with Gen Tony Zinni (Ret) and Tony Koltz, *Battle Ready* (New York: Berkley, 2004), 425–26.

Gen Anthony Zinni's official photograph.
Official U.S. Marine Corps photo

gear, with her M-16 at the ready, looking tough as hell.

I left to conduct my business. Forty-five minutes later, when I came back out, the stir was still at high pitch. It was obvious the Somalis couldn't believe their eyes—an armed woman in Marine battle dress.

On the way back, I turned to Corporal

Watts. "You brought a woman Marine, huh," I said; I knew he'd set this scene up.

He smiled. "She'll kill you just as dead as any man," he said.

I laughed. He loved jerking the Somali tough guys' chains. Back to our headquarters, I drew the woman Marine aside for a quick chat. Corporal Watts was right. She'd kill you just as dead as any man could.[26]

From 1994 to 1996, he served as the commanding general, I Marine Expeditionary Force. During this period, General Zinni served as commander of the Combined Task Force for Operation United Shield, protecting the withdrawal of UN forces from Somalia. He took command of U.S. Central Command in 1997; he oversaw Operation Desert Fox, the 1998 airstrikes against Iraq. In 2000, he retired from the Marine Corps.

General Zinni's career has epitomized the complex difficulties the modern Corps faces; he has been called upon to act as an armed diplomat and peacekeeper as often as he has led his Marines in combat against an enemy. Combining traditional Marine aggressiveness with cultural sensitivity, he showed the way for today's Marines facing diverse, unexpected, operational, and strategic challenges.

[26] Clancy, Zinni, and Koltz, *Battle Ready*, 261.

APPENDIX

HISTORIOGRAPHY FOR MARINES
How Marines Should Read and Understand History

by Paul Westermeyer

Historiography, noun

1 a : the writing of history; b : the principles, theory, and history of historical writing

2 : the product of historical writing[1]

The Marine Corps is devoted to the idea of its history. At recruit training and Officer Candidate School, classes on the history and legends of the Marine Corps are taught to those aspiring to be Marines, and the Corps celebrates its birthday every year with a ceremony and ball. At the National Museum of the Marine Corps, in the Making Marines Gallery, a panel puts a twist on the Corps' hoary "Every Marine a Rifleman" shibboleth and declares "Every Marine a Historian" as it describes the classes taught at recruit training.

Of course, the concept of "Every Marine a Rifleman" does not suggest that all Marines carry rifles, it means that all Marines are trained in basic marksmanship and infantry tactics and techniques and they must complete annual rifle marksmanship qualifications. Similarly, "Every Marine a Historian" should go beyond merely providing Marines with a list of books to read; they should also develop the skills and tools required to evaluate and comprehend historical works. They should do this not with the intent to become professional historians, but rather for basic historical literacy. To understand history, Marines must have a basic understanding of historiography, "When you study 'historiography' you do not study the events of the past directly, but the changing interpretations of those events in the works of individual historians."[2]

The first question that historiography asks is the same question Marines need to ask when looking at any given historical work: What is history? It may seem obvious on its face; history is the story of mankind's past. Why then do we bother, beyond mere entertainment value, to study history; and how is history different from legend, mythology, or fiction?

Primarily, the difference lies in the meth-

[1] *Merriam–Webster*, s.v. "historiography (*n.*)," accessed 5 January 2018.

[2] Conal Furay and Michael J. Salevouris, *The Methods and Skills of History: A Practical Guide* (Wheeling, IL: Harlan Davidson, 1988), 223.

odology and intent. Historians base their interpretations on carefully collected historical facts with the intent to illuminate the past, not invent it. Thucydides, whose work on the Peloponnesian War is one of the oldest histories we have, said, "On the whole, however, the conclusions I have drawn from the proofs quoted may, I believe, be safely relied on. Assuredly they will not be disturbed either by the lays of a poet displaying the exaggerations of his craft, or by the compositions of the chroniclers that are attractive at truth's expense; the subjects they treat of being out of reach of evidence, and time having robbed most of them of historical value by enthroning them in the region of legend."[3] In other words, Thucydides claims he did not base his work on legend nor did he employ poetic license; rather, he based his work on the facts as he could best ascertain them.

This is the basic form of the historical method; as the Father of History, Herodotus of Halicarnassus, wrote: "I am bound to tell what I am told, but not in every case to believe it."[4] Historians examine a great many sources, compare them to each other, and then from these they tease the truth or come as close as they can to it. Thucydides explained in more detail:

> *With reference to the narrative of events, far from permitting myself to derive it from the first source that came to hand, I did not even trust my own impressions, but it rests partly on what I saw myself, partly on what others saw for me, the accuracy of the report being always tried by the most severe and detailed tests possible. My conclusions have cost me some labor from the want of coincidence between accounts of the same occurrences by different eyewitnesses, arising sometimes from imperfect memory, sometimes from undue partiality for one side or the other. The absence of romance in my history will, I fear, detract somewhat from its interest; but I shall be content if it is judged useful by those inquirers who desire an exact knowledge of the past as an aid to the interpretation of the future, which in the course of human things must resemble if it does not reflect it. My history has been composed to be an everlasting possession, not the showpiece of an hour.*[5]

Not every work telling a story from the past is a history, which is based on interpretations of facts from multiple sources. It is important to know when one is reading history and when one is reading something else. Many works look like a history but actually fit more readily into other disciplines, such as political science or journalism. More commonly, memoirs or autobiographies are mistaken for history, when they serve rather as sources that historians work from. Such first-person accounts have value, but the reader should understand their provenance and intent and not accept them as history.

The Commandant's Professional Reading List, for example, includes memoirs, biographies, polemics, and military fiction as well as histories. The easiest way to distinguish these works is to take them at their word. In *First to Fight*, Lieutenant General Victor H. Krulak tells the reader bluntly that his book is "a series of simple vignettes, part history, part legend, and part opinion."[6] Do not stretch a work beyond

[3] Thucydides, *The Peloponnesian War*, trans. Richard Crawley (New York: Random House, 1982), 1.21.

[4] Herodotus, *The History*, trans. David Grene (Chicago: University of Chicago Press, 1987), 7.152.

[5] Thucydides, *The Peloponnesian War*, 1.22.

[6] LtGen Victor H. Krulak, USMC (Ret), *First to Fight: An Inside View of the U.S. Marine Corps* (Annapolis: Naval Institute Press, 1984), xvi.

its author's intent. Other works make the task more difficult, as the author intentionally or unintentionally leaves the book's category unclear. For example, Bing West's *The Wrong War* superficially has the look of a history, describing the war in Afghanistan, but the methodology and intent makes it clear it is a journalistic polemic, advocating policy for the Afghan conflict. It is well worth reading, but not as a history.[7]

Historians generally classify sources as secondary and primary. Secondary sources are histories themselves, they do not purport to be current with the event; they analyze and interpret historical data. These are often used to place a work in context, providing background. Primary sources have direct knowledge of the event, unfiltered by others. These can be contemporary documents, memoirs, oral history interviews, newspaper interviews, photographs, or even archaeological finds.[8] Moreover, the viewpoint and limits of the source must be considered. For example, Colonel Gregory Boyington's memoir, *Baa Baa Black Sheep*, presents his view of World War II, but it was published a decade after the war ended. It is only one man's viewpoint; a more complete understanding of Marine Fighter Squadron 214's war comes from comparing his memoir with interviews and memoirs from the rest of his squadron and commanders, as well as the official records produced during the war itself.[9]

Drawing conclusions from these facts is the primary task of the historian, and historians differ over how they should do this; whether they should accept the past on its own terms, or judge it according to present values and mores. One of the pioneers of modern critical history, Leopold von Ranke, famously rejected the idea that history's job is to judge the past, stating that "History has had assigned to it the office of judging the past and of instructing the present for the benefit of future ages. To such high offices the present work does not presume; it seeks only to show the past as it actually was."[10] Other historians have concluded that history is memory of a particular kind; or as Edward Hallett Carr says in *What Is History?*, history is "a continuous process of interaction between the historian and his facts, an unending dialogue between the past and present."[11] In other words, historians should accept the past on its own terms and judge it as well, constantly reconsidering their conclusions about the past.

The next question is, why do Marines study history? Most famous military thinkers have espoused the need for professional soldiers to study the past. Alfred Thayer Mahan wrote in *The Influence of Sea Power upon History* that "the study of history lies at the foundation of all sound military conclusions and practice."[12] General James N. Mattis laid it out more plainly in an email defending the value of intense reading to Marine officers: "By reading, you learn through others' experiences, generally a better way to do business, especially in our line of work where the

[7] Bing West, *The Wrong War: Grit, Strategy, and the Way Out of Afghanistan* (New York: Random House, 2011).

[8] But always consider photographs with caution. Photographs provide an illusion of objective reality, but they are in truth composed, filtered recordings of the past. A photograph often obscures as much about an event as it reveals.

[9] Col Gregory Boyington, USMC (Ret), *Baa Baa Black Sheep* (New York: Bantam Books, 1958).

[10] Leopold von Ranke, *Geschichte der romanischen und germanischen Völker von 1494 bis 1514* [History of the Latin and Teutonic Nations from 1494 to 1514] (Leipzig, Germany: Duncker & Humblot, 1824), preface; and Ernst Breisach, *Historiography: Ancient, Medieval, and Modern*, 3d ed. (Chicago: University of Chicago Press, 2007), 233.

[11] Edward Hallett Carr, *What Is History?* (New York: Random House, 1961), 35.

[12] Capt A. T. Mahan, *The Influence of Sea Power upon History* (Boston, MA: Little, Brown, 1890).

consequences of incompetence are so final for young men."[13] Indeed, when he founded the Marine Corps University, General Alfred M. Gray Jr. said, "History should be used to teach officers military judgment, not to make academic historians or simply teach facts."[14]

Gray, Mahan, and Mattis make a sound plea for studying military history as a practical measure, learning lessons from history. The wisdom of such an approach is apparent on its face, but less apparent are the dangers of studying history, which are quite real, especially for the unwary. History lends itself to trite sayings and shibboleths, which in turn lead to dogmatic thinking and rote, uncritical analysis; the only antidote for these is careful, critical study. For example, many have heard that those who do not study history are doomed to repeat it. But history cannot repeat itself as each event is unique, springing from specific individuals and conditions and therefore the study of history cannot prevent mistakes. As Geoffrey Megargee explains, "To look back at one historical development and try to draw specific policy conclusions from it is misguided. Such an approach is a leap of faith; it depends on the belief that the historical account is absolutely accurate, and that present circumstances mirror the past exactly."[15]

Moreover, there is a natural tendency for historians to divide and categorize history into different subfields or categories. While understandable, these are essentially false: "There is no military history, political history, social history, African American history, Scots Irish American history, or women's history. While those are useful categories for use in studying ourselves through specific lenses, and for planning conferences, they are, when all is said and done, all history and should be taught as such."[16] Excessive focus on subfields of historical study creates artificial barriers, compartmentalizing historical thought, limiting context, and reducing productive study of history as a whole.

Earlier, the dangers of subdividing history into different fields was mentioned; for our purposes, these topical subdivisions remain useful. When you have identified that a given work lies within one of these categories, it helps to identify the viewpoint of the author, thus providing useful information for analyzing the work. Beyond topical fields (i.e., French, nineteenth century, military, women's, or labor histories), however, histories also can be divided by the author's analytical method and focus. Some historians view history as objective, while acknowledging that the sources, and the historians themselves, are inherently biased and thus subjective. They believe the historian's duty is to strive to be objective, while acknowledging bias. Others view this as an impossible task, arguing that history is inherently subjective and that historians have no duty to objectivity. Some go so far as to argue that historians should be activists, writing histories that support their cause. Not all activist historians openly acknowledge their subjectivity; indeed, many reject the label, claiming to be objective while presenting a subjective historical viewpoint. The history's structure is also important when evaluating the reliability and value of the text. Is it a narrative, tracing the development of the story within a conventional chronological framework? Or is it thematic,

[13] Jill R. Russell, "With Rifle and Bibliography: General Mattis on Professional Reading," *Strife* (blog), 7 May 2013.

[14] Gen Alfred M. Gray Jr. letter to Commanding General, Marine Corps Combat Command, "Training and Education," 1 July 1989, Archives Branch, History Division, Quantico, VA.

[15] Dr. Geoffrey Megargee, "History Cannot Repeat Itself," Facebook, 3 December 2013.

[16] Dr. Glenn T. Johnson, "Subcategories of History," Facebook, 10 July 2016.

covering different concepts in a chronological manner? Is the topic institutional or cultural? Or, if it is a military history, is it operational?

Historians whose work shares an analytical method and focus as well as structure are generally grouped into schools of historical thought. Identifying which school of thought a work falls into is a useful shorthand for evaluating a work.

Examining a small sample of various schools of historical thought illustrates how an author's identification with these schools aids the reader's understanding. Marxist historians posit that all history is economically driven; the most rigid Marxist historians present a deterministic view of history that explains the past and predicts the future. They write from a relativist, often activist, point of view. People's history is a school that generally looks at history from below, attempting to give voice to the voiceless. Because it seeks to examine the history of those who left relatively little in the historical record, people's history is seldom narrative-driven and usually focuses on cultural and institutional topics. Military history often tends toward the "great man" historical school of thought. Often considered old-fashioned and rejected by modern scholars, it focuses attention on the decisions and actions of a few influential individuals and focuses on a narrative view. The French Annales school focuses on long-term change and social rather than political themes. It makes great use of quantification and geographic evidence to examine history from different directions. Recognizing when a given work falls under these schools helps the reader correct for the biases and viewpoints each brings to history.

The history of the Marine Corps specifically has been told in ways that require careful study by the reader. In his seminal work, *Semper Fidelis: The History of the United States Marine Corps*, Allan R. Millet presented an essay on his sources that adroitly summarizes works on Marine Corps history to date:

> *For more than a hundred years, the writing of Marine Corps history has been shaped by internal organizational interest, political controversy, and a perceived public interest in the Corps, the last normally coinciding with the heroics of Marines in wartime. Like most of the writing on military institutions, Marine Corps histories have improved in their scholarly quality, but reflect a bias toward operational narratives and a distaste for either external relationships or internal difficulties. Marine Corps historical writing, which has been largely dominated by Marine enthusiasts in and out of uniform, has had a distinct utilitarian quality, that is, to build loyalty and dedication on the part of serving Marines, create public sympathy and support, present [the] Corps's perspectives on policy issues past and present, and honor the service of former Marines. These characteristics are not unique to the Corps.*[17]

The subcategory of military history can be further divided into differing types. Since at least 1979, Millett has categorized military history into five distinct types. Since most of the history a Marine will read in their career is military history, learning to recognize these five types is an important part of historical literacy:

- Inspirational
- Nationalistic
- Antiquarian/hobby

[17] Allan R. Millet, *Semper Fidelis: The History of the United States Marine Corps* (New York: Free Press, 1991), 768. Millet's summary of Marine Corps historiography remains essentially true 25 years later, though many fine historical works have been published since.

- Military utilitarian
- Civilian utilitarian[18]

Inspirational military histories are designed to highlight military virtues, especially heroism, generally, and for specific military units. Many biographies and unit histories fall into this category. The Lineage and Honors and Commemorative Naming programs run by the Marine Corps History Division's Historical Reference Branch fall under this category, as do many of the exhibits and programs at the National Museum of the Marine Corps. Most of what is taught in the history classes at recruit training and Officer Candidate School fits neatly in this type as well. General Sir William Francis Patrick Napier, who wrote a massive history of the Peninsular War (1808–14), believed all military history should fall within this category: "It is the business of the historian . . . to bring the exploits of the hero into broad daylight. . . . The multitude must be told where to stop and wonder and to make them do so, the historian must have recourse to all the power of words."[19]

Nationalistic military history is very similar to inspirational, but is focused on patriotic or nationalistic themes rather than individual or unit heroism. Perhaps the most widely known example of this type of history is Winston S. Churchill's multivolume magnum opus *The Second World War*, which further proves that nationalistic military history can be quite eloquently written.[20] Other examples include Hans Delbrück's *History of the Art of War* (1920) or R. Ernest Dupuy and Trevor N. Depuy's *Military Heritage of America* (1956). In the last century, much of the history taught in American high schools fell within this type.

Antiquarian military history is focused on historical minutia, such as uniform details or weapon statistics. This type of military history is concerned with the color of a man's jackboots or the type of rivets used on a Panzer VI Tiger tank, but less concerned with the whys and wherefores of warfare and its causes. Model builders, war gamers, and reenactors are typically considered antiquarians, and are often the primary target audience of museums and authors for specialist publishers. The published works produced by enthusiastic antiquarians far exceeds the output of other types of military historians; additionally, the massive amount of data they accumulate is a great boon for historians producing other types of military history.

Military utilitarian histories are usually written by and for professional militaries to educate policy makers and military officers. The primary function of the Marine Corps History Division or the U.S. Army Center of Military History, for example, is to produce military utilitarian histories. The Army's famous *Green Books* and the Corps' *Red Books* produced on World War II represent these types of history.[21] This style is most often narrative, operational history that establishes the basic facts and chronology of an event. The key point is facts not models. Though most government historians are trained professionals, these programs are closely associated with military officers who are

[18] Mark Grimsley, "The Types of Military History," WarHistorian.org, *Blog Them Out of the Stone Age* (blog), 6 January 2012.

[19] John Keegan, *The Face of Battle* (New York: Penguin Books, 1976), 39.

[20] Winston S. Churchill, *The Second World War*, 6 vols. (London: Reprint Society, 1948–53).

[21] The *Green Books* are the *U.S. Army in World War II*, 78 vols. (Washington, DC: U.S. Army Center of Military History, 1946–92); and the *Red Books* are the *History of U.S. Marine Corps Operations in World War II*, 5 vols. (Washington, DC: Historical Branch, G-3 Division, Headquarters Marine Corps, 1958–68).

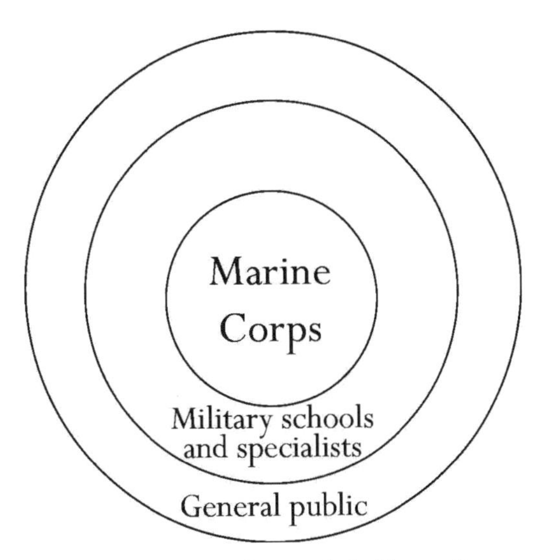

Gen Edwin Simmons often used a bull's-eye diagram to explain the intended audience for History Division's work. Replace "Marine Corps" with "military" and this diagram also neatly illustrates the intended audience of military utilitarian history works.
Marine Corps History Division

usually devoted to one type of military doctrine or another. Government historians have to be particularly vigilant against institutional forces that strive to fit historical events into preferred doctrinal models.

Many academic historians view all official history (the most common form of military utilitarian history) as inherently biased, activist history.[22] Concerning this, Brigadier General Edwin H. Simmons, who led the Marine Corps History Division for more than two decades once said:

> *I frequently use the word "advocacy" and that sometimes puts the academic person's teeth on edge. My point is that anyone working for the Marine Corps Historical Program should believe in the Marine Corps. By the same token, anyone working for the Army, Navy, or Air Force historical programs should be advocates of their respective services. Advocacy does not mean bias, prejudice, or distortion. An advocate can still write objective history.*
>
> *I would indeed argue that official history can be more accurate and objective than that of an independent scholar. The independent scholar is able to first form his hypothesis or premise and then marshal his facts selectively to support that hypothesis or premise. We are not permitted this degree of latitude. We must tell the whole story as best and as completely as we can.*[23]

Civilian utilitarian is the final type, which is defined as academic military history written to help the educated citizen understand war and conflict. It is studied by civilian academics for the same reasons any other historical field is studied—to illuminate our understanding of mankind. John Keegan's *The Face of Battle* is one example, as is Gerhard L. Weinberg's *A World at Arms: A Global History of World War II* (1994). Many of these works examine military institutions, including Millett's *Semper Fidelis* and Aaron B. O'Connell's *Underdogs: The Making of the Modern Marine Corps* (2012). All of these types of history can merge into one another—for example, when a work that is military utilitarian in overall intent includes a sidebar that is clearly antiquarian or inspirational; further, the same historian can produce during their lifetime works of different types.

The Marine Corps determined long ago

[22] The term *official history* refers to a work of history that is sponsored, authorized, or endorsed by its subject. For example, Marine Corps History Division, as a function of higher headquarters, writes and publishes the official history of the U.S. Marine Corps.

[23] BGen Edwin H. Simmons, as quoted by Charles R. Smith, group email to Col Nathan S. Lowrey, Wanda J. Renfrow, and the author, 15 May 2007.

that Marines must study military history, their own most especially, to function more efficiently and effectively. It is not enough to simply read history books; they must be educated readers, understanding the fundamentals of historical method and able to identify different schools of historical thought as outlined above. By attaining this basic historical literacy and understanding historiography, Marines will get the most out of their historical studies.

ABOUT THE EDITORS

PAUL WESTERMEYER is a historian who joined Marine Corps History Division in 2005. He earned a bachelors degree in history and a masters degree in military history from The Ohio State University. He was the recipient of the Marine Corps Heritage Foundation's Brigadier General Edwin Simmons-Henry I. Shaw Award in 2015 for his book *U.S. Marines in the Gulf War, 1990–1991: Liberating Kuwait*. He is the author of *U.S. Marines in Battle: Al-Khafji, 28 January–1 February 1991* and the editor of *Desert Voices: Oral Histories of Marines in the Gulf War* and *U.S. Marines in Afghanistan, 2010–2014: Anthology and Annotated Bibliography*. Westermeyer also serves as the series historian for the Marines in the Vietnam War Commemorative Series. His current projects include a monograph on the Battle of Marjah and an anthology on Marines in the Frigate Navy.

BREANNE ROBERTSON joined Marine Corps History Division in 2015. She earned a bachelors degree in art history from the University of Missouri, a masters degree in art history from the University of Texas, and a PhD in art history from the University of Maryland. Essays drawn from her research appear in *Marine Corps History*, *American Art*, *The Annals of Iowa*, and *Hemisphere: Visual Cultures of the Americas*. She recently authored *Camp Pendleton: The Historic Rancho Santa Margarita y Las Flores and the Marine Corps in Southern California, A Shared History* (2017) in commemoration of the 75th anniversary of that military base. Her current projects include a monograph on Marine Corps activities in the Dominican Republic between 1916 and 1924 and an edited volume examining the history and cultural meaning of the Iwo Jima flag-raisings entitled *Investigating Iwo: The Flag-Raisings in Myth, Memory and Esprit de Corps* (2018).